THE NAVY'S AIR WAR

THE NAVY'S AIR WAR

A MISSION COMPLETED

BY

ALBERT R. BUCHANAN

The Navy's Air War: A Mission Completed by Albert R. Buchanan. First published in 1946.

Annotated edition with footnotes and images published 2019 by The War Vault.

FIRST PRINTING, 2019.

ISBN: 9781092136495.

CONTENTS

FOREWORD

IT HAS OFTEN BEEN NOTED that aviation came of age in World War II. This is true of naval aviation as it is of the air forces of our Army, our allies, and, indeed, of our enemies. This book presents a picture of naval aviation as it was organized and functioned during the war. It not only includes the story of air's part in operations already well known but also sets forth less publicized and less exciting activities.

There are chapters on training, on maintenance, on procurement, and on many other matters not hitherto given much space but vital to the success of the Navy's air war. In a word, it attempts not just to hit the high spots but rather to tell the story of the naval air force as a whole, as a team reaching from a desk in Washington to the fighting areas.

For thirty years I had the honor to play a part in naval aviation. During that time, I watched it grow through two world wars. If this book seems to indicate that many changes in equipment, tactics, and organization were made while the conflict was in progress, that was because aviation was a relatively new weapon and we learned as we went.

That in itself augurs well for the future. When new equipment and new tactics are developed, naval aviation will prepare to meet them in the same spirit and with the same determination that it showed in the war just past.

Admiral Marc A. Mitscher

PART I

1 – PEARL HARBOR

*"I arrived here on October 30, 1940, with the point of view that
the international situation was critical, especially in the Pacific, and I
was impressed with the need of being ready today rather than tomor-
row for any eventuality that might arise. After taking over command of
Patrol Wing 2 and looking over the situation, I was surprised to find
that here, in the Hawaiian Islands, an important naval advanced outpost, we were operating on a shoestring, and the more I looked the
thinner the shoestring appointed to be."*

THIS STATEMENT IS NOT THE POSTWAR REFLECTION of a naval officer
attempting to analyze the factors behind Pearl Harbor, but a quota-
tion from a letter written by Rear Admiral [later Vice Admiral] P. N. L.
Bellinger [Patrick Nieson Lynch Bellinger, of South Carolina, 1885-1962], USN, Commander, Patrol Wing 2, to the Chief of Naval Operations, dated 16
January 1941.

In view of the importance of Hawaii in the scheme of national
defense, this letter revealed a shocking situation. Patrol Wing 2 was
charged with detecting any possible enemy force that might endeavor
to reach this bastion before it could get into a position to attack. The
letter showed that Patrol Wing 2 lacked not only modern planes for
patrol, but also spare planes, spare engines, spare parts, hangar and
beach equipment, bombs, ammunition, stores, adequate base operating facilities, overhaul and repair facilities, as well as qualified
personnel to man sufficient base facilities and shops to ensure continuous operating readiness.

It showed, furthermore, that though PatWing 2's planes and
equipment were obsolete, it was not known whether plans had been
adopted for modernization, or even were there were such plans, the
normal period between request and delivery to this area being nine
months.

The letter went further. It stated that, presumably, "the offices
and bureaus concerned are familiar with the situation in the Hawaiian area over which they have particular cognizance; certainly,
enough correspondence has already been written concerning patrol
plane needs to enable bureaus and offices to take necessary steps to
provide and anticipate such needs."

It closed with specific recommendations for the alleviation of these deficiencies, and listed the requirements in language that could not be misunderstood.

This letter was written almost a year before the raid on Pearl Harbor. The air commanders, both army and navy, aware of the striking power of carrier planes, warned their superiors of the possibility of an enemy attack against the Hawaiian area, and it should not be thought that such warnings were ignored.

There were, furthermore, many factors that had to be considered. It took nine months, in the normal course of official peacetime movement, to act on a specific request from this area, and, if Hawaii was an important point in our defense, there were, in addition, other important areas — the Panama Canal, the west coast, the Philippines, to mention only a few.

The warning might have been considered somewhat exaggerated, since any commanding officer has a normal desire to build up his own establishment and an equally customary tendency to attach particular significance to it. The Navy at this time was heavily committed in the Atlantic to the all-important Neutrality Patrol for the protection of vital shipping to Europe. Finally (and a point not to be forgotten), there was a pervading atmosphere in the United States at that time — much talk of war, but little realization that it would actually come.

In due course of time, therefore, a certain amount of action was taken. Patrol Wing 1, normally based at San Diego, was ordered to reinforce PatWing 2 in the Hawaiian area. Headquarters were set up at the new naval air station at Kaneohe in April and the squadrons followed during the spring and summer.

With the addition of extra planes, it was possible to re-equip four squadrons with the latest type of patrol plane, even though armor and other special gear had to be installed after the planes arrived in Pearl Harbor. Since training was the order of the day, much time was devoted to practice exercises both by aircraft alone and in co-operation with surface units, as well as in excursions to Midway, Palmyra, Johnson and other islands in order to gain experience in advanced base operations.

Because of extensive training activities, with a certain number of old planes in overhaul and a number of new ones in the shops for the installation of the latest gear, and because of an insufficient number of trained crews to maintain a continuous reconnaissance around the islands, patrols were reduced to a dawn, anti-submarine search over the fleet operating areas south of Oahu and to escorting vessels universe to provide their own aircraft as they entered or sortied from Pearl Harbor.

In addition to the big flying boats of PatWings 1 and 2, there were available the planes of three utility squadrons, some of which were capable of overwater search, such fleet aircraft as might be in Pearl at the time, and the Marine Corps aviation units stationed at Ewa, whose fighter planes, in case of emergency, were to operate under army control as part of the defense force.

Meanwhile, the sons of Nippon [the Japanese referred to themselves as the Sons of Heaven] proclaimed their love of peace, while preparing for an adventure in war.

It was almost two o'clock in the afternoon in New York and Boston; a gray and dripping day, just about time to get into the movies which would soon be packed to the doors with the Sunday afternoon crowds. Farther west, in Chicago and the Great Valley [of the Appalachians in the northeast], it was an hour earlier. Those families that customarily ate a midday dinner on Sundays were just rising from the table — stuffed. As yet the radios still carried their usual programs.

Out on the coast, in Seattle, San Francisco, San Diego and Los Angeles, and in all the smaller towns and cities in between, it was going on eleven in the morning. Church bells were ringing, and people were gathering for the morning services.

Here, there was a good deal more discussion than there was in the East, of the Japanese delegation then in Washington [US Ambassador to Japan Joseph Grew and US Secretary of State Cordell Hull negotiating with the Japanese to arrange a peace meeting in Alaska]. New York, Boston, Philadelphia and Charleston had merely noted the President's appeal to the Emperor in the morning headlines, and had forgotten about it. The Coast, always more concerned with the Japanese problem, was inclined to skepticism, but not, as it was to prove in this case, with the skepticism which the situation warranted.

At the great naval base at San Diego [at Point Loma, across from Coronado Island], the big aircraft carrier Saratoga, just arrived from Puget Sound, was warping into her berth, getting over the first lines. Curiously enough a lot of little yellow men in airplanes were pretty sure that at that moment she was somewhere else.

Out in the Hawaiian Islands, in mid-Pacific, two thousand miles still farther west, it was not quite eight in the morning. Directly over the Island of Oahu, with its numerous airfields and defensive facilities, its city of Honolulu, and its great naval base at Pearl Harbor, the rosy clouds hung about the mountains at a base of thirty-five hundred feet.

Visibility was excellent, and in the early light a few civilian planes lazily circled the city and the commercial John Rodgers Airport [now Kalaeloa Airport; John Rodgers (1881-1926) was a US Navy aviation pioneer who died in a plane crash], their pilots getting in solo time or taking private instruction.

They were still in the air, not yet aware that anything was amiss, when the first sleek, gray-silver planes came in out of the north and dived toward the outlying army and navy field, and toward the ships lying peacefully moored at the naval base.

In the residential sections, out at the beach, and even within the confines of the various military and naval reservations, men looked and listened. They saw the lazy planes turn on their sides and dive, heard the scream of their motors, felt the explosions, saw the smoke and flames and counted the splashes, and cursed the Army or Navy, according to their inclinations. This, they allowed, was carrying this war games thing too far. Someone might get hurt.

It was December 7, 1941.

IT SHOULD BE RECORDED THAT on that day our enemy hurled upon us, for the first time in history, a form of attack that was in the end to prove its own nemesis, for the carrier task force, hardened and strengthened by our hands, was to turn against Japan and beat her into submission.

What did naval aviation in the Hawaiian Islands do in the face of this attack? Like our other military forces, it will be shown that, stunned and rendered impotent for the moment by the staggering suddenness and completeness of the blow, it fought back with a heroism that stands out like a bright spot in the dark day.

Kaneohe Bay, the naval air station on the northern coast of the island, was the first to glimpse the shadow of events to come. Patrol Squadron 14 was the ready-duty squadron at Kaneohe that morning, with Patrol Squadron 11 in the stand-by spot. Three planes of Squadron 14 were in the air that day, having taken off at sunrise on a routine patrol over the fleet operating area, PatRon 12 was the third squadron stationed at Kaneohe.

At 0350 that morning, sometime before dawn, the coastal minesweeper Condor was conducting sweeping operations approximately 1¾ miles southwest of the Pearl Harbor entrance buoys when she picked up a submarine contact — the white-feather wake of a periscope where no periscope ought to be. She reported this contact immediately to the duty destroyer Ward, on patrol off the Pearl Harbor entrance, and the Ward immediately took up search.

It was not until 0630, however, that the contact was regained. At that hour, the minesweeper Antares, approaching Pearl Harbor from Canton and Palmyra with a 500-ton steel barge in tow, slowed and turned slowly eastward, awaiting the tug that was scheduled to come out to meet her and pick up her tow. It was as she swung, that the Antares' lookout sighted what appeared to be a small submarine "with upper conning tower awash and periscope partly raised," 1,500 yards off her starboard quarter, apparently tracking her in the direction of Pearl Harbor. The Ward was in sight and Antares notified the destroyer, by visual signals, of what she had seen. The Ward moved in to the attack.

At approximately the same moment, patrol plane 14-P-1, leaving Kaneohe for its patrol, hove in sight and took up its hand in the game about to be played. The plane dropped two smoke pots [sophisticated smokescreen system recently adopted by the US Navy and other military branches, providing a hitherto unavailable camouflage over a very wide area, changing combat strategy forever; also known as The Patterson System, after its inventor Alonzo Patterson] to mark the position of the sub, while the Ward opened fire with guns and depth charges. Simultaneously 14-P-1 dropped one of her depth charges on the suspicious submarine.

Both Ward and 14-P-1 claimed to have sunk the craft, and there seems little doubt that one or the other, or both, succeeded. Ward reported the attack to the Commandant, 14th Naval District, his immediate superior; 14-P-1 reported to the duty officer at Kaneohe Bay. In both cases steps were immediately taken to identify the submarine. There can be little question (whether it was 14-P-1's bomb that sank it or not) that this was the first attack delivered by U.S. naval aircraft upon a Japanese submarine. The crew of 14-P-1 were unaware of it as they proceeded on their patrol, but they had fired the first shots of our naval air war against Japan.

The commanding officer of the air station at Kaneohe was at breakfast at a quarter to eight. From the window of the dining room he could look over the hills to the bay and to the field with its landing strip and hangars. Four Catalina flying boats, PBY's [Patrol Boats] in Navy parlance, were anchored in the bay, and there were several planes in front of the hangars, including the CO's scouting plane, a Kingfisher OS2U-1 [catapult-launched observation floatplane] which was being prepared for his use.

While he was eating, the CO's [Commanding Officer] attention was caught by the persistent hum of a large flight of planes and his suspicion was aroused by the fact that the first ones to come in sight circled to the right rather than to the left as was the rule in the area. Just when he became convinced that these were enemy planes, we don't know, but, as he sped down the hill in his car, he glanced at his watch. It was 0748 and approximately two minutes later the first Jap attack on U.S. soil began.

Service crews were being ferried out to the four big seaplanes anchored in the bay, as the Japanese planes roared in. It was only when the swooping planes began to wink with orange eyes along the leading edges of their wings, and little hissing geysers spouted in straight lines across the ruffled surface of the bay that the men realized that they were under attack.

Then, in the next moment — it was as swift as that — the planes swept on toward the control tower with guns still spitting, leaving behind them four burning planes up the bay and four riddled boats about which splashed startled, angry, frightened men. Some of the men, quite a number of them in fact, did not splash. They simply lay still in the water which softly swirled crimson streamers of their blood about them. They were the first American casualties of the war in the Pacific.

This was only the beginning of the Jap attack. The first wave of raiders, evidently all fighters, swept on to the end of the bay, then turned and came back, concentrating their fire this time upon the planes drawn up on the landing mat.

In this attack they were met by the fire of a single machine-gun rushed from the armory by Aviation Chief Ordnanceman John Finn, who set his weapon up on a covered tin garbage can, directly in the line of attack, and manned it throughout in spite of wounds [later receiving the Medal Honor for his actions that day].

Nor was he left alone to fight. Others followed his example as soon as they were able to get arms and ammunition. The official report of the action submitted by the Commander, Patrol Wing 1, then based at Kaneohe, reads:

"The conduct of all personnel throughout the entire attack was magnificent, in fact, too much so. Had they not, with no protection, deliberately set themselves up with machine-guns right in line with the drop of the attacking and strafing planes and near the object of their attack, we would have lost less men. It was, however, due to this reckless resistance that two enemy planes were destroyed and six more were sent away with heavy gas leaks."

The report, of course, refers to all the attacks which took place on that day at that station. In the first attack no bombs were dropped. It was obviously the intent of the attackers to prevent the planes based there from getting into the air, where they might constitute a threat to the success of the main effort. This, of course, was the strike at the naval base at Pearl Harbor itself.

On the enemy's return, one plane that was singled out for special attention was the wing commander's OS2U-1 on the landing mat. At that time, a chief petty officer was turning over the prop by hand, and the craft was evidently thought to be a fighter preparing to take off. It was thoroughly riddled, as were several of the other planes on the mat.

Following this first strike, the attackers sped away over the hill in the direction of the marine base at Ewa, and there was a brief lull, during which efforts were made to prepare for further action. This came within a very few moments, when a second wave of fighter craft (estimated six to nine), attacked.

This time the planes on the ramp were the objective, but the raiders did not confine themselves to these. Hangars, quarters, cars, in fact anything that moved or chanced to catch the eye of the Jap pilots, was strafed. This attack was followed, some time later, by two waves of bombers, which dropped bombs on the hangars and then swept on. A third and final strafing attack by fighters took place at about ten o'clock, apparently by planes returning to their carriers.

Countermeasures taken were heroic but pitiful. In the main they consisted in getting anything that would shoot into position to repel attack, and then shooting. During the intervals between attacks, every effort was made to clear the vital area of burning wreckage, and to salvage all that had not yet been destroyed. But it was a vain hope to dream of saving much. A survey, following the attacks, showed that all planes actually at base were either destroyed or so damaged as to be useless for the time being. [Ironically, many aircraft had been drained of fuel to prevent sabotage and were sitting-duck targets.]

The three planes of the dawn patrol, not yet returned to base, were not destroyed, and were immediately diverted to a search for the Japanese Fleet. One hangar was destroyed. Seventeen men were killed and eleven seriously injured. During the attack, cars were driven upon the small landing field adjacent to the station in order to prevent any attempted landing by the enemy. No such landing, however, was attempted.

Kaneohe was the first object of attack probably because it was the first reached by the raiders from their ships. The attacks that followed upon the other fields and upon the naval base itself appear to have been delivered simultaneously, or within split seconds of one another.

Several of the first wave of raiders that struck at Kaneohe, as we have noted, seem to have gone over the hill and beyond to fall next upon Ewa, the Marine air base at the other side of Pearl Harbor.

Ewa [pronounced "Evva"] was a comparatively new station. Most of its personnel were still living in tents, and the swimming pool, just begun, was as yet no more than a hole in the ground. It proved mighty useful to quite a number of officers and men that day, for a purpose for which it had not been intended. It formed a convenient revetment in which to take cover while shooting back at the vicious, persistent Jap attackers.

On this morning of December 7, 1941, the officer of the day was sitting in the officers' mess. In five minutes, he would be relieved. It was 0755, and he became suddenly aware of the steady roar of a large formation of planes approaching. He describes what happened then:

"Upon stepping outside, I saw eighteen torpedo planes at about a thousand feet altitude flying down the beach from Barber's Point toward Pearl Harbor. From the northwest, an enemy formation of approximately twenty-one planes was just coming over the hills from the direction of Nanakuli, also at an altitude of about a thousand feet. The first formation of planes continued on down the beach apparently to deliver an attack upon Pearl Harbor. The second formation (single-seater fighters) passed just to the north of Ewa, wheeled right and attacked this camp from a string formation."

Various accounts of this attack testify to the bewilderment of many who witnessed its opening phases. Some have stated that they thought the attacking planes were army aircraft putting on a particularly realistic simulated attack. Only when they saw the little spurts of dust kicked up by Jap bullets did they realize something of what was taking place. A few die-hards have confessed that even then they thought: "Oh-oh! Live ammunition! Someone is going to catch hell for this!"

The officer of the day at Ewa apparently recognized the planes as Japanese, either from their unfamiliar shape or from their markings, and instantly realized what was happening, for he says: "This attack caught me on the way to the guardhouse in an effort to have the camp called to arms."

The pattern set at Kaneohe was repeated at Ewa. There was the same initial shock of surprise followed by the same surge of fury. There was the same almost split-second destruction of planes on the ground, leaving in its wake the same sense of futility. There was after that the same reckless heroism and determination to fight back with whatever weapons might come to hand. There were much the same results which it would be fruitless to chronicle here in detail. Within twenty minutes every plane at Ewa was immobilized.

After that the attackers returned again and again, making innumerable runs in three distinct attacks, strafing personnel and buildings, tents, cars in the parking area and on the roads, even an ambulance and a fire truck on the airstrip, the one picking up wounded, the other endeavoring to get into position to fight the fires which had been started among the planes. At the outset, gunners attempted to bring the free guns in the parked planes to bear upon the diving enemy, and from these positions they fought until the planes in which they stood were fired.

After that they took such guns as could be removed from the planes and fought back from the ground. Despite the heavy strafing here, however, casualties were comparatively light. When the attacks were over, formations of four to six enemy planes, working in two echelons, remained over the airfield apparently for the purpose of protecting the enemy's rendezvous at the assembly point after the return from the main attack, and also, evidently, to hold down any chance aircraft that might have survived the attack and attempt to get into the air.

WHILE EWA WAS BEING ATTACKED, similar strikes were taking place at the army fields, Hickam, Wheeler and Bellows, and the results were devastating. Likewise, at the same time the main naval air base at Ford Island, NAS [Naval Station] Pearl Harbor, was brought under attack.

On Saturday, 6 December 1941, there had been a captain's inspection at NAS Pearl, and in consequence, with the exception of four planes of Patrol Squadron 24 which were out engaging in intertype tactics with friendly submarines in an area considerably removed from the station, all planes were drawn up in neat rows on the field.

They consisted of the big PBY's of PatRons 22, 23, and 24 as well as the SOC's, OS2U's and SBD's belonging to the station and the miscellaneous aircraft of utility squadrons 1 and 2. At 0755 — the exact time was well marked, for the signal for morning colors, hoisted each morning at that exact moment, had just broken out — the first of the attacking planes hurtled down out of the sky upon the unsuspecting station. Unlike the attacks at Kaneohe and Ewa this strike was carried out by dive bombers which both bombed and strafed.

The first bomb is believed to have struck the PatRon 22 parking area, Ramp #4, and the explosion and subsequent fire destroyed six PBY-3 planes, damaged another beyond repair, and put the remaining five out of commission for a period of from one to ten days. Thus, at the very first blow, an entire squadron was reduced to impotence. As the first bombs fell upon the island, the Commander, Patrol Wing 2, at 0758, broadcast the warning:

AIR RAID. PEARL HARBOR. THIS IS NO DRILL!

This was the first official word of the attack to be broadcast, and it was picked up by planes approaching the island, and by ships at sea. A few moments later a similar message was broadcast by Commander-in-Chief, Pacific Fleet.

There appears to have been only a single attack directed at the naval air station on Ford Island. After that the raiders turned their attention to the ships moored in the harbor, or tied up to the docks alongside the island. Battleship Row, as it was called, was close aboard. There lay the Maryland, the Oklahoma, the West Virginia, the Tennessee, the Arizona, and the Nevada. The California lay in a cove on the southeast side of the island.

The Neosho, a navy oiler, was moored to the island dock, having just completed delivery of a cargo of aviation gasoline to the station's tank farm. Over on the other side, between Ford Island and Pearl City lay the target ship Utah, apparently mistaken by the enemy for an aircraft carrier. In addition to the Utah, the cruiser Raleigh and the seaplane tender Curtiss lay upon the northwest side of the island.

Thus, although no further attacks were directed solely at the air station, its location between the two groups of ships, prime targets for the rest of the attacks, left it in a position that was scarcely enviable. The first attack effectively put most of the planes out of action. Thereafter, the damage that was suffered by the station was caused in part by bombs intended for the fleet, but which missed their mark, partly by the hail of falling flak thrown up by the ships in the harbor, and partly by fragments thrown by the offshore explosions on board the Arizona, California and other vessels. Strafing was also carried out by planes crossing the island after attacking the ships.

Here, as elsewhere on that day, there was resistance from hastily improvised positions. But there was more than resistance. As at Kaneohe there were fires to be fought, and there were also rescue operations to be carried out for the crews of the blasted ships in the harbor. Both of these tasks were done coolly and courageously in the face of tremendous difficulties. Smoke from the blazing hangars and burning ships in the harbor drifted across the island, obscuring vision, and the deadly rain of spent flak and debris was almost continuous.

Nevertheless, damage to the station was slight in comparison with that done to the ships. One bomb, apparently aimed at the California, fell short and scored a direct hit on Hangar #6, setting it on fire and killing personnel within. Another bomb fell on Hangar #38, where ammunition was being handed out.

Fortunately, it was a dud. A third bomb struck the dispensary, and another fell in the roadway outside the assembly and repair hangar, but it was the ships that took the worst punishment. The torpedo bombers of the first wave bored in relentlessly and unerringly, with a precision which could only indicate exact information as to the location of their targets.

The Japanese must have been disappointed, however, not to discover the carriers Lexington and Enterprise in port. There can be no doubt that they knew them to be based there. But luck and a secret mission had the Lex and Enterprise at sea and the Saratoga had not yet returned from the west coast where she had been sent for overhaul.

At the hour of attack the Lexington was some four hundred miles south and east of Midway, to which island she was bound with the forward echelon of Marine Scout Bombing Squadron 231, reinforcements for that island's garrison. She received word of the raid at 0822, and at 0900 was informed that hostilities had begun and was ordered to search for the raiding force.

Throughout the day her squadrons flew methodical searches, but without success. A day or two later one of her fliers electrified the force with the announce that he had sighted a Jap carrier, and an attack force was launched. The "Jap carrier," however, turned out to be only a derelict barge. Shortly thereafter the Lex was ordered to give up the search and return to Pearl.

The Enterprise had a somewhat different story to tell. About a week before the attack, she had been suddenly and secretly ordered to transport the forward echelon of Marine Fighting Squadron 211 to Wake Island. She carried out this task under the cloak of strictest secrecy and under war conditions. The situation was acknowledged to be strained, and any potential enemy ships or planes encountered were to be destroyed. The run to Wake was made without incident, and the marine planes delivered. The task force then turned about and headed for home.

On this day, one of the Enterprise's fliers on scout duty thought he caught a glimpse through the mist of three small ships — perhaps three destroyers, he said, or possibly two destroyers and a light cruiser. He could not be sure. Search was made briefly. There was no time for more, and then the planes returned to the ship.

When this incident was recalled later, there was some thought that what the pilot had seen was a part of the Japanese raiding force. In the light of sober afterthought, however, this was impossible. For one thing the ships were considerably out of their way for what followed.

The Enterprise continued on her homeward way, accompanied by her covering destroyers. She was due back in Pearl by Saturday, the sixth, but she encountered bad weather, which prevented the refueling of her destroyers, and this delayed her so that it was impossible for her to arrive before Sunday. Those on board were disgruntled. Many of them had made dates for Saturday night. Others were anxious to be home. By Sunday morning they were two hundred miles west of Oahu, and at 0615, eighteen planes of Scouting Squadron 6 took off for Pearl Harbor — the crews the envy of all their shipmates.

The planes circled and rendezvoused, and then, at 0637, took departure from the task force which made its white-arrow wakes steadily through the dark water below. As far as Oahu the flight was without incident other than the sighting of two or three tankers, which on investigation proved to be American. Apparently strict formation was not observed during the flight. Rather, the planes spread out, in pairs, in a scouting fan to cover a broad lane ahead of the task force to Pearl Harbor.

Shortly before reaching Barber's Point, however, the commander of Scoron (Scouting Squadron) 6 was startled to hear a voice in his radio receiver crying: "Do not attack me! This is 6-Baker-3, an American plane!" There followed a moment's silence, and then the same voice was heard again, this time telling his gunner to break out the rubber boat as he was landing in the water. After that 6-Baker-3 was not heard from again. Back on board the Enterprise they also heard that cry coming blankly out of the air. It was the first warning they received that all was not as it should be. It was also the voice of the first American airman to be shot down in the Pacific war.

Eighteen planes of Scouting 6 approach Pearl Harbor from over Barber's Point that morning. They saw planes swarming and diving over Ewa. There were more planes over Pearl Harbor itself.

But the fact that the commercial radio stations in Honolulu were still playing popular music tended to lull any suspicions that might otherwise have arisen immediately. The air group commander himself, in his report, states that he believed the planes circling over Ewa to have been U. S. Army planes, and he gave them a wide berth.

As he approach Ford Island, however, he first noticed bursts of antiaircraft fire, and an instant later was attacked by Japanese planes.

Those eighteen planes from the Enterprise were the only U.S. carrier-borne aircraft to engage in action at Pearl Harbor. Taken by surprise and attacked by superior numbers, they were unable to beat off the enemy. There is no doubt, however, that they left their mark upon him. One Japanese plane was definitely shot down, and at least one more was considered probably destroyed. Scouting 6's losses, however, were not light. Six of the eighteen planes that left the Enterprise were shot down or missing when the smoke cleared.

It is not altogether clear where each of the Enterprise planes landed. Some appear to have come in at Ewa. Others, including the air group commander and the Commander, Scoron 6, managed to land at Ford Island, which in spite of everything remained usable. As soon as possible these planes were refueled and rearmed, loaded with bombs, and sent away to search for the enemy's attack force, but without success. This operation became increasingly risky because of the tendency of the gunners on the ships, their nerves now worn threadbare by the attack, to open fire on anything that flew. This state of tension continued throughout the following night, long after the attack had ended, and more than one American plane was shot down as it came in for a landing.

Before the Enterprise planes could be refueled to take off, Utility Squadrons 1 and 2 at Pearl, and Utility Squadron 3 stationed at the Maui airport, a commercial field only recently taken over by the Navy and now known as NAS Puunene, Maui, were sending out their lightly armed or even completely unarmed planes in search of the enemy.

By eleven o'clock, when the enemy had cleared the area, more aircraft were able to get off and planes from the stricken ships in the harbor were transferred to the air station to form an impromptu squadron which commenced operations that same afternoon. In addition to the two carriers, cruisers operating outside the harbor launched their aircraft, and such army planes as were available joined in the search.

Although false radio calls, probably disseminated by an enemy ship stationed south of the islands for that purpose, drew many of the searchers off in the wrong direction, two aircraft from the cruiser Northampton made contact with a Jap fighter at 1120 and succeeded, after twenty minutes, in shooting him down.

About noon, a Catalina of PatRon 14 was damaged but was able to continue its patrol, and a plane of Utility Squadron 1 penetrated apparently to within fifty miles of the attacking fleet before enemy fighters forced it to turn back.

PatRon 21 from Midway, two Catalinas of Utility Squadron 2 at Johnson, and the planes from the Enterprise and Lexington had no luck at all in their efforts to track the Japanese carriers. As darkness fell over Hawaii, the naval air force found more than 100 of its 156 aircraft destroyed and quite a few more damaged.

Of those left, almost half were noncombatant types, but even they were usable for patrols as the next few days showed, when pilots, whose principal activity had recently been towing target sleeves, took off to guard against the enemy's return with only the lightest of armament, sometimes with only a rifle slung across the knee.

These, in brief, were the actions of naval aviation on December 7, 1941. The unprovoked assault on Pearl Harbor was an initial victory for the Japanese, and permitted them to go far in their first sweep across the Pacific. On the other hand, no other single action of the enemy's could have welded the American people as firmly together in a common will for victory. As far as naval aviation was concerned, no one had to argue any longer concerning the striking power of a carrier force. We now had evidence, costly evidence, that this was a key to the defeat of the Japanese Empire.

It is proposed in the second part of this account to go to some length into the background of naval aviation. This examination may throw some light on the factors that caused us to be caught flat-footed at Pearl Harbor. More important, we believe it will show the significant strides that naval aviation had already made by December 1941, in building up a force that was to be so instrumental in crushing the enemy.

PART II: INVENTORY

2 – Background

WHEN THE BOMBS BEGAN FALLING on Pearl Harbor, naval aviation had over thirty years of experience behind it. Starting in 1908, when a few naval observers witnessed a demonstration put on by the Wright Brothers for the Army, it had developed slowly until World War I. As a result of that conflict, the value of aviation to the Navy had been recognized and, in 1920, air commands were created within the fleet. Even more significant was the establishment the following year of Aeronautics as one of the co-ordinate bureaus of the Navy Department.

Although aviation, like other branches of the service, was hampered throughout the twenties and early thirties by lack of funds, the first chief of the new bureau, Rear Admiral William A. Moffett, [William Adger Moffett, of South Carolina, 1869-1933] by careful handling of the monies at his disposal, was able to carry on considerable experimentation. From this period dated the power catapult used for launching planes from battleships and cruisers, the aircraft carrier, and considerable work with large, rigid airships.

Of these developments, the carrier was easily the most important. In 1922, the Langley, converted from the collier Jupiter, joined the fleet and was used largely for experimental purposes in working out the equipment and techniques for efficient carrier operation. The next two carriers, the Lexington and Saratoga, converted from battle-cruiser hulls upon which construction had been halted as a result of the Washington Treaty, were commissioned in 1927. The Ranger was the first American ship designed as a carrier from the keel up and was ready in 1933.

The same year that saw the Ranger commissioned also represented the low ebb of aviation during the postwar period. The number of students at the great training station at Pensacola sank to thirty in May, but the turn was already in sight.

First, the Navy benefited from funds available for public works to commence construction on existing air stations and even to begin work on new installations at San Pedro, California, and Sitka, Alaska. With the deterioration of the international situation, it was decided to build upon our defenses and, in 1934, the Vinson-Trammel Act provided a five-year expansion program for the Navy in which aviation was granted 1,910 additional planes.

To obtain pilots for the growing air force, Congress passed the Aviation Cadet Act of 1935.

Because the inadequacy of this legislation soon became apparent, further expansion of the Navy was called for in 1936 and 1938. The rapid increase in the size of naval aviation and the constantly changing nature of the program gave rise to numerous problems. Each jump in the number of planes meant that more pilots would be required, and the need for greater pilot recruitment taxed the existing training facilities.

By April 1938, there were 605 students at Pensacola compared to the 30 of some five years before, and one of the ever-present boards was recommending that pilot personnel be doubled. The Naval Aviation Reserve Act of June 1939 fixed the maximum number of reserve pilots at 6,000.

In the meantime, another board, meeting under the chairmanship of Admiral [Arthur Japy] Hepburn [Vice Chief of Operations, overseer of U.S. Naval logistics], had studied shore station development and recommended among other things the construction of the giant training base at Corpus Christi, Texas.

Late in 1939, the Horne Board investigated the relationship between the various phases of the expansion in naval aviation and worked out a percentage system whereby the training program could be expanded to keep pace with plane production.

For the regular officer who entered aviation, training had been based on the theory that he should be proficient in flying all plane types used by the Navy. With the rapid increase in personnel after 1935, such all-around training gradually gave way to specialization in a particular type. Late in 1938, the President allocated certain emergency funds to the Civilian Pilot Training Program to be administered by the Civil Aeronautics Administration and to be expended in giving preliminary ground and flight training to college students.

Both the Army and Navy shared in this program that was given formal sanction by Congress in June of the following year when it was provided that 10,000 students in 460 colleges should be given instruction at a cost of $4,000,000. At the other end of the program was operational training given to pilots after they had gained their wings and joined the fleet.

In July 1941, this phase of the training program was set up independently so that pilots would henceforth receive, at shore establishments, instruction in such subjects as advanced fighter tactics, additional gunnery, carrier qualification, and advanced training in the particular type of plane in which they were specializing. The purpose of these innovations was to save time and to ensure that pilots reporting to the fleet would be ready for immediate employment.

An augmented air force required not only fliers but also a large number of enlisted men for air and ground crews. The Navy had always trained its own mechanics, ordnance, and radiomen largely by the apprenticeship system at air stations. Under the stress of expansion some quicker method had to be found, and by 1941 there were two large schools for aviation ratings, one at Chicago and the other at Jacksonville, with a capacity of 10,500.

Another 4,600 were under instruction at the air stations at Norfolk, Pensacola, San Diego, Alameda, and Seattle. These were in addition to specialized schools for aerographers' mates at Lakehurst, photographers' mates at Pensacola, and parachute riggers at San Diego, Corpus Christi, and Lakehurst. Even with these prodigious efforts, the supply was running behind actual needs, partly because the program was constantly being stepped up and partly because each new plane type proved more complicated than its predecessor and hence required more specialists to care for it.

On 1 February 1941, a course for nonflying aeronautical engineers drawn from civilian life was created at the Massachusetts Institute of Technology. After three months of indoctrination the students were sent out to air stations to relieve pilots as engineering officers. This commissioning of civilian specialists recalled a practice of World War I and was to be greatly expanded after Pearl Harbor. By 7 December 1941, the main lines of the training program had been laid down and the necessary techniques learned for the even more rapid expansion that lay ahead.

At the same time the Bureau of Aeronautics was struggling with the problems of procurement. It was one thing to obtain authorization for ever-increasing numbers of aircraft, but it was quite another to obtain delivery.

Some of the explanation lay, of course, in the difficulties inherent in planning and in operating the mechanics of procurement. Much more serious were the small size of the aircraft industry and the competition between Army, Navy, and foreign buyers for the limited output. The first was in part outcome by providing government funds for plant expansion.

The second problem, that of competition, led to the establishment of the National Advisory Committee under Mr. William S. Knudsen in the summer of 1940 to co-ordinate the entire aircraft production program. At the same time every effort was made within the Navy Department to improve and speed up planning and procurement methods.

The rapidity of the expansion, including sums for new plants, is reflected in appropriations for naval aviation; for the fiscal year 1940, (1 July 1939 to 30 June 1940), $131,459,000; for the fiscal year 1941, $982,320,200; and appropriated prior to 7 December 1941 for the fiscal year 1942, $1,016,596,500.

Not only was the number of aircraft increased but also new types were introduced. By the summer of 1941, although the Brewster Buffalo (F2A) was still in use, both navy and marine squadrons were being equipped with the improved Grumman Wildcat (F4F). The Douglass Dauntless (SBD) dive bomber was in use throughout the fleet, and, even though a top-notch torpedo plane was not yet available, one was under development.

For patrol planes the Navy had the Consolidated Catalina (PBY), while the Martin Mariner (PBM) and Consolidated Coronado (PB2Y) were already in production. The old SOC, a Curtiss float plane used on battleships and cruisers, was giving way to the Vought-Sikorsky Kingfisher (OS2U). The latter plane was also used widely both for operational training and, in the early part of the war, for anti-submarine patrols. In addition, plans were well advanced for the Grumman Avenger (TBF), the Curtiss Helldiver (SB2C), and the Chance-Vought Corsair (F4U).

Along with development of new plane types went the construction of additional ships. In 1936, the Langley, which had been frankly experimental, was retired as a carrier and converted to a seaplane tender. Before Pearl Harbor four other carriers — the Yorktown, Enterprise, Wasp, Hornet — were added to the fleet, and in June 1941, a merchant vessel was converted with the idea of using it for an auxiliary carrier, especially to escort convoys.

This ship, the Long Island, proved to be the forerunner of over a hundred similar vessels built and converted in American yards for the United States and British navies. As it became obvious that war was drawing closer, the prospect of advanced base operations became more and more certain and the need increased for tenders for patrol planes. Again, there was some resort to conversion both of merchant vessels and of old four-stacker destroyers laid up since the First World War.

Because the latter were regarded as makeshifts to meet an emergency, work was also begun on a new class of small, especially designed tenders, the first of which — the Barnegat — was commissioned on 3 July 1941.

The steady addition of operating planes created a problem of adequate shore bases. Such older stations as the one at Norfolk began early to suffer from overcrowding and, as naval aviation expanded, there appeared the danger of reduced efficiency from lack of space and repair facilities. Ever since 1923, the Navy had possessed reserve air bases near leading cities for the purpose of giving reserve aviators the opportunity to fly planes whenever possible. Although most of these were relatively little developed beyond the barest essentials for the relatively primitive flying of the twenties and the early thirties, they were capable of development.

More important was the great air station at Alameda, California, authorized in 1937. Earlier air stations had grown in a rather hit or miss fashion as aviation developed, but Alameda was designed from the start to serve as a model station, and, although many of its standards were later modified in the light of experience, basic principles of construction and certain other problems of aviation shore establishments were first worked out there.

By 1 July 1940 there were already 26 naval and marine air facilities in operation and by 7 December 1941 there were 41 air stations in the continental United States, not including bases in Alaska, overseas or on sites acquired from the British in Newfoundland, Bermuda, and the West Indies.

While the situation was far from ideal when hostilities began at Pearl Harbor, naval aviation had already made great strides in men, equipment, and bases. Even more important, plans had been made to continue expansion at an accelerating rate. When war came, the groundwork had been laid for the rapid building of the most powerful naval air force in history as an integral part of the most powerful Navy in history.

FOUR-STAR ADMIRAL FREDERICK J. HORNE

3 – NAVAL AIR ORGANIZATION

FOR TWENTY YEARS BEFORE PEARL HARBOR, there had been much discussion of the best organization of the Navy for war. As generally stated, the problem had two parts. In the first place, there was the question of actual combat operations, which required a flexible system that could be readily adapted to the constantly changing needs of the strategical situation. In the second, there existed a need for a relatively permanent administrative organization to provide logistical and material support, to handle personnel, to insure the combat readiness of men and equipment, and to furnish repair and maintenance facilities.

Although it might at first sight appear difficult to reconcile these two, and much discussion was devoted to the subject, in practice it turned out to be relatively easy and even before hostilities began, the Navy had achieved the essentials of the system it was to follow. War conditions brought surprisingly few changes.

Operational control was based on the task force principle, the basis of which was that composition of forces and the nature of command should be determined by the mission assigned. For example, a convoy was in essence a task force to which certain naval vessels and even an escort carrier might be assigned. Obviously when the convoy reached its destination, the escorting ships were available for reassignment and the task force was automatically dissolved.

Other tasks might be more complex. A carrier raid on the coast of Norway would require the presence not only of carriers but also a protective screen of destroyers, cruisers, and, perhaps even battleships. The invasion of the Marianas called not only for air and surface units, including carriers, but also for amphibious forces and ground troops, while for the invasion of Normandy carriers were unnecessary because of the proximity of flying fields in England. A task force, then, was simply an assemblage of military power for the accomplishment of a specific objective; it was, further, a method by which units drawn from the several services and even from different nations could be combined under a single command.

Before considering how naval aviation was fitted into the task force pattern, it is necessary to look briefly at the United States Fleet. During the war its commander-in-chief, Fleet Admiral E. J. King, maintained his headquarters in Washington where he represented the Navy on the Combined, and on the Joint Chiefs of Staff.

These two bodies, the first of which included British as well as American officers, planned the global strategy of the western allies. Once a campaign had been decided upon, the assignment of naval vessels and aircraft and the necessary over-all planning of the Navy's share in the operation was the responsibility of the Commander-in-Chief, United States Fleet.

There were two principal subordinate fleets — the Atlantic and the Pacific — and also a number of smaller fleets that were in reality permanent task forces. Such were the Eighth Fleet in the Mediterranean, the Twelfth Fleet with headquarters in the United Kingdom, and the Seventh Fleet in the southwest Pacific area. The commanders of the subordinate fleets might create such task forces within their commands as they deemed desirable.

Under the Pacific and Atlantic fleets, these subordinate forces sometimes reached such size and importance in themselves as to be recognized as independent fleets. This was the case with the Third and Fifth fleets in the Pacific and with the Fourth Fleet in the Atlantic, to all of which considerable space will be devoted in the narrative of operations that follows.

When a specific operation was contemplated, air units were assigned as needed. For example, when the Fifth Fleet went into the Marshall Islands, it was assigned fast carriers of the Essex and Independence classes, which were organized with their supporting destroyers, cruisers, and battleships as Task Force 58, and the whole was placed under the command of a senior naval aviator, Vice Admiral [later Admiral] Marc A. Mitscher.

This was regarded as a striking force whose first function was to destroy enemy air power by wide-ranging raids over a vast area; second, if an attempt were made by surface vessels to interfere with our landing forces, the fast carrier force was to engage, and, if possible, destroy the Japanese Fleet, or at least drive it off; and third, it was to assist the landings by close support if required. Such a force was self-contained and could be used either in co-operation with other Fifth Fleet elements or operated independently.

Usually, close air support was the function of escort carriers assigned to the amphibious force and organized into task groups within that force. This was true in all major landings. Again, the conduct of air operations was under a senior naval aviator who carried out orders received from the commander of the amphibious forces.

In the Marshalls, still a third form of naval air power played its part. For weeks before the assault, navy search and photographic planes, based in the neighboring Gilberts, had reconnoitered enemy shipping, carefully noting ship movements, checking on the condition of beach defenses and airfields, and generally serving as the eyes of the fleet. In this they had co-operated with marine dive-bomber squadrons and army heavy and medium bombardment groups. The units of three services had been placed under a single command and given the designation of Task Force 57 to indicate their connection with the Fifth Fleet.

Independence-Class Carrier (CVL). Built on cruiser hulls as a wartime conversion, teamed with the Essex-class CV's in the Pacific offensive. With functions similar to those of Task Force 57 but of a more permanent nature, were the sea frontiers. These were coastal defense areas set up in such a way that they were coterminous with similar zones created by the Army and permitted a coordinated defense of the United States and its possessions. There were the Eastern, Gulf, Caribbean, and Panama sea frontiers in the Atlantic and the Western, Northwestern, Hawaiian, and Philippine, in the Pacific.

Each was organized as a task force with air units fitted into the structure as subordinate task groups. Because the sea frontiers early proved their value in anti-submarine warfare, additional ones were set up in Morocco and Alaska, and the Philippine Frontier was revived in November 1944. Building a task force was a game played with pre-fabricated units. If the air support plan called for half a dozen carriers, they were assigned from the units available, it being assumed that all carriers were alike. Although individual differences did exist, of course, actually there was great uniformity among units of a like type and they could be substituted for one another with a minimum of difficulty. Task forces in operation had to be supplied and their equipment maintained.

The commander wanted to be certain that when he gave the order, planes would rise into the air and would not just sit idle while someone looked around for a few spare parts, or sent an urgent dispatch for the gasoline that hadn't arrived. Should some scientist develop a gadget to destroy parked planes more effectively, it was important that it be installed in the fleet right away, not next year. As men and machines began to show signs of combat fatigue, they had to be replaced.

Meeting the needs of the operating forces was the work of administrative commands all the way from Washington to Guam or Morocco. To attempt a complete description of what they were and all they did would take a volume in itself, but some knowledge of their structure and function is necessary to an understanding of naval aviation. In the Navy Department were the bureaus — essentially procurement agencies — that supervised the development and manufacture of equipment. Especially important to aviation, of course, was the Bureau of Aeronautics. All the others also supplied equipment and services that were needed by the air organization. The work of the bureaus was coordinated by the Chief of Naval Operations upon whom fell the main burden of logistical planning. As of August 1943, there exists a Deputy Chief of Naval Operations (Air) who performs CNO's aviation functions.

Within the various fleets, organization was by types, i.e., there were commands for battleships, cruisers, destroyers, etc. Type commanders performed numerous functions including the very important one of seeing that all vessels under their command exhibited uniformity of equipment, training, and doctrine. Before the war, aviation was not concentrated into a single command in either the Atlantic or Pacific Fleet, a distinction being made between carrier and patrol planes. This situation caused some confusion in the early part of the war, and on 1 September 1942 Rear Admiral (later Vice Admiral) A. W. Fitch became Commander, Aircraft, Pacific Fleet.

Summarized briefly, his functions included the allocation and distribution of all planes, material, and aviation personnel through the Pacific area, the making of recommendations concerning types, numbers, and characteristics of aircraft required for current and projected operations, the advanced training and combat readiness of squadrons, and the preparation of tactical instructions and doctrine for all Pacific Fleet aircraft. He also had cognizance of all carriers, tenders, and other vessels assigned to the aeronautical organization and served as aviation adviser to the Commander-in-Chief, Pacific Fleet.

The success of ComAirPac in carrying out his numerous duties led to the formation, on 1 January 1943, of a similar command in the Atlantic. Because new carriers of the Essex and Independence classes were built and commissioned on the east coast, ComAirLant trained their air groups and gave the ships their initial shakedown. The very great success of our carrier aviation in 1944 was in real measure the result of what the boys had learned along the Atlantic coast.

Obviously, there existed numerous commands subordinate to ComAirPac and ComAirLant. The basic unit of naval aviation was the squadron, which exercised both operational and administrative control of its members. Although the majority of escort carriers (CVE's) had only a single squadron, a few of the larger ones and all fast carriers had two or more squadrons organized into air groups. All land-based squadrons, however, were assigned to fleet air wings.

Originally patrol wings, their designation was changed on 1 November 1942, when the wings' functions were extended from patrol planes to include all naval aircraft habitually flown from shore bases or tenders.

Beginning in September of the same year there was established in each wing a headquarters squadron whose function was to take over routine maintenance from the individual squadrons. This was done because the early months of the war had placed a great premium on mobility, and it was desired to streamline squadrons so that they could move all their personnel and gear in their own aircraft.

Maintenance was another responsibility of ComAirPac and ComAirLant. The same principle that was applied in wing headquarters squadrons was widely used elsewhere. There was a tendency to deprive operative squadrons of routine maintenance and to turn it over to specially created units. The most widespread were the carrier air service divisions (CASDiv's), which provided for the upkeep of planes on carriers, and the carrier aircraft service units (CASU's) which performed the same functions at shore bases.

Ultimately, CASU's in the forward areas had their responsibility enlarged to cover repair of all types of navy planes and were consequently renamed Combat Aircraft Service Units (F). Since present-day planes possess not only many thousands of parts but also a lot of highly technical gadgets, special mobile maintenance units were created to take care of certain instruments and other special items. Cursed when something went wrong, rarely thanked when everything functioned as it should, never sharing the glory of the headlines, members of these units struggled along frequently with inadequate equipment, improvising solutions to the most serious problems, fighting jungle heat and arctic cold and occasionally, on recently captured islands, enemy survivors. It was a hard, dull way to make war, but it contributed mightily to victory.

Marine Corps aviation had a parallel organization closely integrated with that of the Navy. The Director of Marine Aviation was a subordinate of the Deputy Chief of Naval Operations (Air), and elsewhere various marine commanders reported to their navy counterpart. For example, the Commanding General, Marine Aircraft Wings Pacific — later renamed Commander, Aircraft, Fleet Marine Force, — was under ComAirPac for logistical and material support. The internal organization of the marine air was based upon squadrons, two or more of which formed a group. Two or more groups were then put together to form a wing.

The above description of the aeronautical organization necessarily leaves out much. No mention is made of utility squadrons that provided target tow and pilotless drones for antiaircraft practice and performed other miscellaneous services as well. Nor does it consider the battleship and cruiser planes that were under the control of their respective ship commands. Since the Coast Guard in peacetime operates under the Treasury, it continued to be organized separately. All of these various units, however, had one thing in common — they depended for logistics on ComAirPac and ComAirLant.

MARC ANDREW "PETE" MITSCHER

PART III: THE ATLANTIC THEATRE

4 - IT TAKES PRACTICE TO WIN A WAR

IT IS CUSTOMARY FOR COLLEGE FOOTBALL TEAMS to engage in spring practice to prepare for the coming fall season. During this time fundamentals are stressed. The players perfect their knowledge of blocking, tackling, passing, and running with the ball. Basic plays are run over and over again until they become routine. Techniques of offensive and defensive action are studied on the field, through skull sessions, and blackboard talks.

The Navy had its preliminary practice period for World War II, the Neutrality Patrol, from September 1939, to December 1941. The analogy, of course, does not hold true in all respects. The Navy did not carry out its practice behind a high board fence on its own practice fields. It secured it training on a field on which a bitter struggle was already in progress. On the wide sweeps of the Atlantic, the Allied navies were making a desperate attempt to cut down the serious inroads that Nazi submarines and raiders were making on Allied supply lines. There was the constant possibility that the United States might be drawn into the fray as a combatant, and there were times when the practice seemed dangerously like the real thing.

On the last day of August 1939, word flashed across the Atlantic that the German military juggernaut had rolled into Poland, and war, so long feared and expected, became an actuality. Declaration of war by Britain and France two days later made it clear that the conflict would engulf most of Europe. The scene was set for a re-enactment, on a grander scale, of the great World War of 1914-1918.

Many factors were similar, if not identical. England, an insular nation, unprepared for war, would depend heavily upon supplies and munitions which could reach her only by way of the shipping lanes of the Atlantic. Most of these goods would have to come from the United States. Germany, lacking a surface navy large enough to challenge the British on the seas, would send out hundreds of U-boats in an attempt to send all shipping bound for the Allies to the bottom of the Atlantic. Surface raiders would add to the destruction, and the Atlantic would again become a crucial battleground. The United States, as in 1914, was neutral.

That fighting in the Atlantic would endanger our neutrality became evident, not only from our experience in World War I, but also from reliable information immediately at hand. On the very first day of the war, the Chief of Naval Operations in Washington informed our forces afloat that German submarines were set to operate on Atlantic trade routes and that a dozen German merchant vessels would operate as armed raiders.

The dispatch further stated that neutral merchantmen might expect Great Britain to institute similar practices as in the last war. It became the duty of this nation, as a neutral, to see to it that none of these activities was carried on within our territorial waters, and that none should interfere with our rights on the high seas.

Even before this edict was given to the world, the United States Navy, upon which most of the burden of enforcing our neutrality would inescapably fall, had swung into action.

On 4 September 1939, the Chief of Naval Operations sent a dispatch to Rear Admiral A. W. Jackson, commander of the Atlantic Squadron, ordering him to establish as soon as possible a combined air and ship outer patrol for the purpose of observing and reporting in cipher the movements of warships of the warring nations. The patrol was to extend east from Boston to latitude 42-30, longitude 65, then south to latitude 19, then around the seaward outline of Windward and Leeward islands to the British island of Trinidad, near the shore of South America.

The following day a second dispatch ordered the patrol "to observe and report in a confidential system the movement of all foreign men-of-war approaching or leaving the east coast of this country or approaching or leaving the east coast of the Caribbean."

The limits of the patrol were set about three hundred miles off the eastern coast line of the United States and along the eastern boundary of the Caribbean Sea. On the same day a surface patrol of destroyers and Coast Guard craft was ordered established to patrol across the steamer lanes to the southward of Grand Banks, operating directly under the Chief of Naval Operations.

Thus, was envisioned a sentry line of ships and planes along our Atlantic frontier from the northernmost approaches southward to and around the islands bridging the entrances to the Caribbean, ending at the island of Trinidad off the northeast coast of South America.

That such a line would be thin was evident in view of the forces at that time assigned to the Atlantic squadron. Most of our naval strength, undersea, surface, and air, was in the Pacific.

Forces available were, briefly, as follows: Battleship Division 5, composed of USS Texas, USS New York, USS Arkansas and USS Wyoming, together with their air unit, Observation Squadron 5 (this nine-plane squadron was not commissioned, however, until 16

October 1939); Cruiser Division 7, composed of USS San Francisco, USS Tuscaloosa, USS Vincennes, USS Quincy, and USS Wichita, together with their air unit, the sixteen planes of Scouting Squadron 7 (Wichita did not report until 25 September); the aircraft carrier USS Ranger, at that time engaged in training its own air group and that of USS Wasp, which had not yet been commissioned; approximately forty destroyers of all types and ages in commission; an undetermined number of old destroyers to be recommissioned; about fifteen old submarines; and five squadrons of patrol planes, totaling about fifty-four. This was exclusive of the small force assigned to the Panama Canal Zone.

The patrol planes mentioned were Patrol Squadrons 51 (12 PBY-1 planes), 52 (6 P2Y-2 planes), 53 (12 P2Y-2 planes), and 54 (12 PBY-2 planes), all in Patrol Wing 5 and Patrol Squadron 33 (12 PBY-3 planes) of Patrol Wing 3. Four seaplane tenders were attached to the wings.

Immediately upon receipt of the orders from Washington, Rear Admiral Johnson put his forces in action. Cruisers, destroyers, and seaplane tenders put to sea from Norfolk and other Atlantic ports. Lumbering patrol seaplanes churned the waters of eastern bays taking off for assigned bases or patrol stations. On September 6, Admiral Johnson informed the Chief of Naval Operations that the patrol was beginning to function, and by 20 September, when Commander, Atlantic Squadron's Operations Order No. 20-39 became effective, our Atlantic coastal waters from Nova Scotia in the north to the southern tip of the Lesser Antilles were under daily surface and air surveillance.

A brief summary of the patrols indicates the close co-operation between navy sea and air forces. Four destroyers together with Coast Guard units patrolled the steamer lanes south of Grand Banks, and two destroyers worked out of Boston. Patrol Squadron 54 flew daily air searches from Newport, Rhode Island, in connection with a destroyer surface patrol. Patrol Squadrons 52 and 53 teamed up with destroyers to cover the waters adjacent to Norfolk.

Two destroyers based at Key West, Florida, patrolled the Florida Straits, Yucatan Channel, and neighboring waters. The Guantanamo-San Juan area was covered by surface forces consisting of destroyers and the two cruisers USS Tuscaloosa and USS San Francisco, together with air patrol composed of Patrol Squadron 33 based at Guantanamo and Patrol Squadron 51 based at San Juan.

Two cruisers, USS Quincy and USS Vincennes patrolled the general sea approaches between Newport and Norfolk. Battleship Division 5 and USS Ranger, together with shore-based air detachments made up a reserve group based at Norfolk and engaged in training and outfitting.

This, then, was the original deployment of forces for the Atlantic patrol. Within a month it had begun the expansion which was to continue until the United States entered the war. In October, Patrol Squadron 52 moved to Charleston, South Carolina, and a detachment of Patrol Squadron 33 moved to Key West to fill gaps in the coverage of our southern Atlantic coast line.

USS Ranger joined the ships of Cruiser Division 7 to form a striking group capable of long-range searches. In November 1939, a surface patrol of destroyers was established in the Gulf of Mexico to watch suspicious craft in that area. The U. S. Coast Guard conducted inshore patrol operations with both surface and air craft, and co-operated fully with the Navy by consultation and exchange of information.

Whatever the Neutrality Patrol became in the months immediately preceding the attack on Pearl Harbor, it was not, in the beginning, a warlike or offensive project. Its original mission was purely one of observation and reporting, and every precaution was taken to minimize hazards to personnel, planes, and ships in this peacetime patrol. At the same time, every effort was made to avoid performing unneutral acts or giving the appearance of performing such acts, and ships and planes were instructed to exercise care, when approaching foreign vessels, to avoid any action which might be interpreted as being of hostile intent.

Early overall orders to the patrol were necessarily general and lacking in clear detail, due to the speed with which operations were set up and the lack of experience upon which definite procedure could be based.

In the beginning, only foreign men-of-war were to be reported, and no orders were issued for trailing ships observed by our forces. Commander, Atlantic Squadron's Operation Order No. 20-39 directed the units of the patrol to investigate reports of submarine operations, perform missions of mercy on the contiguous high seas, prevent engagements between hostile belligerents within territorial waters of the United States (fortunately no such engagements occurred), exchange such information among task groups as became necessary for co-ordination, train and indoctrinate personnel in the requirements of neutrality patrols, and conduct gunnery and other training as conditions permitted.

Operations were to be modified during approach of bad weather to insure safety of personnel and material, and normally not more than one-third of air units and one-half of surface units were to be employed on patrol at a time.

These instructions were soon expanded. The commander of the Atlantic Squadron, on 16 October 1939, issued another order (No. 24-39) that expanded patrol activity. Not only were foreign men-of-war to be reported, but "suspicious" vessels were to be noted, and both types of vessels were to be trailed until their actions were satisfactorily accounted for. As time went on the patrol expanded and increased the extent and intensity of its activities, but the basic method remained the same, and the purpose did not change until the months immediately preceding the entry of the United States into the war.

All units of the Atlantic Squadron were included in the patrol task organization, but the destroyers and patrol planes bore the major part of the actual patrol operations.

The battleships were not placed on the patrol line at any time, and while Cruiser Division 7 actually began the period by patrolling between Newport and Norfolk, it was soon withdrawn, the San Francisco going to the Pacific and the other cruisers going to duty in the San Juan-Guantanamo area.

What patrolling it did subsequently was broken up by Naval Reserve cruises, fleet practice exercises, and special missions. Battleship Division 5 was occupied for the most part with Reserve cruises and intensive training.

In general (except in areas where no aerial patrol existed, such as Grand Banks and West Gulf), patrol planes were used for observation and search, while destroyers patrolled centrally in areas or sub-areas prepared to develop fully any air contacts.

THE YEAR 1940 WAS A YEAR OF STEADY OPERATION, training, and expansion in the Atlantic Neutrality Patrol. Assigned areas were covered constantly, but due to our position as a neutral, and the restriction of the patrol to the waters near our coast, little of excitement occurred. For the most part it was dull, arduous duty for seamen and aviators.

This is not to say that the patrol did not perform a vital function. Its mission was an essential one, and it was performed with fidelity and skill. Extended overwater flights day after day and month after month, and long stretches at sea without rest were exhausting to our men, but they kept the war at a distance and protected our shipping.

Moreover, the patrol trained our forces for the struggle to come. Concurrently with the patrol, the intensive training program advanced our readiness for war, the best training being obtained on the flight tracks and patrol lines. When war finally came to us, our patrol organization had the advantage of more than two years of intense activity under conditions closely paralleling those of war.

During 1940, the patrol was augmented as rapidly as practicable. Old destroyers were recommissioned, and new patrol plane squadrons were added. On 1 August 1940, Patrol Squadron 55 was commissioned at Naval Air Station, Norfolk, and on 1 October 1940, Patrol Squadron 56 was commissioned at the same station. Both began intensive training preparatory to joining the patrol.

On 1 November 1940, Atlantic Squadron became Patrol Force, U. S. Fleet, and an additional division of cruisers was assigned. On 17 December 1940, Rear Admiral (now Fleet Admiral) E. J. King relieved Rear Admiral Hayne Ellis as Commander, Patrol Force. (Rear Admiral Ellis had relieved Rear Admiral Johnson as Commander, Atlantic Squadron on 30 September 1939.)

New duties loomed ahead. Events in Europe during 1940 had altered the whole war outlook for the United States. The fall of France in June followed by the aerial blitz on London and other English cities reduced British fortunes to such an ebb that her defeat and the surrender of her fleet to Germany became distinct possibilities. Dangers to America inherent in such an event were too apparent to be ignored. Our concern in the Atlantic became not only the preservation of a precarious neutrality, but self-protection against possible aggression.

In September 1940, the famous destroyers-for-bases transaction was made between the United States and Britain, by which the United States obtained long-term leases on eight bases in the Atlantic and Caribbean areas in exchange for fifty old destroyers. Included in the bases thus obtained were Argentia (Newfoundland) and Bermuda, both destined to play major parts in Atlantic Patrol activities in 1941.

On 1 February 1941, the United States Atlantic fleet was established under command of Admiral E. J. King, USN. It was composed of patrol force units greatly augmented and completely reorganized. With these forces the patrol of the Atlantic was extended greatly to give more complete protection to the supplies which were flowing in ever-increasing quantities from our eastern and southern ports to the United Kingdom. In addition to the coastal patrols which had been in operation since late 1939, long-range patrols and convoy escorts were set up.

For purposes of these operations the Atlantic was divided into three sectors. The trade routes to northern Europe were patrolled by Task Force 1, composed of battleships, cruisers, and destroyers. Task Force 2, composed of cruisers, aircraft carriers and destroyers covered the central North Atlantic. Task Force 3 composed of cruisers, destroyers, and mine vessels based at San Juan and Guantanamo patrolled the South Atlantic, pushing ever farther south as trade and war operations moved in that direction, until, by the time the United States entered the war, this force was patrolling South American waters, basing at Recife.

One of the most important developments in the new organization of Atlantic forces was the creation, on 1 March 1941, of Support Force, U. S. Atlantic Fleet under command of Rear Admiral A. L. Bristol, USN. Support Force (designated Task Force 4) was composed of destroyers of Support Force, and Patrol Wing of Support Force, the latter initially under command of Commander (later Rear Admiral) H. M. Mullinix. Patrol Wing of Support Force consisted originally of Patrol Squadrons 51, 52, 55, and 56, and the tenders USS Albemarle and USS Geo. E. Badger. On 5 April 1941, Patrol Squadron 53 was added.

The original directive ordering establishment of Support Force stated in part: "It is desired the Force be prepared for distant service in high latitudes and that it be given intensive training in anti-submarine warfare, the protection of shipping, and self-defense against submarine, air, and raider attack." Support Force was intended to be developed into a powerful, well-trained, highly mobile force to operate from bases in the North Atlantic to prevent Germany from cutting communications between the United States and Great Britain.

The original offshore patrol, now extended to include air surface forces operating from Coco Solo, Canal Zone, was regrouped as Task Force 6, and that portion of it based at points north of the Caribbean and Gulf areas was designated Northern Patrol. The mission of the Northern Patrol was to "investigate reports of potential enemy vessels and movements and of other non-American activities in the North Atlantic. Operating bases at Norfolk, Bermuda, Narragansett Bay, and, in due course, Argentia (Newfoundland)."

Support Force, Atlantic Fleet, was assigned as a part of this Northern Patrol, and it was in this assignment that Patrol Wing, Support Force spearheaded the advance of naval aviation to the strategic islands to the north and east, and helped insure the safe passage of materials of war to Britain.

The advance northward did not begin at once, however. The early months after the creation of the wing were devoted to intensive training in co-operation with other units of Support Force. This training involved convoy and escort operations under conditions simulating war, and it involved periods of arduous duty with routine patrols along the Atlantic coast. Patrol Squadron 51 was established at Naval Air Station, New York (Floyd Bennett Field) on 8 April 1941. Patrol Squadrons 52 and 53 were established at Naval Air Station, Quonset Point, R. I., on 3 April 1941 and 24 May 1941, respectively.

Next came the movement to Newfoundland, to the strategically located base of Argentia, recently obtained from Britain in the destroyer-for-bases agreement. With this movement began a series of advances more important, perhaps, than any other development in the Atlantic prior to Pearl Harbor. From Newfoundland our planes could patrol farther out over the major Atlantic shipping lanes than from any continental base; and, later, from Iceland, to which Newfoundland was a steppingstone, aerial patrols could, if necessary, reach the British Isles.

On 15 May 1941 the wing flagship, USS Albemarle, arrived at Argentia, laid thirteen heavy plane moorings, obtained weather analyses, and prepared for the commencement of seaplane operations. Three days later, on 18 May 1941, planes of Patrol Squadron 52 arrived at Argentia and commenced air operations. Argentia remained the principal base of the wing until July 1943, when operations were largely transferred to bases in the United Kingdom.

Operations at Newfoundland were rendered difficult and hazardous by the rugged contour of the island, rudimentary local communications, bleak and severe climate characterized, according to the season, by severe storms, gales, high winds and/or dense fog. Icebergs, in season, and local sheet ice which forms in winter on all quiet bodies of water, presented additional hazards. Land base facilities were in early stages of construction, and the wing was, therefore, largely dependent upon equipment which USS Albemarle brought with her. It was rough, hazardous duty, lacking entirely the glamour of open combat, but calling for skill, courage, and stamina in the constant battle with the elements.

On 24 May 1941 came a dramatic assignment. The German battleship Bismarck had engaged and sunk HMS Hood in the Strait of Denmark. In the general search which followed, planes at Argentia were ordered to make immediate aerial reconnaissance over a sector five hundred miles southwest of Cape Farewell, Greenland. Despite foul weather and extremely bad flying conditions, the search was flown. Neither the Bismarck nor any other vessel was found, however. It was learned later that the Bismarck had turned southeast after her engagement with the British and did not pass through the area covered by the patrol.

In the meantime, the eleven PBY-5's of the patrol came to grief. Buffeted about and thrown off course by the rough weather, the planes became separated and lost, and none made it back directly to Argentia. After extensive flights, the planes came down separately in various bays in Newfoundland, Labrador, Quebec, and adjoining islands. None was destroyed or seriously damaged, and all were ultimately flown back to base.

On 1 July 1941, in a general reorganization of patrol planes, Patrol Wing, Support Force became Patrol Wing 7 (remaining a unit of Support Force), and its squadrons 51, 52, 53, and 55 became Patrol Squadrons 71, 72, 73, and 74, respectively. Squadrons 71, 72, and 73 had 12 PBY planes each, and Squadron 74, after receiving three planes transferred from former Squadron 56, had 12 PBM-1's. Remaining planes of former Squadron 56 became a training unit under Commander, Patrol Wings, Atlantic Fleet.

At the same time, Patrol Wing 8 was formed, composed of redesignated squadrons and new squadrons yet to be commissioned. Patrol Wing 8 was initially attached to Support Force, but saw little actual duty, other than training, with that organization. With the outbreak of war with Japan the new wing was ordered west for Pacific duty.

The summer of 1941 was a busy period for the planes stationed in Newfoundland. In addition to routine patrol and escort missions, our fliers conducted long-range reconnaissance flights, surveys of Iceland, Greenland, and Labrador, ocean rescue operations, and various flights for other forces, services, and governments.

In August 1941, President Roosevelt and Prime Minister Churchill held their historic six-day conference at Argentia aboard USS Augusta, from which emerged the Atlantic Charter. Before, during, and after this meeting the planes of Patrol Wing 7 flew a very heavy reconnaissance schedule guarding Argentia and its sea approaches to the limits of plane endurance. No enemy vessels were sighted.

From 6 August to 20 August 1941 a three-seaplane detachment of Patrol Squadron 71, in conjunction with U. S. Army representatives, carried on aerial surveys of Greenland, basing on the USS Lapwing in Tungdliafik Fjord on the west coast.

In this month came another long step in the advance of our patrol forces to the north and east. On 6 August 1941, six PBY-5's of Patrol Squadron 73 and five PBM-1's of Patrol Squadron 74 landed and came to rest on the waters of Skerja Fjord, near Reykjavik, Iceland. Moorings were borrowed from the Royal Air Force, which also operated seaplanes there, and operations were begun at once, the planes basing initially on USS Goldsborough. Convoys were covered as far as five-hundred miles from base, and air patrols were maintained in the Denmark Strait to Greenland.

If conditions at Argentia had been bad, those in Iceland in this pioneer period were infinitely worse. Living accommodations and other base facilities were practically nonexistent. At the outset as many as possible of the squadron personnel crowded aboard the Goldsborough for berthing, while the balance slept in the planes, in tents ashore, and a few in huts loaned by the Royal Air Force.

Base construction was begun at once, but was kept at a slow pace by lack of material. Iceland itself afforded practically nothing in the way of building supplies, and harbor congestion at Reykjavik was such that ships bringing vital material from the States frequently were forced to lie offshore for over a month awaiting their turn to unload.

Flight operations, while pressed energetically, were hampered by this same lack of supplies and made hazardous by "the worst weather in the world," characterized, even in the less severe seasons, by high, frequently shifting winds and quick, unpredictable changes.

By this time the functions of the Neutrality Patrol had expanded greatly, and the United States was drawing near the "shooting war" soon to be upon us. For several reasons this was especially true in the northern Atlantic.

In the first place, the vast bulk of lend-lease material going from America to Britain was transported by the northern route. Our policy of defending America through lend-lease and other support to Britain and the other nations fighting Germany could be effective only if most of these supplies reached their destination. Expansion of the patrol activities of the United States forces helped to assure safe delivery of goods and eased the burden on British forces protecting Atlantic sea lanes.

In the second place, development of sea and air bases in Newfoundland, Greenland, and Iceland, together with maintenance of air and sea patrols therefrom, were intended to prevent the Germans from developing a strong position in the North Atlantic or seizing any part of the strategic islands on the northwest approaches to the

United States. Attacks on these islands without warning or declaration of war were within the realm of probability, in the light of German activities of the past. The fall of Denmark underlined this danger and made the islands of Iceland and Greenland potential springboards for a grand invasion of America.

These and other considerations, although we were technically not at war and were ostensibly protecting our neutrality, led, in 1941, to great expansion of the geographical limits of our patrols and sharp intensification of their missions.

An operation plan issued by Commander, Argentia Air Detachment on 12 September 1941 and revised 23 October 1941 defined the western Atlantic area in which the planes of the force were to operate as "that area bounded on the east by a line from the north along 10 degrees west as far south as latitude 53 degrees north, thence by thumb line to latitude 53 degrees north, longitude 26 degrees west, thence south, and extending as far west as the continental land areas but excluding Naval Coast Frontier and Naval District land and water areas and Canadian coastal zones, and the territorial waters of Latin-American countries."

The mission of the force was, within this area, to

"(a) Protect United States and foreign flag shipping other than German and Italian against hostile attack by:

(1) Escorting, covering, and patrolling for the defense of convoys, and by

(2) Destroying German and Italian naval, land, and air forces encountered.

(b) Insure the safety of sea communications with U. S. strategic outposts.

(c) Support the defense of U. S. territory and bases, Iceland, and Greenland.

(d) Trail merchant vessels suspected of supplying or otherwise assisting the operations of German or Italian naval vessels or aircraft.

Maintain constant and immediate readiness to repel hostile attack. Operate as under war conditions, including complete darkening of planes when on escort duty during darkness, varying plane altitudes as necessary."

On 1 October 1941, four PBY-5 planes of Patrol Squadron 71 arrived at Kungnait Bay, Greenland, based on USS Gannet (on temporary duty from Patrol Wing 5) for patrols of the Greenland coasts. Due to the uncertainties and violence of the Greenland weather, regular convoy escort patrols were not feasible, and the detachment returned to Argentia on 18 October 1941.

Such was the story of our Neutrality Patrol, convoy coverage, and preparation for war in the North Atlantic.

At the same time ships and planes assigned to other areas of the Atlantic were equally vigilant and active. From bases extending down our Atlantic and Gulf coasts to the Canal Zone, and from West Indian bases as far south as Trinidad, surface craft and patrol planes of the Navy and Coast Guard were combing our coastal waters daily. Suspicious ships were being searched out, trailed, and prevented from fueling or tending submarines.

Bermuda, one of the bases obtained from the British, had become a highly strategic outpost and trans-Atlantic steppingstone. From it our planes were flying daily patrols in co-operation with surface forces. Long-range patrols of heavy units were roaming the high seas, observing and gathering information.

Thus, by the time we were called upon for full participation in the war, the western Atlantic was, in effect, an American ocean. The Neutrality Patrol was an operation devoid of the glamour of combat, but in its twenty-six months of existence it had developed into a vast coordinated project which exerted a profound immediate and long-range effect on the war. In the early days it safeguarded our neutrality and guaranteed the sanctity of our home waters. Later it safeguarded our northern and southern flanks from encroachments of a potential enemy and assured delivery of supplies which kept our future ally in the war. Throughout, it advanced our state of readiness by training personnel and making ever-increasing demands for procurement and improvement of ships, planes, and other matériel.

In the Neutrality Patrol, from beginning to end, naval aviation played a major role. Without land-based patrol planes, carrier planes, and the scouting and observation planes aboard our cruisers and battleships, such an effective patrol could not have been developed. On the other hand, without the stimulation and training of Neutrality Patrol activity, naval aviation in the Atlantic would not have been in a state of readiness when war came.

The Jap attack on Pearl Harbor ended the Neutrality Patrol as such. But only the name died. The forces of the patrol, unhampered by the word "neutrality" entered into a new, more dangerous task — the final defeat of the German submarines.

5 – ANTI-SUBMARINE WARFARE

THE NEUTRALITY PATROL IN THE ATLANTIC from 1939 to 1941 had given considerable training in certain basic techniques of air war. Unfortunately, this training was available only to the comparatively small forces we had during that period. When war actually came and Neutrality Patrol became anti-submarine warfare in earnest, our forces, experienced though they were becoming, were forced on the defensive by virtue of their small numbers.

Consequently, anti-submarine warfare in the Atlantic fell into two main phases. From December 1941, to December 1942, we were without question on the defensive. During this period our shipping losses on the average exceeded our construction figures. The second phases, extending roughly from December 1942, to December 1943, will be taken up later and was a period of gradually mounting offensive.

There was no doubt which opponent was better prepared for the struggle in December 1941. Within five weeks after Pearl Harbor, German submarines moved into American coastal waters and sank an American tanker 150 miles south of Montauk Point, Long Island. In the ferocity of the opening attack, in the month of January, 14 ships were sunk off the east coast and 12 in Canadian waters.

Unfortunately, in spite of the training afforded by the Neutrality Patrol, the United States was ill prepared to counter the first blow. There was nothing to do but extemporize and try to hold out until the huge training program and production schedules could get under way, and start an adequate flow of men and materials to the field.

The American northern air defense consisted of Catalinas of Patrol Wing 7 based at Argentia, Newfoundland, and Reykjavik, Iceland. The pilots braved the terrible North Atlantic winter for hours on end to try, with their inadequate anti-submarine tools and techniques, to protect shipping on the northern convoy route.

To the south, Commander, Patrol Wing 3, based at Coco Solo, Canal Zone, sent his patrol planes sweeping over both eastern and western approaches to the Panama Canal in search of enemy surface raiders, invasion forces, aircraft, and submarines.

In the center of the line, planes of Patrol Wing 5, based at Norfolk, Virginia, attempted the overwhelming task of trying to cover from various hastily arranged advanced bases the whole eastern coast line of the continental United States.

As can be imagined, the mad scrambling of helpless merchant vessels and the blind punching of the pitifully few air and surface craft had little deterring effect upon the German U-boat commanders who sent seventeen merchant vessels in continental waters to the bottom. By February, a large-scale U-boat assault, which saturated the defenses of the Eastern Sea Frontier, carried into the Caribbean and Gulf and resulted in the destruction of forty-two merchantmen with practically no damage inflicted on the enemy.

In March, the main weight of the U-boat attack shifted north off the continental United States, and despite all that navy planes of Patrol Wing 5 could do, twenty-eight merchant vessels were sent to the bottom. These attacks were marked by cool, deliberate ferocity on the part of U-boat commanders. Submarines, apparently completely without fear of reprisal, surfaced and sank ships by shellfire even within sight of the coasts.

Two moves were instituted by the Navy in April in an attempt to stem the wave of almost unrestricted sinkings in American waters: the system of convoying coastal shipping, and the establishment of the sea frontier commands. Every available type of patrol vessel, including converted yachts and civilian craft, was pressed into service.

Likewise, the numerically small Anti-Submarine Air Squadrons composed of navy Kingfishers and Catalinas of Patrol Wings 3, 5, and 7 were augmented by Coast Guard, civilian, and army aircraft. The I Bomber Command, whose squadrons flew high-speed, land-based Liberator bombers, was assigned to anti-submarine duty in co-ordination with the Navy.

Eastern Sea Frontier, Panama Sea Frontier, Caribbean Sea Frontier, and Gulf Sea Frontier commands were set up to co-ordinate the efforts of these diverse branches of the service. Theirs was the task of directing the flow of men and materials, soon to come, into channels where they could best be employed to drive back the U-boat and win the crucial Battle of the Atlantic.

The effect of these two moves was not, of course, immediately apparent. Sinkings continued at a high rate in the coastal and Caribbean areas, but the U-boat was beginning to pay a price for its successes. Nineteen attacks, resulting in three U-boats damaged and one sunk, were made by navy, Coast Guard, and army aircraft in U. S. coastal waters from Newfoundland to Florida.

By May, new anti-submarine measures and killer teams of air-surface forces made it so difficult for U-boats to operate off the east coast that the enemy once again shifted his attention to the lightly defended southern approaches to the United States. In that month, the Caribbean and Gulf of Mexico were turned into a flaming hell of burning tankers and merchantmen shattered by U-boat torpedoes and gunfire.

In all, seventy-two American and Allied ships were destroyed in that theater by U-boats which suffered negligible damage themselves.

Following up their success of May, Hitler's U-boat fleet moved deeper into the Gulf and sank thirteen merchant vessels in the eastern approaches to the Canal Zone. This region was being patrolled at the time by Catalinas and Kingfishers of Patrol Wing 3 and by Mitchells of the Panama Army Command. These aircraft made a few sightings but were unable to attack the enemy or to prevent him from torpedoing (mostly at night) the helpless merchant vessels.

At the same time, squadrons operating on the Caribbean and Gulf Frontier regions were again unable to cope with the aggressive U-boat campaign which resulted in the sinking of forty-eight merchantmen and tankers bound for New Orleans, Galveston, and the Canal Zone. Air opposition continued light; only seventeen attacks were made by United States aircraft resulting in four U-boats damaged.

As time passed, however, and as the production of air and surface craft increased, it became possible, after convoy requirements for transatlantic shipping were met, to discover more and more attention to the protection of coastwise sea lanes which could be continuously swept by air and surface craft.

By the middle of July, convoying of merchant vessels in the Caribbean and Gulf, as well as in east coast waters, was instituted. This move was followed by a considerable decrease in shipping losses in southern waters.

No sooner had pressure been released from this part of the defense line, however, then the enemy struck the northern convoy routes between Iceland and Newfoundland. In this windy, fogbound region where planes of Patrol Wing 7 were flying their interminable patrols, seven merchantmen were torpedoed and sunk. At least one submarine responsible for this activity was discovered, however, by a Catalina of VP-73 and sunk under rather unusual circumstances. One of the PBY's depth charges landed on the U-boat's deck and stuck in the grating.

The bomb, of course, would have been harmless until it reached a fixed depth under water, but instead of attempting to disarm it or float it off on a raft, the charge was rolled over the side by an inexperienced seaman and promptly blew up the U-boat. Prisoners picked up later by a surface craft evinced our first definite U-boat "kill" of the war.

In August, the enemy who had pulled back to regroup his forces in the preceding month lashed out viciously with the greatest offensive yet mounted in the West Indies area. Availing themselves of the innumerable bays, inlets, and harbors of the Caribbean for concealment, and the rather localized channels through which Allied convoys passed for points of attack, Nazi U-boats succeeded in sending thirty-three merchant vessels to the bottom. Most of these attacks were made by surfaced U-boats against merchant ships proceeding unescorted.

An American countermove, which had been some time in the making, was the establishment of Patrol Wing 11 under the jurisdiction of the Caribbean Sea from Command at San Juan, Puerto Rico. The purpose of this step was to intensify the war against the U-boat in the area from Cuba to Natal, Brazil, and to afford greater protection to convoys passing through the islands.

Six days after the arrival of navy patrol squadrons at San Juan, the first attack on a U-boat was made. This was followed by other attacks from time to time, though few were assessed as damaged and none as sunk or probably sunk. To make these contacts with the enemy, navy planes with the supporting army units had to fly patrols to the coast of South America and far to the east of the Windward and Leeward islands.

At approximately the same time, a detachment of Patrol Wing 11 was sent to Trinidad to deal with a group of submarines attacking shipping off the coast of northern Brazil. Before the air patrols could be effectively set up, seven ships had been sunk.

On 16 September 1943, in order to provide closer supervision of anti-submarine air coverage under the operational control of Gulf Sea Frontier, Patrol Wing 12 was commissioned. Administrative headquarters of the wing was set up at Key West. Elaborate block patrols of the Florida Straits, Bahama Channel, and the Yucatan Channel were flown by aircraft of the wing from Key West and San Julian, Cuba bases, to prevent enemy U-boats from using southern coastal waters for their depredations.

September saw a great decline in U-boat activity in American waters, the only regions under attack being the northern and southern approaches to the United States. This happy state of affairs continued throughout October when little if any U-boat effort was exerted in the American theater. Except for a sudden offensive against shipping in the lightly defended Brazilian coastal zone, out of range of planes of Patrol Wing 11, in November and December 1942, no enemy U-boat fleets ever thereafter successfully invaded American waters, patrolled by aircraft of the patrol wings (later called fleet air wings) 3, 5, 9, 11, and 12.

This securing of the approaches to the Western Hemisphere from the deadly menace of the German U-boat did not satisfy the American government. Plans which called for the reconquest of Africa and operations in the Mediterranean required the establishment of land bases for American aircraft farther afield. The blow was to fall early in November.

Through waters infested by German U-boats, as we shall see later in more detail, a huge American invasion fleet pushed its way, and on 8 November, in spite of all the enemy could do, our troops poured ashore at Mehdia [or Mehdya], and Safi in French Morocco.

As soon as hostilities ceased three days later, two navy patrol squadrons, VP-92 and VP-73, began anti-submarine operations out of Casablanca and Port Lyautey. These operations entailed close-in patrols off the three beachheads in order to frustrate desperate German attempts to disrupt the landing and supplying activities of the fleet. Actually, the invading forces encountered relatively light opposition from French shore establishments and German submarines.

During the whole operation, 9 ships and 2 escort vessels were sunk, and 5 merchant vessels damaged by U-boats. The worst loss occurred on the night of 11 November when 4 merchant vessels and 1 destroyer were torpedoed off Fedala. Counterattacking American surface, and British air forces in the invasion period sank 2 U-boats and made seven damaging attacks.

During December, after the urgency of the invasion began to give way to a consideration of more long-range anti-submarine problems, the U. S. Navy squadrons based at Casablanca and Port Lyautey instituted a series of "canned" patrols designed to cover thoroughly the area within three hundred miles of Casablanca. Six attacks were made before the end of the year.

These anti-submarine activities were soon to be directed and administered by Commander, Fleet Air Wing 15, whose unit was commissioned in Norfolk on 1 December 1942.

Thus, as 1942 came to an end, what at first had seemed a picture of utter chaos gradually took on a semblance of shape and order and the outlines of a definite plan of campaign began to appear. An umbrella of aerial protection, spread over the western Atlantic from Brazil to Iceland, had effectively driven the U-boats far out to sea.

Wherever a land base could be established, the United States Navy put up an aerial barrier impenetrable to more than an occasional undersea raider. This cover now stretched from Iceland to Newfoundland, down the Atlantic coast, through the West Indies, and along the coast of South America from Panama to Brazil. Thence it leapt the South Atlantic, pausing at Ascension Island, and passed up the West African coast to the Straits of Gibraltar.

This changed U-boat situation was due in part to rapidly completed production schedules of ships and planes and to an accelerated pilot training program. But more important were the improved anti-submarine weapons and techniques which had been developed since the hectic days of the spring of 1942. Doctrine for attacks on U-boats by aircraft was gradually evolving as a result of the successes and failures of army, navy, Coast Guard, RAF, and British Coastal Command aircraft in more than two years of anti-submarine warfare.

New and improved aircraft were coming into operation: powerfully armed and long range PB4Y's, fast maneuverable PV-1's, very long-range Mariner flying boats, and navy blimps. Moreover, the Navy was beginning to use each anti-submarine aircraft in a specialized capacity designed to wreak the maximum damage on the enemy.

Also, these planes were going into action armed with far more efficient U-boat detection devices and more powerful weapons than had been available a few months earlier.

Land-based aircraft, however, did not supply the whole answer to the problem, effective as it was within its limitations. Though the enemy was driven offshore, he was not forced off the seas. He still sat astride the mid-Atlantic sea lanes, out of range of land-based aircraft, and by operating in wolf packs he succeeded in exacting a heavy toll of merchant ships, even in convoys.

The answer to this was the escort carrier — small aircraft carriers converted from former merchant hulls or built from the keel up. They were designed to carry enough planes, generally Wildcat fighters and Avenger bombers, to afford protection for the convoys passing through the U-boat hunting grounds.

In addition, CVE's with a supporting group of destroyers or destroyer escorts were to be used eventually as hunter-killer teams to root out and destroy enemy submarines wherever they might be. By the end of 1942, United States shipyards were beginning to turn out escort carriers in sufficient numbers to bode ill for future enemy operations in the Atlantic theater.

With two notable exceptions, United States aircraft carrier operations in the Atlantic, European waters, and the Mediterranean were the exclusive responsibility of small escort carriers. The exceptions were the first USS Wasp which was later sunk off Guadalcanal in September 1942, and the USS Ranger. Both ships made their important contributions to victory over the European end of the Axis by a series of aircraft ferrying missions in addition to anti-submarine patrols and offensive strikes against the enemy.

The stories of Great Britain's battles against a vastly superior enemy from the fall of France to the entrance of the United States into the war are legion. But none are more impressive than those of the little island of Malta and the convoys sent to its relief. Malta's importance in preventing complete Axis control of the Mediterranean cannot be overemphasized. With pitifully few obsolete planes for its defense, the problem of keeping the island supplied with food, munitions, fuel and aircraft was acute.

There were no British bases within range of Malta from which fighter aircraft replacements could be sent, while Graziani and Rommel controlled the nearest points on the North African coast. The only way fighter planes could be delivered was by ship. But the fact that every Malta convoy was subjected to severe submarine, surface, and air attacks made shipment by cargo vessel too tenuous a thread upon which to hang the island's defense. Aircraft carriers which could launch Malta-bound planes equipped to defend themselves against enemy air attack, were the only solution.

Because Britain's few carriers were desperately needed elsewhere to support a fleet already spread thin by operations in waters all over the world, the United States Navy lent its support by sending the Wasp together with other naval vessels to join the British Fleet for a time.

The Wasp left Casco Bay, Maine, on 26 March 1942 and arrived at Scapa Flow, Scotland, on 4 April. After receiving Royal Air Force Spitfires and personnel aboard at Glasgow, she departed for the Mediterranean accompanied by the American destroyers Madison and Lang and other screening British warships, on 14 April, and cleared the Straits of Gibraltar five days later. Early in the morning of the twentieth the Wasp launched Fighter Squadron 71 for combat air patrol and then launched forty-seven Spitfires which proceeded to Malta.

The Wasp and her escort immediately began the return trip to Scapa Flow where they arrived on April 26. On May 2 she again left Scotland on a similar ferrying mission and on the ninth launched another forty-seven Spitfires in the Mediterranean for Malta, returning to Scapa Flow a week later en route to the United States. Not every British plane delivered by this carrier to within range of Malta reached its destination safely. But enough got through to insure considerable strengthening of that base which ultimately withstood its grim siege by enemy bombs until the Axis was finally driven from the Mediterranean.

In a report on Malta to the House of Commons on 2 July 1942, Prime Minister Churchill said, "Fighter aircraft have been flown in from aircraft carriers by the Royal Navy and we were assisted by the United States Navy whose carrier Wasp rendered notable service on more than one occasion enabling me to send them the message of thanks, 'Who says a Wasp cannot sting twice?' "

The Ranger, the first United States naval vessel designed and constructed, from the keel up, as an aircraft carrier, was also given important ferrying duties in the early phases of the war when she took United States Army Air Forces fighter planes and pilots to West Africa. The first of these trips was begun on 22 April 1942 when sixty-eight P-40's (Warhawks) destined for the Tenth Army Air Force in India were carried to the African Gold Coast where they were launched and flown to Accra.

These planes and their pilots all landed safely at Accra on May 10 and in a series of hops finally reached Karachi, India. A second P-40 cruise to the Gold Coast was made in June, after which the Ranger returned to the United States in time to pick up its air group and take part in the Casablanca landings where that group performed so capably. Again, loaded with P-40's and pilots, she left Norfolk on 8 January 1943 and launched her cargo off the coast of French Morocco.

The planes landed at Casablanca on January 19. An identical cruise with Casablanca-bound P-40's was made in February 1943, with the planes landing on the twenty-third.

Her ferrying duties completed for a time, the Ranger provided air escort for the SS Queen Mary carrying Prime Minister Churchill to the Quebec Conference with President Roosevelt, after which she joined the British Home Fleet at Scapa Flow.

During this tour with the Home Fleet she operated in Norwegian waters for several months guarding the northern convoy routes to Russia against attacks by larger German surface raiders and the Luftwaffe. It was here that the Ranger, commanded by Captain Gordon Rowe, USN, made the first American naval air strike against German forces in Norway. On 4 October, Ranger aircraft made two strikes against German shipping — one in and around Bodo Harbor, the second from Alter Fjord to Kunna Head, Norway.

Thirty-one thousand pounds of bombs were dropped during these attacks, resulting in the destruction of at least six enemy ships including a 14,000-ton tanker. In addition, two other ships were damaged by bombs or strafing, and the Ranger's Combat Air Patrol shot down two enemy aircraft.

The Ranger was detached from British command in November 1943, and returned to the United States. She made a final Atlantic ferrying cruise leaving New York on 24 April 1944. This time the cargo was P-38 Lightnings, the Army's famous twin-engined fighters, plus Allied military personnel which were delivered to Casablanca in the first week of May.

On the return voyage, she carried battle-worn army aircraft and Allied personnel. In July 1944 the Ranger was transferred to the Pacific where she was operating at the time of the Japanese surrender.

6 - PROTOTYPE OF INVASION

IN ANY UNCRITICAL, OVER-ALL VIEW of the exploits of naval aviation in World War II, the Atlantic is likely to fall into a position of relative unimportance. The heroic accomplishments of the fast carrier task forces which ranged the Pacific during the last two years of the war, their overwhelming size and power, and the dramatic climax of their final triumph in the air over Japan, so completely overshadow earlier efforts in the Atlantic as to make the latter seem puny by comparison.

Yet it must be remembered that the Atlantic was a proving ground for Pacific operations; there, men, ships, and plane squadrons received the training which later carried them to fame in the war with the Japs. It must be recalled, too, that in the early years of the war, Germany, not Japan, was considered our Number

One enemy, that the defeat of the oriental foe had to wait until the Atlantic mission was accomplished — the mission of carrying overwhelming quantities of matériel [supplies] to be thrown against Germany, and of transporting and supporting the legions which finally forced surrender.

It might be profitable to remember, too, that one of our largest amphibious operations was carried out in the Atlantic a full year before the Marines stormed the beaches of Tarawa, and that the assault and capture of Casablanca, however imperfect it may have been in execution, pointed lessons which were definitely useful in subsequent Pacific operations. Here carrier-based aircraft proved their value in the protection of fleet and transport units, support of initial landing assaults, and direct support of subsequent advances inland by ground forces. Here the escort carrier established its worth as an adjunct of the fleet in offensive operations. Here cruiser and battleship planes demonstrated under fire their versatility as reconnaissance planes, fire spotters, and anti-submarine guards. Here was tragically demonstrated the need for thorough training of aviators, especially in navigation.

When our high command reached the decision to attack North Africa, the task of transporting troops, putting them ashore, and supporting them in the capture of Casablanca, was entrusted to Task Force 34, a vast naval armada organized under command of Rear Admiral H. K. Hewitt, Commander, Amphibious Forces, Atlantic Fleet. Specifically, the mission of Task Force 34 was "establishment of Western Task Force (U.S. Army) on beachheads ashore near Mehdia, Fedala, and Safi, and support of subsequent coastal military operations in the capture of Casablanca, French Morocco, as a base for further military and naval operations." D-Day was to be 8 November 1942.

The air group attach to Task Force 34 was made up of the large carrier USS Ranger, the escort carriers USS Sangamon, USS Suwannee, and USS Santee, the light cruiser USS Cleveland and nine destroyers. The escort carrier USS Chenango was attached to the task force, but its mission was the ferrying of seventy-eight Army P-40-1 planes, to be launched and flown to the airport at Port Lyautey after the capture of that field. The air group was under command of Rear Admiral E. [Ernest] D. [Doyle] McWhorter, USN, Commander, Carriers, Atlantic Fleet.

When word of the coming operation was first received by Rear Admiral McWhorter, his forces were far from ready. USS Ranger and her air group were well trained and seasoned, but some of the carriers and many of the squadrons had just been commissioned. For example, USS Santee reported to Chief of Naval Operations and sailed from the navy yard on 13 September 1942, with yard workmen still aboard. During the ensuing month she returned to the yard twice, and at no time during that period was she free of the yard workmen. There were only five experienced aviators aboard, and a bare handful of officers and men who had seen salt water before. In the case of USS Sangamon, 50 percent of the ship's company had never been to sea in any capacity.

Little time for preparation remained, however, and with more experienced forces unavailable, Rear Admiral McWhorter was compelled to do the best he could with what he had. In an effort to overcome the "greenness" of ships, crews, and squadrons, the carriers were ordered to sail for Bermuda early in October 1942, to carry out intensive training.

At Bermuda, classes were conducted, lectures given, flight operations carried on, all adding up to one of the shortest, but most highly concentrated training periods on record for an operation such as the one about to be undertaken. Three and one-half tons of papers, maps and photographs were used in study and briefing. Finally, a full dress rehearsal was held, using Bermuda as the coast of Morocco. There was not time for another.

On 25 October 1942 the air group sailed from Bermuda under conditions of complete radio silence. Meanwhile a covering group consisting of one battleship, two heavy cruisers, five destroyers, and one oiler had sailed from Casco Bay, Maine, on 24 October, and detachments of battleships, cruisers, destroyers, transports and other auxiliaries had sailed from Norfolk on 23 and 24 October. The Casco Bay and Norfolk groups rendezvoused at sea on 26 October, and on 28 October the air group joined the disposition.

The combined fleet thus formed was the largest ever to sail the Atlantic up to that time. It spread out over the ocean to a length of twenty-five miles, and an equal breadth. The voyage to Africa was a busy one for the air group. In addition to training and briefing, which continued unabated, the carriers set up aerial anti-submarine and scouting patrols to function in co-operation with the destroyer screen for the protection of the formation.

Task Force 34's plan called for a general covering group of one battleship, two heavy cruisers, and four destroyers, and three separate attack groups composed of battleships, cruisers, destroyers, and transports. The Northern Attack Group was to go ashore near Mehdia, the Center Attack Group at Fedala, near Casablanca, and the Southern Attack Group at Safi, to the south.

After beachheads were secured and neighboring towns and airfields were captured, all forces were to converge on Casablanca and effect its capture. For the assault, the air group, too, was divided into the Northern, Center, and Southern air groups, each to operate in support of the appropriate attack group. The assignment of the air groups was to provide protection for warships and transports from air and submarine attack, bomb and strafe enemy coastal guns, troop concentrations, airports, and warships, and support troop landings and subsequent extension of beachheads.

The air group, and indeed the whole task force, was hampered in its planning by uncertainty as to the amount of fighting that would be required. Nobody knew how strong the French resistance would be. Reports were varied and contradictory, and estimates varied all the way from immediate, bloodless capitulation to full-scale, all-out defense. However, the strength of French forces was known, and no chances were taken. The operation was planned as a complete surprise to the defending forces, and while it was the intention of our leaders to make peaceful landings if possible, preparations included provisions for dealing with determined and continued resistance at all points. On the morning of 7 November (D-1 Day) the task force was divided, and the various groups proceeded to their stations, preparatory to making simultaneous landings at the three predetermined points at 0400 the following morning.

USS Sangamon and her destroyer screen accompanied the Northern Attack Group, and took station approximately thirty miles off Mehdia; USS Ranger, USS Suwannee, USS Cleveland, and screen accompanied the Center Attack Group to a point off Fedala, near Casablanca; and USS Santee with her screen proceeded to Safi with the Southern Attack Group.

At dawn on the morning of 8 November the carriers launched fighter planes and established combat air patrols and anti-submarine patrols over warships and transports at the three attack points. These patrols were maintained from dawn to dusk during the entire operation, and the carriers and their screens dispersed to sea at night to return to station at dawn. At no time did an air attack develop against our shipping, although submarines appeared sporadically, and on the twelfth made a concentrated attack, narrowly missing our carriers and sinking several transports.

At approximately 0500 on the morning of 8 November, landings originally scheduled for 0400 were made at all three points. Resistance developed, particularly in the Casablanca area, and within two hours Rear Admiral Hewitt had ordered a general offensive.

Torpedo planes, scout bombers, and fighters from the carriers went into action against airfields, coastal defense guns, and shipping in the harbor of Casablanca, and the fight was on.

There followed three days of intense aerial activity.

In the north, on the first morning, enemy planes appeared and strafed our troops on the beaches at Mehdia. Fighter patrols from USS Sangamon ranged the beaches all day, but still the troops complained of strafing. It developed that one enemy plane would sneak in low over a ridge, almost invisible to our fighters above, make one run, and disappear over the ridge.

Finally, with the aid of a naval air-ground liaison officer, our planes were able to catch the lone enemy and shoot it down. In the meantime, other planes from the Sangamon spent the day in attacks on the Port Lyautey airfield and other objectives.

In the center, bombers from USS Ranger and USS Suwannee struck at Casablanca, and Ranger fighters followed with heavy strafing. (Fighters from the Suwannee, meanwhile, were flying combat air patrol over the Northern Group, releasing Sangamon's planes for offensive sweeps.) Intense antiaircraft fire was encountered by the planes over Casablanca. Initial bombing was not too good, due to excitement and uncertainty as to character of targets.

Hits were scored on the French battleship Jean Bart in the harbor, and at least one of three French submarines attacked was sunk, but the Jean Bart's guns were not silenced, nor were the harbor defenses neutralized. Throughout the day, the batteries of El Hank and the 15-inch rifles of the Jean Bart ranged the beaches at Fedala.

During the day a French light cruiser and three destroyer leaders sortied from Casablanca Harbor and headed north toward Fedala. They were attacked by surface forces, together with Ranger and Suwannee planes, left burning, and the next day were observed to have been beached.

In the south, at Safi, the first day was quieter than in other areas, but was not without its mishaps and confusion. Two torpedo bombers went into the water on the first take-off, due to extremely light wind across the deck.

The first flight of six fighters, led by Lieutenant Commander Blackburn, was sent aloft to establish combat air patrol. One plane was lost from undetermined cause, and the pilot listed as missing in action. The remaining five planes became lost from their ship due to faulty navigation.

Lieutenant Commander Blackburn ran out of fuel first, due to excessive gasoline consumption occasioned by failure of landing gear to retract completely. Before ditching, he ordered the other four pilots to head for shore and attempt to land on a suitable airfield. The four landed at Mazagan airfield, and were promptly captured, to be released four days later when the U. S. Army took over the field. Lieutenant Commander Blackburn was picked up by a destroyer after fifty-six hours in the water.

In the meantime, four Santee scout bombers had established anti-submarine patrol early in the morning, and torpedo planes were aloft awaiting orders to proceed to the support of ground forces on the beach. Not until 1030 was communication established with the naval liaison station ashore, and then only one plane was called for.

Accordingly, Lieutenant Commander Joseph A. Ruddy, in a torpedo plane, spent eight hours in the air as a one-man force, carrying out reconnaissance and photography missions. He landed only once for refueling during that time.

At dusk, planes of all groups were taken aboard, and the carriers with their screen steamed away from the coast to avoid the ever-present danger of submarine attack in concentrated shipping areas.

In this first day of operation, the carriers and their planes had established immediate air superiority and control, maintained combat air patrol and anti-submarine patrol over all American shipping, given direct support to ground forces, and in co-operation with battleships and cruisers, inflicted damage on enemy shipping and coastal defense installations.

In addition, scouting and observation planes from battleships and cruisers had been in the air throughout the day, spotting for ship's gunfire, carrying on reconnaissance, search, photography, and anti-submarine patrol missions.

Fedala had surrendered to ground forces at 1430, and Safi was captured at approximately the same hour.

At dawn the next morning (9 November) the carriers returned to their stations and continued operations as on the first day. By now there was a semblance of organized resistance in the air, and the guns of Casablanca were definitely giving trouble.

In the north, Port Lyautey was reported captured at 1430. The airfield, however, which was to be used by the Army P-40's brought over by the Chenango, was still in French hands.

In the center, French planes bombed the beach early in the morning, and were engaged by our combat air patrol. At 0915 Commander of the Center Attack Group asked for fighters to stop enemy strafing of the beach. Pounding of shore batteries and harbor defenses of Casablanca went on all this day, and many planes were destroyed on the ground at Rabat, Cazes, and Marrakech.

In the south the Santee, which had been forced to cease flight operations on the afternoon of D-Day because of insufficient wind across the deck, was again having trouble on the morning of the second day. Because it was thought that the Santee's planes would not take the air, the heavy ships of the Southern Attack Group launched all planes for anti-submarine patrol and reconnaissance.

Later, when the Santee planes appeared on station, the cruiser and battleship planes were recalled, not, however, before they had reported bombing a submarine near Cape Kantin.

On this second day the French air force in the Safi area dropped its apparently nonhostile attitude of the day before. In the morning a plane came though the overcast over Safi Harbor, to drop a bomb near the USS Lakehurst which was unloading heavy tanks to be used in the assault on Casablanca. The plane was shot down by antiaircraft fire from the transports.

At 0800 a reconnaissance plane from the Santee was fired upon by antiaircraft guns at Marrakech airport. The pilot dropped two bombs on the hangars and returned to his ship. Later a bombing strike was flown against the field, destroying 20 planes on the ground and damaging 5 others. En route to and from the field, the flight supported an attack by an American armored combat team, destroying 14 trucks loaded with enemy troops.

The airfield at Safi was reported captured at 1655.

The third day of the attack, 10 November, was the last day of actual fighting. It was much like the second, except that the fighting was more intense. Submarines, which had been sighted and engaged throughout the operation, now became much more numerous. One, apparently headed for Dakar, was beached after heavy strafing.

In the north, Port Lyautey airfield was captured in the morning, and the army P-40's from the Chenango were requested. The first plane to land damaged its landing gear. Forty-three planes were landed, then the operation was temporarily discontinued because of the poor condition of the field. The remainder of the 78 planes were flown to the field the next day, too late to participate in the fighting.

In the center, our battleships, together with planes from the Center Air Group, worked on the guns of Casablanca, while fighter planes strafed and bombed installations, and shot down enemy planes that came up to oppose them. In the afternoon the order came to silence the Jean Bart. Nine bombers with 1,000-pound bombs attacked, and twenty minutes later the leader radioed, "No more Jean Bart" — a pardonable exaggeration.

In the south, Santee planes, supporting the Army, destroyed three trucks, strafed gun emplacements and machine-gun nests, shot down two twin-engined bombers near Bon Gehdra. A plane from the Philadelphia bombed a beached submarine at Cape Blanca. By night the landing field at Safi was in operation for emergency use.

Again, on this day, six Santee planes failed to return to the ship, due to faulty navigation and crowded communications. One was abandoned in the air, one landed in the water, and four landed at Safi.

The final air, sea, and ground assault on Casablanca was planned for 0715 on 11 November. It was halted, however, on a report from General Patton that the French would capitulate. The "Cease firing" order was given to the task force, and the battle for Casablanca was over. The carriers reduced their flight to anti-submarine patrols and began recovering stranded pilots.

The land struggle was over, but the sea battle continued. On 12 November, enemy submarines attacked in force. The Ranger was narrowly missed by four torpedoes, and several others were observed to pass close aboard other carriers. Fortunately, no warships were hit, but four transports were destroyed. On 12 November, units of the task force began leaving the area and within a few days practically the whole force was en route to the United States.

All fighting had gone off on schedule. Not a man or ship was lost on the voyage to Africa, none of our planes was shot down in aerial combat. Twenty-six enemy planes were shot down by our fighters in the air, and over a hundred were destroyed on the ground.

There were valuable lessons to be gained from the Casablanca operations for both future training and operation of naval aviation. We had seen a sample of some of our strength, and, what was fully as important, had discovered some our weaknesses. The comments of some of the leaders of the venture indicated both these points. On the credit side of the ledger, the Commander-in-Chief, U.S. Atlantic Fleet, noted:

"The value of aircraft on cruisers and battleships was clearly illustrated. The use of these aircraft for spotting, close anti-submarine patrol, and search relieved an equal number of carrier aircraft for combat missions, thus increasing the striking power of the air group by the full number of VO-VS aircraft involved."

The worth of escort carriers converted from tankers was realized. As Rear Admiral McWhorter, commander of the air group, phrased it, "When a vessel of this class is in the striking force, tanker requirements can be eliminated at least in part." This advantage outweighed a weakness of this type of vessel. Because of its low speed, as has been seen, it was sometimes difficult to launch planes in light winds.

Important lessons were learned concerning armament. The devastating character of .50 caliber machine-gun fire was seen in its ability to put a destroyer out of action. On the other hand, the ineffectiveness of light bombs was made apparent, and it was considered that 1,000-pound bombs were needed to inflict real damage on heavy ships and gun emplacements.

As to training, there was much on both sides of the ledger. On the credit side, Cominch wrote:

"It is considered that the planning for the use of the Air Group, and the execution of the tasks assigned, were outstanding. Even on the part of the most inexperienced aviators there was a strict adherence to activity in support of the general task, which was impressive evidence of thorough indoctrination by the Air Group Commander."

On the debit side, there were occurrences which, though regrettable, directed and stimulated future training, thereby preventing recurrences on a larger scale. Outstanding in this category were plane losses through faulty navigation and poor plane procedure. The commander of the air group wrote, "Excessive loss of planes on Santee due to faulty navigation and inability to bring these planes in with equipment provided, indicates necessity of adequate training." Cominch also drew attention to this matter and at the same time gave the explanation for such losses. He wrote:

"Increasing evidence has shown that proper training of pilots is, up to a certain point, of more value than numbers. . . . It is considered that training is the most vital need at the present time in naval aviation. . . .

"Three-fourths of the carriers involved in this operation had less than one-half of the shakedown training usually considered necessary, prior to the operation. . . ."

Another weakness of training that showed up in the engagements was the unfamiliarity of our forces with our own plane models and those of our allies. As a result, there were instances of friend firing upon friend.

All in all, the Casablanca venture formed an important piece in the building up of the Navy's air strength. The prototype of invasion had been built. Its improved descendants were to invade the islands of the Pacific in the years to come.

USS RANGER (CV-4), PANAMA CANAL, 1945

7 – THE NAVY'S AIR WAR ON THE "WOLF PACK"

WITH THE ESTABLISHMENT OF FLEET AIR WING 15 at Port Lyautey, French Morocco, in March 1943, the ring of American land-air bases encircling the Atlantic was complete. During that same month, sinkings of Allied shipping in the Atlantic reached the total of nearly 600,000 tons—the third highest monthly figure of the war up to that time—yet the tide was beginning to turn in our favor. The defensive phase of the war was over and the Allied offensive to destroy the enemy both on the high seas and in his home waters was well begun.

As the ring of American air bases became tighter, preventing the U-boats from launching more than occasional attacks in coastal waters, the Nazi high command began to utilize a different mode of attack. Although group tactics had been sometimes used as early as 1940, the U-boats now formed "wolf packs" designed to operate against convoys out of the range of land-based air cover. While huge wolf packs struck savagely at convoys carrying supplies and troops to Britain, American aircraft based in Iceland, French Morocco, Newfoundland, and the West Indies flew hundreds of patrol hours with negligible results in terms of U-boat sightings and attacks.

Our answer to the new German tactic was to assign several small carriers of the converted type to anti-submarine duty in the Atlantic. The first of these "baby carriers" to enter the struggle was the Bogue, which got under way for Argentia, Newfoundland, on 24 February 1943, escorted by the Belknap and the Geo. E. Badger. During its fight against the sub the Bogue compiled an impressive record.

Of twenty-five U-boats attacked prior to the end of November 1944, she is credited by the Navy with 9 sinkings, 3 probable sinkings, and the inflicting of damage on 7 others. Shortly after the Bogue began operations, her efforts were augmented by those of the Card. That the enemy was completely unprepared for this development is indicated by the fact that these CVE's achieved astonishing results in terms of U-boat kills.

During the grim days of February, March, and April, several attacks were made by TBF's flying from the carriers and by destroyer escorts accompanying the carriers. The pattern of organization and the tactics of the hunter-killer group were becoming well established. The group was normally composed of one CVE and four to six destroyer escorts, with a composite squadron, composed of fighters and torpedo bombers, based aboard the CVE.

So successful was this unit in combating the underwater menace that other CVE's as they became available, were assigned to Atlantic duty. These were the Core, Croatan, Block Island, Mission Bay, Guadalcanal, Solomons, and the Tripoli. Although operating in constant danger, only one of them, the Block Island, was sunk. This fine record reflects credit on the vigilance and ability of the personnel, nonflying and flying alike.

In April 1943, three moves in the American anti-submarine campaign took place. First, in order to facilitate the shift of American air power to the East for use against the enemy in home waters, control of operations in Canadian coastal waters was turned over to the Canadian Air Command, and Commander, Fleet Air Wing 7 prepared to move his main units to Iceland.

Second, anti-submarine air forces in the northern and southern approaches to the European theater were reinforced by the arrival of Ventura and Liberator aircraft units in Iceland and French Morocco. As will be seen, this event marked the beginning of greatly increased anti-submarine activities by land-based units. Third, the establishment of the Anti-Submarine Development Detachment at Quonset Point, R.I., gave great impetus to the already active effort on the part of the United States Navy to provide its men with the best possible tools and techniques to use against the U-boats.

In April and May 1943, U-boats again made a determined effort to break down the supply lines between America and Britain. Some wolf packs operating on the Atlantic convoy lanes sank merchant vessels, mostly stragglers, but were so roughly handled by long-range aircraft flying from United Kingdom, Iceland, Newfoundland, Bermuda, and Africa and by three CVE hunter-killer groups ranging the central Atlantic, that ship sinking began to decline. The fiercest convoy battle of the month of May and probably of the war to that time, involved a slow west-bound convoy, ONS-5, which ran into heavy gales, snow, hail, icebergs, and an estimated twenty-five U-boats.

During the night the U-boats attacked in twos and threes using regular tactics. During the day they approached from ahead of the convoy center and fired torpedoes from between columns. Little air coverage was available because of bad weather, but surface craft did outstanding work making forty counterattacks in which at least eighteen U-boats were sunk or damaged.

Shortly thereafter, on 21 May 1943, a major engagement between carrier-based aircraft and enemy submarines took place in the mid-Atlantic out of reach of land-based aircraft. Within twenty-four hours six promising attacks, one resulting in the destruction of a U-boat and capture of twenty-four survivors, were made by Avenger aircraft flown from the USS Bogue. The convoy, though endangered on every side, passed through the area safely.

In June the enemy apparently had withdrawn from the North Atlantic theater to lick his wounds. His main concentration appeared to be shifting to the southwest of the Azores where about thirty U-boats were estimated in mid-June to be patrolling the convoy lanes to Africa.

While in the Moroccan Sea Frontier area U-boats were still clearly on the defensive, a new German technique which had been developing in the spring became increasingly evident: the use of snooper aircraft for spotting convoys moving up or down the coast of Spain.

Ever since the fall of France yielded Bordeaux and other major airport facilities in southern France to the Germans, Focke-Wulf 200's and Dornier 217's were occasionally seen by convoys passing between the Azores and the Iberian Peninsula. It was not, however, until May that the enemy aircraft did more than locate the convoy, radio its position to enemy U-boats, perhaps attempt homing procedure, then drop its bombs halfheartedly and return to base.

Beginning on the twenty-second of May, however, viciously aggressive attacks were made on practically every convoy moving through the area. Generally heavy antiaircraft fire thrown up by the convoy and its escorts drove off the attackers; yet on some occasions merchant vessels were sunk or damaged by bomb hits or near misses.

This sudden increase in air activity apparently was designed to accomplish important results. First, the work of the submarines would be vastly simplified if convoy positions were frequently and accurately reported to them; moreover, there is nothing like a straggler or a convoy slowed down to a snail's pace by a damaged merchant vessel to please the eye of a U-boat commander. Secondly, strong aircraft activity most certainly would tend to divert the attention of British and American anti-submarine aircraft from their primary duties.

The relative ineffectiveness of the German U-boat fleets apparently led Admiral Doenitz, in July, to order his submarines to adopt aggressive tactics and to stay on the surface and fight back if they could not safely submerge. The first evidence of this change of policy was felt by aircraft of Fleet Air Wing 15 when, on 2 June, the approaches to the Straits of Gibraltar were invaded by enemy submarines.

Army and navy squadrons between 5 June and 15 June, made 16 attacks which resulted in 5 U-boats sunk and 6 U-boats damaged. British aircraft operating out of Gibraltar at the same time sank 5 and damaged 1 undersea raider. In all but one case the U-boats remained surfaced and poured a hail of shells at the attacking aircraft, damaging three B-24's and one PBY-5A. At the end of that period, the battered remnants of Doenitz' flotillas limped north along the Spanish coast and into the bay ports.

While aircraft of Fleet Air Wing 15 were blasting the enemy off Africa, British and U. S. Army anti-submarine squadrons were mounting an offensive of their own against U-boats entering and leaving their lairs in the Bay of Biscay.

Meantime, the USS Bogue transferred from the North Atlantic to United States-Gibraltar convoy duty, broke up a large assembly of U-boats lying directly across the path of a convoy. Four separate attacks were made in forty-eight hours. Three days later a submarine only ten miles from the convoy was attacked and badly damaged by air attack. Before three more days had passed, a U-boat was discovered and sunk by coordinated air-sea assaults. No further incident occurred, and the convoy reached port safely.

Allied air successes in the Bay of Biscay and off Morocco, however, were achieved in the face of great enemy air opposition. During this period practically all flights from the United Kingdom encountered groups of two or more JU88's, long-range German fighters, while clashes between FW-200's and Liberators and Catalinas of Fleet Air Wing 15 were commonplace.

A most unusual combat took place on 12 June 1943 between two navy Catalinas and two FW-200's. The navy planes were sent to provide air and anti-submarine protection to the Port Fairy, a British merchant vessel en route to Casablanca, French Morocco, with survivors of a previous air attack. Shortly after reaching the ship, Lieutenant (jg) Drew, pilot of one of the Catalinas, sighted FW-200's apparently on a bombing run.

Ordering the second pilot to provide close anti-submarine coverage against the two U-boats believed to be shadowing, Lieutenant Drew climbed as quickly as possible but could not prevent the completion of the run. One bomb struck the stern of the Port Fairy. The navy pilot, however, stayed in the fight and succeeded in breaking up all other attacks by flying collision courses and forcing the enemy to change course whenever bombing runs were attempted. Many shots were exchanged between the Catalina and the FW-200's, no damage, however being sustained by the former. After the enemy craft were driven off, the Port Fairy continued on its way to port.

The new German submarine tactic of fighting back on the surface was the occasion for a flare-up of American air successes in the Caribbean where planes of Fleet Air Wing 11 made several successful attacks both day and night against moderate to heavy antiaircraft fire. In the Eastern Sea Frontier area, aircraft of Fleet Air Wing 9 did not encounter submarines fighting on the surface until August. Heavy antiaircraft fire through which American planes plunged to drop their depth charges marked the final phase of active anti-submarine warfare in American waters.

SUMMER IN ENGLAND BROUGHT WITH it increased anti-submarine operations in the Bay of Biscay. It was believed that by compelling U-boats, based in French bay ports, to make a large part of their passage through the bay in a submerged condition the length of the effective patrol of each U-boat would be reduced materially. Besides, the necessity for U-boats to run the gauntlet of heavy air attack for several days both on leaving and on returning to base was sure to react unfavorably on the morale and general aggressiveness of U-boat crews. That intensified bay operations were feared by the enemy was indicated by strenuous attempts to break up air patrols by long-range fighter interception.

To aid the hard-pressed coastal command aircraft and to relieve army Liberator squadrons for duty over Europe, plans were made to send several navy squadrons of Fleet Air Wing 7 to England. On 23 July 1943 the first navy squadron of MAD-(Magnetic Airborne Detectors) equipped Catalinas, VP-63, arrived at Pembroke Dock, England, and began an intensified training and familiarization program in this type of submarine detection. Operations were begun in the bay at the end of July and one day later the first VP-63 plane was shot down by enemy fighter action.

Shortly thereafter, two navy Liberator squadrons, VP-103 and VP-105, landed in England and were based temporarily at St. Eval, Cornwall. Commander, Fleet Air Wing 7 who was to direct their operations under over-all British control, established his headquarters at Plymouth, England, on 21 August.

Though the Navy's commitment was at that time only a temporary reinforcement of Allied anti-submarine effort in the bay, long-range planning envisaged the throttling of the U-boat menace in the western Atlantic preparatory to invasion of the Continent. Ten days later, the full-fledged navy patrol squadron got under way with the beginning of the operations of PB4Y-1 Squadron, VP-103.

August was the most successful month of the war for Allied anti-submarine forces in the Atlantic. Only two merchant vessels were sunk by submarine action — off the coast of Brazil — in areas patrolled by U.S. or Allied aircraft. Meantime the bay offensive netted four U-boats sunk out of ten attacks, and CVE's escorting convoys or operating as separate task forces destroyed 10 submarines in 12 attacks west of the Azores. Scattered contacts with enemy U-boats were made by aircraft of Fleet Air Wing 9 and Fleet Air Wing 11 — one of the latter resulting in the sinking of a U-boat in the Curacao-Aruba area.

By the end of the month almost all enemy U-boats were en route to French bases with practically none left to operate offensively on main shipping lanes. The tactics also had by that time become much less aggressive than in July. The CVE's operating in the central Atlantic reported that U-boats were now more prone to submerge at the first favorable opportunity than to remain on the surface and fight it out with attacking aircraft. Whenever, however, a U-boat was surprised on the surface, the aircraft always could expect to receive a burst of heavy aircraft fire. Another change in tactics brought about by the effectiveness of aircraft operations was becoming apparent in the Biscay area. Previously, the U-boats proceeded to and from their French bases in groups of three to five for mutual protection; but this practice was later abandoned in favor of individual passage, the U-boat remaining submerged during daylight hours.

September saw U-boat concentrations move out to the North Atlantic convoy lanes. There a wolf pack of approximately twenty U-boats attacked the ONS-18 and ON-202 convoys for six days. Land-based coverage from Iceland and from Newfoundland was provided for four days but the convoys were without support for forty-eight hours. Three escorts and six merchant ships were sunk before the battle was over. Convoy escort craft destroyed three U-boats and damaged several.

U. S. Navy aircraft sank no submarines in the Atlantic during September despite intensified efforts by planes of the air wings ranging that ocean and by the CVE escort and "killer groups." This can be attributed in part to the reduced activities of U-boats early in the month but mainly to the use of superior defensive search radar by the submarines.

Although it is not at all certain that this decided fall-off in aircraft sinkings was due essentially to the use of radar rather than to a combination of circumstances, it does illustrate well the ever-changing aspects of the anti-submarine war and the need for continuous effort to overcome the temporary gains made by the enemy through introduction of new equipment or tactics. Countermeasures to minimize the effectiveness of U-boat radar were, however, well under way.

New equipment designed and developed by the Anti-Submarine Warfare Operations Research Group (ASWORG) and the Anti-Submarine Development Detachment, Atlantic Fleet (AsDevLant) was becoming available to aircraft for the detection of U-boat radar. Aircraft tactics were revised and undergoing tests through exercise against our own submarines.

Likewise, two detection devices, Magnetic Airborne Detector and Sono-buoy, were already coming into operational use. Rocket projectiles were soon expected to be used against the enemy in the Battle of the Atlantic.

Most of September was spent by navy squadrons of Fleet Air Wing 7 in training for the specialized warfare of the Bay of Biscay. This involved a substantial amount of what was known as fighter affiliation training, for the enemy was at that time using large numbers of ME-110, 210, 410, and JU-88 fighters. The latter, most commonly encountered, were long-range two-engined fighters which often operated in groups of six to twelve and seldom attacked PB4Y-1's unless they outnumbered them by at least six to one.

The RAF endeavored, with some success, to intercept enemy fighters with Spitfires, Mosquitoes [DH.98], or Beaufighters [Bristol Type 156]. But since navy anti-submarine patrol planes were to operate alone and principally in the day, encounters were definitely expected. Each squadron, therefore, when it arrived in England, was given a stiff course in anti-fighter tactics and steps were taken to place the gunnery operations of the planes on the highest level.

The principal U-boat activity was again in the North Atlantic in October. On 4 October a pack of twenty U-boats, while attempting to intercept two westbound convoys, moved within range of Iceland-based squadrons of Fleet Air Wing 7. No sooner had they done so than a series of spectacular air attacks drove them off with at least three of the submersibles sunk and several damaged. Not a ship of either convoy was touched.

On 8 October the remnants of the same concentration attacked a small eastbound convoy about six hundred miles west of Iceland.

This time a merchant vessel and an escort craft were torpedoed and sunk, but at heavy cost to the enemy.

Again, Allied shore-based aircraft were sent out to aid the convoy from Iceland and British bases. In the area of the convoy several U-boats were located and attacked so aggressively that three of them were sunk and a number damaged.

While absorbing this beating from navy and Allied land-based aircraft, the U-boat fleet in the central Atlantic received an even worse whipping from American CVE-based aircraft. Of twenty attacks made by TBF's flying from the ubiquitous baby flattops Card, Core, and Croatan, twelve resulted in assessments of submarines sunk or probably sunk.

Besides the highly successful work of land-based aircraft from Iceland and the United Kingdom, there was little glamour attached to the operations of navy aircraft of Fleet Air Wings 7 and 15. Long, exhausting patrol missions were flown during October in the Bay of Biscay and off the Iberian Peninsula with few sightings and no successful attacks rewarding the efforts of navy pilots.

They did, however, keep the enemy on the defensive and prevented him from attacking the convoys which sailed in a steady stream between the United Kingdom and Africa. These aircraft were, during the same period, subjected to numerous attacks from German fighters and long-range bombers operating out of French air facilities. In all but two instances, however, the navy craft escaped by skillful evasive tactics.

Toward the end of the month, navy planes of Fleet Air Wing 15 were attacked from another quarter. For almost a year, navy and French patrol aircraft had operated anti-submarine sweeps among the Canary Islands [archipelago which forms the southern tip of Spain] in order to reduce the refueling activities of German submarines in that area. By the fall of 1943, the Spaniards had apparently become quite used to the daily or weekly "Canary Sweeps" of Allied aircraft. Sometimes, if Allied planes passed inside of the three-mile limit [imposed by Spain's ruler Francisco Franco who officially stayed out of the war, but was sympathetic to Germany as an enemy of communist Russia and sent volunteers to assist German troops on the Russian front], a Spanish antiaircraft gun would fire warning shots, but to all intents and purposes the gun crews did not have orders to "shoot to kill."

Suddenly, on the afternoon of 26 October, a Catalina on a routine patrol off Las Palmas was attacked from a cloud by a Spanish fighter plane. The attack was pressed home vigorously, and the American plane returned the fire and dove for the water. This maneuver enabled the Catalina to escape with minor damage — a fact which attests to the poor quality of Spanish planes and marksman-ship.

When the same thing happened the following day, two Venturas of the wing were sent to an advanced base at Agadir, French Morocco, to fly the patrol. Needless to say, the two fighter planes that took off when the navy patrol was picked up on the shore radar were disagreeably surprised to find two fast bombers stripped down to fighting trim instead of one slow lumbering Catalina. As they approached the Venturas, about seven miles offshore, the navy pilots peeled off and attacked vigorously. The Spaniards immediately broke off and fled for the beach where both made forced landings.

After this incident enough diplomatic pressure was brought upon the Spanish government at Madrid to cause a cessation of fighter attacks on Fleet Air Wing 15 planes.

October also witnessed a most damaging blow to German hopes for successful operations against Allied convoys in the eastern Atlantic — the establishment of Allied air bases on the Azores. Under cover of an old Anglo-Portuguese agreement, diplomatic entreaty and pressure caused the Portuguese government to grant to Great Britain the use of two airfields, Lagens Field on Terceira Island and Santa Anna Field on San Miguel Island, for anti-submarine bases. Little imagination is needed to visualize the tremendous significance of this move in the U-boat war.

Almost at once, long-range aircraft, Fortresses and Liberators, were smashing at the U-boat concentrations, north, west, and south of the Azores. Convoys were given coverage when necessary 1,000 miles farther west than before. British planes based on the Azores began to escort northbound convoys to the point where United Kingdom aircraft could take over.

And most important, CVE groups which hitherto had escorted the huge Gibraltar-bound convoys to within eight hundred miles of the African coast were freed a larger part of the time for independent killer operations in the mid-Atlantic.

The beginning of a new Allied move against the fleet of Admiral Doenitz was indicated by the sinking of two U-boats in the Straits of Gibraltar by British surface craft. For some months an occasional U-boat had sneaked down the coast of Spain, sometimes inside territorial waters, and passed the Straits into the Mediterranean through the gauntlet of British searchlight-equipped Wellington aircraft at night, and American army and, later, navy Liberators by day.

Because of the large number of convoys passing through Mediterranean waters to supply and reinforce the various land operations in that theater, it became imperative that the U-boats already in the

Mediterranean be isolated and destroyed and that others be prevented from passing through the Straits of Gibraltar.

At first, surface patrols were instituted in the Straits at night, and British Swordfish and Hudson aircraft swept the Straits and its approaches by day. When the enemy, not deterred by a strong British defense, persisted in his efforts, planes of VB-127, a PV-1 squadron of Fleet Air Wing 15, were brought into the struggle. Continuous daily patrols were set up off the immediate approaches in order to attack U-boats approaching the Straits on the surface or compel them to submerge too far out to avoid surfacing again in the Straits themselves.

In November the over-all picture was most encouraging. The only merchant vessel sinkings occurred in the Panama Sea Frontier where intense air effort by planes of Fleet Air Wing 3 yielded numerous disappearing radar blips but no sightings. The same condition prevailed in other areas where U-boats were known to be operating — the Bay of Biscay, and the Gibraltar-Morocco area. This disappointing state of affairs was probably due to excessive caution on the part of U-boat commanders and to the use of improved radars and special gear for the detection of our radars.

Early in November, the enemy, discouraged by fruitless patrolling, moved the main U-boat concentration from the Newfoundland area to the region northeast of the Azores astride the Gibraltar-United Kingdom convoy route. There, by surfacing only at night and by employing long-range FW-200 aircraft to spot the convoys, the U-boats hoped to avoid damage to themselves and to achieve some ship sinkings. In both those expectations they were disappointed. Land-based Coastal Command and U. S. Navy aircraft from the Azores and England seconded by an aggressive CVE killer cruise destroyed five U-boats and damaged several. No merchant ships move east-west or north-south were attacked successfully.

Since in the main navy aircraft were limited during this period to day operations, they were compelled to devote most of their time to "hold down" patrols. It was decided, therefore, to send over to the European-African theater navy PB4Y-1 squadrons equipped with a new type of searchlights. Further, it was hoped that availability of additional CVE's and spare carrier squadrons would soon permit the use of CVE's regularly for night work also, using specially equipped TBM's.

The enemy, by December, had apparently given up the wolf pack tactic which had enabled U.S. and British killer groups to destroy several U-boats at a time. A thinly spaced patrol line in the northeast Atlantic brought no results. A second line along the United Kingdom-Gibraltar convoy route was kept on the defensive by constant attacks from Navy CVE-based aircraft and night-flying [a Vickers] Wellington [long-range medium bomber] squadrons from the Azores. The only ships sunk by U-boat action were destroyed by lone submarines operating in four American sea frontier areas.

Submarine sinkings were likewise at a low ebb, during December, except in the Azores area, where a destroyer escort of a CVE group sank a U-boat, a TBM destroyed another, and a third was hounded to its death by a coordinated air-sea attack.

From this point on, the U-boat war became more and more one-sided. Floods of new equipment, perfected tactics, thoroughly trained replacement crews, faster and more powerful aircraft all played their part. In the mid-Atlantic, U-boats operated singly and were at all times threatened by an increasing number of CVE groups whose TBF's were equipped with new types of secret weapons, sonobuoys, and rocket projectiles, and whose escort craft employed every available type of detection gear.

Not only did the small carriers account for several sinkings, but, under the command of Captain Daniel Gallery, a hunter-killer group produced one of the most dramatic and daring achievements of recent naval history.

On 4 June 1944, in the vicinity of the Azores, the group made a positive contact which a determined attack by the destroyer escort Chattelain forced to surface. Boats from the carrier Guadalcanal and the destroyer escort Pillsbury reached the U-boat before the scuttling charges or flooding mechanisms could accomplish their purpose, and the United States Navy, in its first successful boarding attempt since 1814, found itself with a real German submarine to study.

For the enemy sub commander who operated close to shore there were larger numbers of patrol planes outfitted with all kinds of complicated, but none the less deadly gadgets. Beginning in February, Catalinas of VP-63, equipped with Magnetic Airborne Detectors, blocked the Straits of Gibraltar by day, and after 1 June, when the first blimps began arriving, the patrol was extended around the clock. As Doenitz could not reinforce his dwindling Mediterranean fleet with additional submarines, U-boats became a virtually nonexistent menace by the time of the landings in southern France.

At the same time, the continuing offensive in the Bay of Biscay and the bombing of German submarine bases and navy yards by army and Allied Aircraft, greatly decreased the effectiveness of the French ports as supply and repair bases and hindered new construction and the refitting of obsolete and damaged U-boats.

Although the anti-submarine war continued to the last day of hostilities with both sides developing new equipment and tactics, and although vigilance could not for a moment be relaxed, the issue was no longer in doubt. It was for the most part a hidden war of nerves and men. On the Allied side it required the co-operation of scientists and naval leaders, of surface and air components, of British, French, Americans, Brazilians, and others.

All put in long hours of tedious, unrewarding labor, only very, very rarely alleviated by a few moments of thrilling action. Naval aviation — both ship- and shore-based — played its part, and it was a distinguished one, but it was not by itself sufficient. The converted French trawler, the British corvette, the little American destroyer escort, the sleek destroyers of many nations, were all just as essential to getting the vital convoys through as were the American and British carriers, the little inshore and large offshore patrol planes or the great silver blimps.

It was the triumph of science and technical skill operating through the surface and air forces of half a dozen nations, in which the United States Navy and its air arm did its share.

From this point on, the U-boat war became more and more one-sided. Floods of new equipment, perfected tactics, thoroughly trained replacement crews, and faster and more powerful aircraft made it impossible for the U-boat to approach any Atlantic shore with impunity.

The establishment of an air-sea blockade of the Straits of Gibraltar prevented Doenitz from reinforcing his Mediterranean fleet. Intensification of the bay campaign did much to render over costly the use of St. Nazaire, Lorient, and Bordeaux as repair and supply bases. The bombardment of those ports and the German sub bases and navy yards by Allied and army aircraft, cut down the supply of new U-boats and hindered the refitting of antiquated and damaged ones.

Those U-boats which cruised in the mid-Atlantic in 1944, in general, operated singly and were at all times threatened by the increasing numbers of CVE groups whose TBF's were equipped with secret bombs, sono-buoys, and rocket projectiles and whose escort craft employed every available type of new detection gear. Naval Aircraft had played a stellar role in the Battle of the Atlantic.

8 – The Battle of the South Atlantic

The struggle against Nazi submarines and blockade runners in the South Atlantic continued throughout the course of the war against Germany. The diplomatic aspects of activity in this theater, coupled with the fact that this warfare came to a peak during 1943, make it desirable for purposes of clarity to deal with this phase of the war as a unit, rather than to break the story into chronological segments.

Naval aviation's development in Brazil for the immediate purpose of waging anti-submarine warfare in the South Atlantic was so tied up with long-range diplomatic objectives that any attempt to separate the two would lead to confusion. To the average observer, the winning of the war in the strategic waters between South America and Africa was an obvious triumph, but fighting U-boats was neither the beginning nor the sole result of our efforts in that area.

Antecedents of U. S. naval aviation in South America, particularly in Brazil, go far back, but the first definite step toward obtaining air bases came in November, 1940, when the United States Government, through the War Department, negotiated a contract with the Pan-American Airport Corporation (subsidiary of Pan-American Airways) for the purpose of creating certain additional land airports and seaplane bases, improving existing bases, and providing other specified facilities.

For many years, developments in South America had been a matter of grave concern to those entrusted with our international relations and our national security. Fascist commercial, ideological, and political infiltration, too well known to require discussion here, presented a very real threat to the political integrity of the Latin American republics, and in turn to the safety of the United States itself.

It was well recognized that if fascism overran South America, we in the United States could bid farewell to security. It was recognized, too, that in the event of actual hostilities between the United States and the German-Italian Axis, Brazil would occupy a position of great strategic importance as a steppingstone for invasion of North America, or, if that did not develop, as a site for bases to be used by one belligerent or the other in the South Atlantic submarine war.

By 1940 the situation, from our point of view, had become critical. Our State and War departments increased their co-operative efforts for the dual purpose of combating Fascist infiltration in South America (at the same time building up United States influence in that area) on the one hand, and securing bases for the operation of American armed forces in Brazil on the other. The contract with Pan American was the first and most important act in attaining the latter objective.

The first bases provided under the contract were built for the U. S. Army, actual construction being carried out by ADP (Airport Development Program, subsidiary of Pan-American) under supervision of the U. S. Engineering Department. They were used, originally, for ferrying war planes to Africa and the Far East.

In 1941 and 1942, bases for this purpose were established at Amapá, Belem, Fortaleza, Fernando de Noronha, Natal, and Ibura Field (Recife). A glance at the map will indicate the strategic location of these bases in forming a chain of stopovers along the northeast coast of Brazil on the plane ferry route from the United States to war areas across the South Atlantic.

In the meantime, naval aviation was arriving in Brazil. In early 1941, Task Force 3 (later 23) under command of Rear Admiral Jonas H. Ingram [later Admiral Jonas Howard Ingram, 1886-1952], was assigned the duty of long-range patrol in the southernmost area covered by our Neutrality Patrol in the Atlantic. As the area of the U-boat depredations extended southward, Admiral Ingram's force moved south ahead of it, extending protection to shipping, and building up our naval and diplomatic relationships with Brazil. Based originally on San Juan and Guantanamo, Task Force 3 visited Recife in April 1941, and shortly thereafter was using both Recife and Bahia as replenishing stations.

At this time the only U. S. naval aviation units operating in this area were the ship-based scouting planes of Scouting Squadron 2, operating from cruisers of Cruiser Division 2 (Cruiser Division 2 comprised the major part of Task Force 3). Patrol squadrons of land-based planes, however, were soon to come into the picture.

Because obviously we could not establish military installations and activities within the territory of a friendly power without the consent of its government, the problem in its early stages was a diplomatic one in which the Navy was concerned only in so far as it could be of assistance to the State Department. Although relations between the United States and Brazil had usually been friendly and an American naval mission at Rio de Janeiro had fostered mutual confidence between the navies of the two countries, the Brazilian authorities can hardly be blamed for feeling a little hesitant about admitting our soldiers and sailors to their territory.

As long as the war in Europe favored the Axis, the possibility existed that friendliness toward the United States and Britain might ultimately bring reprisals. Strong, carefully fostered German and Italian interests were there to remind the Brazilians of what might happen in case of an Allied defeat.

As the danger of open conflict between the United States and the Axis became more evident, the inadequacy of the program carried on by Pan American Airways through the Airport Development Program was revealed. This was one reason for sending Task Force 3 under Admiral Ingram to Belem in April 1941; it indicated to the Brazilians both our desire and ability to help in their defense. At the same time our Embassy was instructed to obtain air bases in Brazil as an open governmental matter and to win permission for American military men to operate in the open rather than surreptitiously as the employees of a private company.

Success in gaining the desired concessions was conclusive evidence of the friendly disposition of the Brazilian government and people, even in the days when things were not going so well for our side, and of the skill and tact of our diplomatic and military representatives.

So well did the latter do their work that in 1942, when Brazil declared war, President [Getúlio Dornelles] Vargas delegated operational command of the Brazilian Army, Navy and Air Force to Admiral Ingram, giving him full authority and responsibility for the protection of Brazilian territorial and sea areas.

It has been noted that the first U.S. aviation bases established in Brazil were built for the use of the Army. At the outset, the Navy did not intend to use landplanes in this area, and as a consequence confined its recommendations to seaplane bases. In April 1941, these recommendations were revised to include five seaplane bases. On one of these, the Natal seaplane base, construction was started in March 1941, and the others followed at intervals. Since this work was being done by ADP under U.S. Army supervision, and since the Army had construction of its own going on at some of these places as well as at many others, the navy program was necessarily slow and irregular.

The seaplane base at Belem, for example, was not begun until September 1943, although it was one of the first authorized. In the meantime, naval planes used army bases and facilities at some points pending construction of their own.

Construction of U. S. naval bases in Brazil was under Admiral Ingram, who served successively under the titles of Commander, Task Force 3; Commander, Task Force 23; Commander, South

Atlantic Force; and, beginning in March 1943, Commander, Fourth Fleet. The setup at the outset was not too satisfactory because of the devious channels through which recommendations and requests had to be routed.

Fourth Fleet made requests or recommendations to Naval Operations who, after approval, forwarded the request to the Bureau of Yards and Docks. The bureau then passed the word to the chief engineer in the War Department, and, after it passed from him to the division engineer, it finally reached the contractor who was to do the work.

In October 1942, a fleet civil engineer and several Civil Engineer Corps officers were added to Admiral Ingram's staff to represent the Navy at construction locations, plan and supervise construction, and thus simplify and expedite the carrying through of the air base program. The result was salutary to a high degree, and to the Fourth Fleet and its Fleet Engineer Office belongs much of the credit for the rapid expansion of our air power in Brazil.

As war operations became more intense in the South Atlantic in the early days of 1943, it became apparent that the seaplane base program was inadequate. Naval land-based planes were using army fields and facilities as bases for convoy coverage and anti-submarine patrol, causing congestion of fields and personnel facilities.

In the interest of efficiency, and in anticipation of further expansion, it was highly desirable that the Navy have landplane facilities of its own. Accordingly, Admiral Ingram, on 9 February 1943, recommended a program of landplane base construction for the Navy. This program, with increases recommended by Commander, Air Force, Atlantic Fleet, was approved by the Vice Chief of Naval Operations on 22 May 1943.

In the meantime, Commander-in-Chief, U. S. Fleet, had recommended, on 26 February 1943, the diversion of two airship docks with facilities, and twelve operating blimps to northeastern Brazil. After approval by the secretary of the Navy, the Vice Chief of Naval Operations appointed a board to make recommendations for the lighter-than-air program in Brazil, including location of the blimp bases.

The board made its report on 8 April 1943, and on 17 May the Vice Chief of Naval Operations directed that the board's recommendations be acted upon. Accordingly, lighter-than-air facilities were installed at ten sites along the Brazilian coast. The main overhaul base (for the operation and maintenance of twelve airships) was established at Santa Cruz, near Rio de Janeiro, where an old German hangar built for the Graf Zeppelin was made available for navy use.

Additions were made from time to time to the programs outlined above, with the result that by late 1943 an unbroken chain of naval air bases extended from French Guiana in the north, 2,500 air miles along the Brazilian coast to Santa Cruz, near Rio de Janeiro, in the south. Beginning at Amapá, near the French Guiana border, and extending to Rio de Janeiro, the bases supported landplanes, seaplanes, and blimps for a complete coverage of adjacent waters in convoy escort and anti-submarine patrol.

In 1944, the United States extended aid to the government of Uruguay in constructing a landplane base on Laguna del Sauce, also in the general area of Montevideo. This marked the most southerly limit of the base support for land-based naval planes in the Western Hemisphere.

The first squadron of naval patrol planes to come to Brazil was Patrol Squadron 52, which arrived in Catalinas at Natal in December 1941, as a part of the Neutrality Patrol. These planes used the Pan American facilities at Natal, and their work consisted largely of patrol flights, since no submarines had appeared in the Brazilian area up to that time, and Brazil was still at peace with Germany and Italy. The pilots and crews lived ashore in the town or in tents and other makeshift facilities near the Pan American ramp. They remained at Natal until relieved by Patrol Squadron 83.

On 7 April 1942, six Catalinas of Patrol Squadron 83 landed at Natal, to be followed, on 13 June 1942 by the other six planes of the squadron. On the night of the arrival of the first planes, the men were surprised to hear the story of their arrival on the German nightly news broadcast. The news, it was later established, had been radioed to the Nazis by personnel of the German Condor Air Line quartered nearby.

With the coming of Patrol Squadron 83, naval aviation's part in the war in the South Atlantic began in earnest. The first attack on a submarine was made on 2 May 1942 off the island of Fernando de Noronha. While submarines in great numbers had not yet begun operations in Brazilian waters, sinkings had become alarming in the Trinidad area, and it was expected that the undersea boats would soon move south in force to attack shipping along the entire coast of Brazil. Accordingly, a convoy system for important shipping was established between Trinidad and Bahia starting in June 1942, and Patrol Squadron 83 started protective convoy sweeps at that time. By the year's end all important shipping was convoyed south of Trinidad as far as Bahia.

It is interesting to note the manner in which the squadron, with no base facilities to speak of, and little equipment beyond its twelve planes, managed to cover convoys over the 2,200-mile line from Cape Orange to Bahia. It must be remembered that Patrol Squadron 83 was at that time the only naval squadron in Brazil, and that practically the only bases available outside of Natal were Pan-American fields or partially completed army bases consisting mainly of runways, and having absolutely no facilities for quartering or messing naval personnel.

When word came from Fourth Fleet that a convoy was due to come into the squadron's area (a Trinidad-Bahia convoy for example) the squadron, based at Natal, would send a detachment of two or three planes north to Belem and Amapá. On arrival at one of these sites, a camp of tents for the plane crew would be set up adjacent to the runways (or the landing area, if a water landing had been made), and all hands would cook their own meals and spend the night.

The following day one or two planes would go out and cover the convoy, while the crew of the third plane broke camp, packed up the tents and mess gear and proceeded to the next night's stopping point where they would have the camp established in time for the return of the crews of the two planes out covering the convoy. The following day the plane that had made the camp would take the coverage, and the process would be repeated until the convoy had reached Bahia. When day and night coverage was required, four planes instead of the usual two or three would be sent.

Patrol Squadron 83 carried the entire burden as the sole U. S. naval plane squadron in Brazil until 7 November 1942, when Patrol Squadron 74 arrived at the Natal seaplane ramp and joined in the work.

The next to arrive was Patrol Squadron 94, the first detachment of which landed at Natal on 7 January 1943. These three squadrons were the only ones operating in Brazil up until the forming and reporting of Fleet Air Wing 16, and they became the original three squadrons of the wing.

It should be noted at this point that Patrol Squadron 83 made important contributions to the technique of anti-submarine warfare during its pioneer months in South America. The whole game was new, and the squadron, through trial and error, built up its methods and passed them on to the squadrons which came later and were assigned to Natal for familiarization.

In March 1943, the Germans began to intensify their submarine operations in the South Atlantic. In anticipation of this development, Fleet Air Wing 16 had been commissioned at Norfolk on 16 February under command of Captain R. D. Lyon, USN. Its function was to control and co-ordinate under the Fourth Fleet all air operations and training of U. S. Navy squadrons and other aircraft assigned to the Fourth Fleet. Captain Lyon reported to Commander, Fourth Fleet, on 14 April 1943, and wing headquarters were set up in Natal.

At this time Patrol Squadrons 83 and 94, with Catalina amphibian planes, were based at Parnamirin Field, Natal, using the field jointly with the U. S. Army; and Patrol Squadron 74, with Martin Mariner seaplanes, was located at the Natal seaplane ramp. Detachments of planes from these squadrons were kept at the Pan-American and army fields and seaplane landing areas at six other spots along the Brazilian coast.

The arrival of the wing in Brazil was a part of the expansion program designed to combat the greatly increased operations of submarines in the area, and occurred at the time when approval was being granted for construction of extensive landplane base facilities. The coming intensification of the anti-submarine effort, entailing as it would the assignment of many additional squadrons and the carrying out of coordinated surface, land-based and carrier-based air attacks, made necessary a coordinating command such as the wing would afford.

The task force principle was applied in this as well as in other theaters of war. The operational phase of Fleet Air Wing 16 was, therefore, designated as Task Force 44 and assigned a basic mission. This mission included a patrol of the seas off the coast of Brazil, air coverage for convoys as needed, intensive anti-submarine operations in conjunction with surface vessels or independently when needed, and co-operation with surface vessels in attacks on blockade runners or raiders. At a later date, the training of Brazilian squadrons and detachments was included as an additional duty.

From the date of the arrival of the wing, the anti-submarine warfare was stepped up in the South Atlantic in opposition to the submarine offensive which developed as expected. New squadrons were thrown into the fight as they arrived. Bombing Squadrons 127 and 129, flying new Venturas, arrived on 14 May and 1 June 1943 respectively, both in time to participate in the battle which reached its peak in July. During that month at least fifteen submarines were operating along the east coast of South America, nearly six times the average number that had appeared there in the preceding four months.

Of these, eight were probably sunk, for a ratio of one submarine probably destroyed for every 1.75 ships lost. This is in comparison with a record of one submarine sunk for every 3.3 ships lost in all areas during the first half of 1943. U. S. Navy planes made twenty-four sightings, leading to sixteen attacks, which in turn resulted in the probable destruction of eight submarines.

On 27 September 1943, aid in the submarine war arrived in a new form when a navy blimp landed at Fortaleza. It was the first unit of Fleet Air Wing 4, whose commander, Captain W. E. Zimmerman, USN, had arrived and reported on 2 August. The blimps concentrated on convoy coverage and rescue work, and in the latter category, particularly, they distinguished themselves by the record they made. Beginning in September 1943, they took a sizable portion of the daily convoy coverage load off the backs of the heavier-than-air squadrons.

New squadrons continued to arrive, from August 1943, to February 1945. These squadrons operated from the various bases along the Brazilian coast, being shifted from one to another, or splitting into detachments as the tactical situation required. Fourteen squadrons in all were assigned to the Fourth Fleet, but at no time were there more than nine squadrons in the area simultaneously.

In addition to its own planes, Fleet Air Wing 16 had assigned to it for training and operation, three Brazilian squadrons. It should be noted that throughout its stay in Brazil, the wing engaged in training Brazilian pilots and teaching them the necessity and techniques of plane care and maintenance. In this task, particularly in the maintenance phase, the Americas had to start practically from scratch. It is to their credit as well as to that of the Brazilians that the latter developed into effective units which fought efficiently in co-operation with United States forces.

In addition to its routine duties of covering convoys, conducting regular anti-submarine patrols, and training Brazilian fliers, Fleet Air Wing 16 engaged upon occasion in special operations involving co-ordination with other forces. Chief among these were barrier sweeps against submarines or surface blockade runners conducted in co-operation with surface ships and in some cases with U. S. Army planes.

The technique of barrier operations was fairly simple. When the plotted position, course, and speed of an enemy vessel, together with intelligence as to its destination, indicated that it would necessarily pass over a given line or through a given area within a certain time period, a plan was drawn up whereby the area would be covered by air and surface craft in such a manner that the enemy could not make its passage and escape detection. Aircraft, in most cases, developed the contacts, and their reports brought patrolling surface craft to the scene. The attack might be made by aircraft, surface craft, or both, depending upon the tactical situation.

The barrier operations against German blockade runners in December 1943, and January 1944, provide a typical example. During these two months, five German blockade runners — Osorno, Weserland, Rio Grande, Burgenland, and Alsterufer — attempted to pass through the Atlantic, returning to Europe from the Far East with important cargoes of rubber, tin, and other strategic war materials.

When Commander, Fourth Fleet received intelligence that the ships were to pass through his area, he immediately drew up plans for an air and surface barrier to extend across the South Atlantic Narrows from Natal, at the point of the bulge of Brazil, east to Ascension Island, thence north-northeast to within four hundred miles of the coast of Africa, which was as far as available ships and planes could extend an effective barrier line. Allied patrols from Africa partially closed the remaining gap near the African coast.

The longest and most important sector of the barrier line, from Natal to Ascension Island, was assigned to planes of Fleet Air Wing 16 for air patrol. Regular patrols would fly east from Natal and west from Ascension Island, meeting in mid-ocean (within radar observation distance, that is) before turning back to base. These flights were so timed that the blockade runners, with their known maximum speeds, could not traverse the belt covered by the planes' radar in the interval between patrols.

At the same time, cruisers and destroyers patrolled the area, ready to rush to the scene when a plane reported sighting a ship. The effect was like that of a vast net stretched across the ocean, in which the enemy was almost sure to become entangled.

The result of this far-flung hunt was that three of the five blockade runners — Weserland, Rio Grande, and Burgenland — were caught and sunk by planes and ships of the Fourth Fleet. The Alsterufer was subsequently sunk by aircraft in the North Atlantic. Only one ship, the Osorno, succeeded in reaching the French coast, and even this one had to be beached because of previous damage.

Such was the story of land-based U. S. aviation in the South Atlantic. Hundreds of tales of individual heroism could be told, of attacks pressed home by slow patrol planes in the face of devastating antiaircraft fire from surfaced subs, of plane crews downed at sea and rescued by their comrades, of men dying in operations too prosaic ever to achieve even the brief fame of newspaper mention.

A good yarn can be made of "hold down" operations, in which planes watched over submerged U-boats, like a cat watching a rat hole, forcing them to remain under water until they could stand it no longer, and then destroying them when they came up for air. Tales could be told of the opposite "baiting" tactics in which the planes left the area, lulling the sub into a sense of security, only to return and pounce upon him when he came unsuspectingly to the surface. Incidents such as these, in infinite variation, will make good fireside stories for years to come.

Between the moments or days of excitement, however, came the weeks and months of tedious and fatiguing routine — patrol and convoy canvas, a duty which, while never free of danger, was absolutely lacking in glamour. The historian of Fleet Air Wing 16 sums it up in these words:

"Mock heroics have not been engaged in down in this area but, as the record shows, the aviation personnel has been willing to accept death in action when the chips were down, and a job had to be done. Many of these unsung heroes have not received medals, but they have all done their bit. In this area it has been a war where patience and steadiness have counted for as much as brilliance and dash in other theatres where there has been more shooting."

It has been noted that the peak of the submarine attacks on shipping in the South Atlantic came in July 1943. They never again presented a threat of equal seriousness. The wholesale destruction of U-boats during that month by our air and sea forces, plus the continued building up of our air strength during the remainder of the year (five squadrons were attached, two detached, and two blimp squadrons added) ushered in a decline in submarine activity which continued except for occasional flare-ups, until the end of the war.

The vigilance of our fliers was not relaxed in any degree and, indeed, sinkings by submarines continued in lesser numbers until the end of the war. But the Battle of the South Atlantic had been won. In May 1944, Fleet Air Wing 16 was reduced by two squadrons, and by the end of the year, plans were well advanced toward closing out aviation activities in Brazil. The wing was decommissioned in June 1945, its mission completed, and its few remaining squadrons returned to the United States.

Carriers did not take part in Fourth Fleet operations in the South Atlantic until the spring of 1944. On 31 March 1944 the USS Solomons, an escort carrier, with Composite Squadron 9, embarked, together with assigned destroyer escorts, became Task Group 41.6, and carried on long-range submarine hunts until 15 August, on which date the Solomons departed Recife for Norfolk.

Two days before the departure of the Solomons, USS Tripoli with Composite Squadron 6, embarked, and later arrived at Recife in company with a screen of destroyer escorts. The Tripoli and her escorts became Task Group 41.7, and operated in the South Atlantic until 15 November 1944.

On 21 September 1944, the USS Mission Bay, carrying Composite Squadron 36, departed from Dakar, French West Africa with five destroyer escorts for anti-submarine operations southwest of Cape Verde Islands. From 28 September to 2 October the Mission Bay engaged in joint operations with the Tripoli. On 15 November 1944, both the Tripoli and the Mission Bay departed Recife for Norfolk, ending carrier operations in the Fourth Fleet area.

Any final estimate of the accomplishments of naval aviation in the South Atlantic during this war must take into account both military and political objectives.

From a military point of view, the purposes of the South Atlantic campaign were to protect the Western Hemisphere from attack, protect shipping, insure the delivery of war material to Europe, Africa, and the Far East, and accomplish the final defeat of the U-boat in the South Atlantic area.

These purposes were achieved in full. Beginning with relatively untrained personnel, planes poorly equipped and too few in number, no tried and proved doctrine for anti-submarine warfare, and practically no base facilities, Commander, Fourth Fleet, and his aviation forces established air bases from one end of the Brazilian coast line to the other, developed an organization of skillful and seasoned airmen with a complete combat doctrine based on experience, and finally reduced the U-boat to impotence in the South Atlantic.

In the political field, to the continuing policy of fostering amicable relationships with Brazil was added, in the early days of the war, the immediate and urgent objective of obtaining bases in Brazil and permission to operate them, as well as that of obtaining the full co-operation of the Brazilian government in the active effort against the enemy.

These objectives were attained in full measure. But this achievement does not tell the whole story of our diplomatic success in that area. The mutual confidence, good will, and reconciliation of basic interests engendered during our wartime operations will have salutary effects upon hemisphere solidarity long outlasting the achievement of the immediate objectives of this war.

To what extent naval aviation was responsible for our diplomatic success is open to debate. It can be safely stated, however, that the reassuring presence of American naval air might in the skies over Brazil, protecting Brazilian shipping and coastal areas, the co-operation of our Navy in training Brazilian fliers and furnishing them with planes, and the fair and statesmanlike attitude of Commander, Fourth Fleet, in negotiations over the establishment of the air bases, all played their part in sealing the friendship between Brazil and the United States. In the South Atlantic, naval aviation's mission was accomplished, and more.

TYPE 156 BEAUFIGHTER MK X WITH ROCKETS

9 – Naval Aviation in the Italian Invasion

IF THE NATURE OF NAVAL OPERATIONS in the Atlantic and in European waters relegated the Navy's air power to a less spectacular role than it played in the Pacific, that role was by no means unimportant. Wherever the Navy functioned in those theaters, it was almost always accompanied by air units and practically every form of naval air operation was undertaken at one time or another, from offensive carrier plane strikes to air-sea rescue.

Until the landings made in North Africa in November 1942, naval aviation in the Atlantic theater, as we have seen, devoted itself primarily to protecting the sea lanes between America and our European allies, that is, to anti-submarine warfare. Beginning with the African landings, American naval air power played an ever-increasing part in the crescendo of combined operations which finally precipitated the collapse of Germany.

In each landing the Navy as a whole performed two major functions: the first was the strictly amphibious operation which ferried assault troops and equipment to the beaches; the second was gunfire support. Here the Navy took up positions offshore and supplied artillery support needed by the assault troops until a beachhead was secured and until mobile ground artillery could be landed and brought to bear against the enemy.

In this latter phase, naval planes performed vital tasks of spotting and directing the fire of naval vessels upon shore targets. In the African landings and, as we shall see later, in the final European landings made on the southern coast of France in August 1944, a third important function was undertaken by American naval forces with escort carriers. Because carriers were able to operate close inshore, they provided immediately available air bases from which long-range fighter planes were able to strike at the enemy far inland until the Army Air Forces and Royal Air Force were established at land bases.

Traditionally viewed as "the eyes of the fleet," naval observation and scouting planes operating from battleships and cruisers emphasized the accuracy of that view with distinctive credit throughout the war. Normally, each ship carried from one to four float-mounted scout-observation planes. They were launched or literally "shot" into the air from deck catapults.

When they returned for recovery they landed in the relatively smooth water of the ship's wake and were hoisted aboard by aircraft cranes usually on the ship's fantail. The men who flew these planes received dual training in aviation and naval gunnery and were under administrative command of the ship's gunnery officer.

Because of their relatively slow speeds, older types of scout-observation planes could almost hover above the target area correcting their ship's main battery fire and indicating new targets by radio. The Curtiss SOC's, called "Socks," and the newer "Kingfishers" did remarkable work from the outbreak of war until they were finally replaced, in the latter part of 1944, by faster, more heavily armed SC-1 Curtiss "Seahawks," which were not so vulnerable to enemy aircraft. While they were never intended for such tasks, all these types were used at various times for bombing, strafing, anti-submarine patrols with depth charges, and similar types of combat missions.

With the exception of the Casablanca naval operations, none of the landings made in the European-Mediterranean theater was strictly a navy "show," even though our Navy usually provided the greater number of ships. For the most part each landing was a true combined operation with American, British, French, Netherlands, Canadian, Greek and Polish units, plus units of other nations taking part in varying arrangements according to the availability of their sea and air power.

The major naval strength was furnished by the two greatest sea powers, the United States and Great Britain, and naval command for each operation was shared or alternated between the two. This was the case when Sicily was invaded some eight months after the African landings.

The invasion of Sicily on 10 July 1943 was the largest amphibious operation ever attempted to that date. In size and scope, it was exceeded in the European war only by the invasion of northern France almost a year later.

It was the first time we were able to strike a major blow against the enemy in his home territory, his vaunted Festung Europa. General Eisenhower commanded the expeditionary forces, and top naval command went to British Admiral Sir Andrew Browne Cunningham.

The over-all plan called for landings at five different places on the island, three of which, Scoglitti, Gela, and Licata on the south coast of Sicily, were primary American objectives. Transports, cruisers, and destroyers were assembled at Oran and Algiers, and landing craft were assembled at Tunis and Bizerte. More than fifteen hundred United States naval vessels participating in the operation began to leave North African ports on 5 July, and on the morning of the tenth they were all in position for the assault, which was preceded by a heavy naval bombardment of shore targets.

The light cruisers Brooklyn, Birmingham, Philadelphia, Boise, and Savannah supplied the heaviest gunfire support for the landings, most of which was directed by their spotting planes. By the afternoon of D-Day all three beaches were secured by our troops who began the job of unloading supplies to consolidate positions. However, the enemy began a series of strong counterattacks by air and land that kept our cruisers seriously occupied for the next few days.

One of the most effective jobs undertaken by our cruisers and destroyers was in stopping counterattacks by enemy tanks which threatened to drive our forces into the sea. Naval gunfire was directed against shore batteries, roads, bridges, and other targets of opportunity throughout July and August until attention was shifted to the coast of Italy proper.

It was during the early hours of the D-Day assault on the Scoglitti area that a Navy pilot in one of the Philadelphia's "Socks" performed one of the most remarkable exploits in the annals of naval scout-observation aviation. The pilot, Lieutenant (junior grade) Paul E. Coughlin, while spotting for his ship over the assault area, reported enemy troop activity a short distance inland from where our forces were unloading. Since he carried two 100-pound bombs on his plane he requested permission from his ship to bomb the enemy.

Permission was granted by the Philadelphia's commanding officer to bomb all troops on the beach provided they were positively identified as enemy. He made a dive-bombing attack on the enemy troops dropping his first bomb which failed to explode because of incorrect fusing. Nevertheless, it had the desired effect of dispersing the enemy troops and keeping them away from their gun position. As he pulled out of his dive, Coughlin circled past an American forward patrol on the beach at the foot of a cliff. This group waved to him and pointed to the hill position from which it was evidently being held up by the enemy.

Circling low over the hill, Coughlin spotted a group of enemy soldiers running into a hedge, whereupon he immediately dropped his second bomb. This bomb also failed to explode because the low altitude from which it was dropped did not give it sufficient time to arm itself, so the pilot and his radioman began to strafe the position until white flags and Italians began appearing from the hedge. Coughlin then flew low and motioned for them to move in the general direction of the American patrol on the beach.

The radioman, Richard Shafer, ARM2c, emphasized the pilot's instructions by well-placed machine-gun bursts at their heels. Four more white flags appeared over entrenchments as the prisoners moved toward the shore. As the first group of Italians passed one of these entrenchments, the pilot flew over and guided these new groups in the direction of the main party, again punctuating his instructions with gunfire which effectively convinced the hesitant few.

Using tactics of the cattle roundup, the two navy men in the SOC herded their prisoners into the open and directed them to the beach, discouraging by machine-gun fire those who indicated a desire to stray. At one point, when the prisoners were scattering, the radioman's free machine-gun jammed.

The pilot circled low over the group and the radioman kept the gun pointed at the men, none of whom realized the situation, for when one did try to break away and run for shelter, the radioman turned him back by firing his .45 pistol along the machine-gun barrel. The prisoners needed no further convincing and about a hundred of them reached the positions on the beach where they were taken by American troops.

Coughlin returned to the hill and "flushed" an additional thirty or more Italians by judicious strafing with the pilot's fixed gun, but as he was directing them to the crest of the hill, he suddenly saw a burst of antiaircraft fire on his starboard beam and two German Messerschmitts closing on his tail.

He went into a steep dive and at that moment 5-inch shells from the Philadelphia began breaking between him and the enemy planes, forcing them to turn away from the pursuit. Men on cruiser had been watching the entire show, and when they saw German planes some five or six miles from the SOC shoot down a plane spotting for a British ship, they anticipated the attack on Lieutenant Coughlin and the gunnery officer ordered immediate fire on the enemy.

As soon as the German planes disappeared, Coughlin returned to his prisoners who had taken advantage of his difficulty and dispersed. Nevertheless, he again rounded up most of them by strafing near-by buildings and other possible hiding places. When the Italians were within firing range of the Americans on the beach, he returned to his ship, having brought in approximately 150 prisoners with an expenditure of 1,000 rounds of .30 caliber ammunition.

The Sicilian campaign was scarcely ended on 17 August 1943 when British forces crossed the two-mile Straits of Messina and landed on the Italian mainland. On 9 September the Allies moved in force to secure the important harbors of Naples, Gaeta, and Salerno with their main effort directed against the last of the three in the Bay of Salerno.

A combined American-British fleet of more than six hundred ships and large landing craft under Vice Admiral [later Admiral] H. K. Hewitt [Henry Kent Hewitt, 1887-1972] supported the landings. This fleet was divided into two groups, one predominantly American and the other mainly British, with the American Southern Attack Force covering the Salerno operations.

From D-Day until they bombarded Naples on 1 October, our naval forces were engaged primarily in supplying the beach forces, in fighting off enemy attacks, and in destroying shore targets by naval gunfire. Together with British battleships and lighter units of both nations, the United States cruisers Boise, Philadelphia, and Savannah which supported the Sicilian invasion, continued their offshore bombardment of enemy installations and concentrations.

Here again, heavy spotting responsibilities fell to fleet aircraft. New considerations were introduced, however, which greatly altered former concepts of aerial spotting and observation. The hazards of spotting naval gunfire from the then available cruiser planes had been conclusively proved at Casablanca and Sicily. "Socks" and Kingfishers were designed as adjuncts to the gunnery functions of battleships and cruisers, not as fighter planes. The very nature of naval gunfire support for an amphibious invasion meant that these planes were now operating close inshore and therefore well within range of land-based fighter aircraft.

Their vulnerability to these fighters because of their relatively slow speeds was obvious, and consequently it was decided that because of anticipated enemy fighter strength in the Salerno operation, the use of cruiser-based planes for spotting would be of limited value.

The Commander, Western Naval Task Force, Vice Admiral Hewitt, in his report on the Salerno operations explained the steps taken to overcome this problem as well as the future planning for naval air spotting which was so successful in later operations:

"A conference of representatives of the Royal Air Force, Royal Navy Fleet Air Arm, United States Naval Aviation, and staff gunnery officers (British and American) . . . was held to determine the plane most suitable to perform this mission. The P-51 (Mustang) was selected. As time did not permit training cruiser pilots to fly P-51's, it was decided that U. S. Army Air Force pilots would be trained to spot naval gunfire. Navy pilots trained the Army pilots in spotting procedure. Aircraft operated in pairs, one spotting, the other furnishing cover. (This procedure doubled spotting effectiveness in that both pilots were trained spotters, and either could take over if the other was rendered ineffective by mechanical failure or enemy attack.)

Contrary to general belief, the naval gunfire spotting from P-51 aircraft proved exceptionally successful. Unfortunately, only four P-51 aircraft were made available for spotting. In future planning, when it is necessary to use Army Air Force reconnaissance for this work, the number of aircraft needed for spotting should be laid down as a naval requirement, and the pilots trained to 'spot' rather than to 'sense.' . . . for immediate future operations in this theater the U. S. Navy in order to have suitable aircraft for spotting naval gunfire must use high performance aircraft, and Army pilots trained to use naval procedure. Steps have been taken to train immediately the cruiser pilots in this theater to fly fighter aircraft, and to learn fighter tactics and gunnery."

Vice Admiral Hewitt's recommendations were heeded, and naval pilots trained in fighters proved their exceptional value in the invasion of northern France.

Despite the dangerous nature of such missions, SOC's did fly from American cruisers during the opening phases of the Salerno landings and provided valuable spotting-observation information. One plane from the USS Savannah directed cruiser fire on a group of about twenty enemy tanks on D-Day. The pilot reported four more tanks moving on a road in the direction of the main group, plus thirty-five additional vehicles as well as troops in the area.

After ascertaining that the vehicles and troops were not friendly and that there was no Naval Shore Fire Control Party in the vicinity, he effectively directed cruiser fire at the target. After 9 September, spotting was taken over completely by Shore Fire Control Parties and the P-51's.

The only carriers in this operation were British; nevertheless, the U. S. Navy controlled other fighter aircraft. The USS Ancon flagship for the Commander, Western Naval Task Force, also served as the fighter director ship for all land-based fight aircraft of the XII Air Support Command throughout the operation.

The famous "end run" terminating at the Anzio-Nettuno beachhead was the final amphibious invasion on the coast of Italy. Designed to break the stalemated military situation and hasten Allied capture of Rome, it was a direct flank attack on supply and communication lines fifty-five miles behind the front, which forced the enemy to turn about and fight and then withdraw to new defense positions.

The operations began in the early morning hours of 22 January 1944, supported by a naval task force composed of British, American, and Greek naval units, which were later supplemented by French and Netherlands ships for gunfire support. The entire naval task force of 243 ships was under the command of Rear Admiral F. J. Lowry, USN.

The principal American naval gunfire support during the early phases of the campaign was furnished by the light cruiser Brooklyn and the destroyer Edison, but other American naval units continued to support the beachhead until it was expanded with the Fifth Army move on Rome in June 1944. Here again the primary function of navy aviation was spotting. Because the landings caught the enemy by complete surprise, there was little initial opposition and most naval gunfire was directed at whatever targets were located by either spotting aircraft or Shore Fire Control parties rather than to previously selected objectives.

In addition, Army Air Forces planes spotted fire for all Allied ships. Our cruisers engaged in long-range duels with enemy artillery, but the most important spotted naval gunfire was interdiction fire on roads, junctions and bridges, disrupting enemy troop and supply movement.

10 – Naval Aviation in the Invasion of France

Amphibious invasions of hostile territory by combined naval, ground, and air forces were one of the outstanding military characteristics of World War II. The beginning date of these operations was normally called D-Day and the precise time of the attack H-Hour. Although there were many such days for our armed forces during the war, to most Americans D-Day meant 6 June 1944 — the day of the assault on northern France in the Normandy area. It was here that the major Allied offensive effort against Germany was directed.

To transport the bulk of the assault forces and their equipment to the attack area, the Allies assembled the greatest and most diversified naval amphibious and support forces ever engaged in a single operation. The invasion fleet composed of American, British, Canadian, and French units was under the command of Admiral Sir Bertram Ramsay, RN. It was divided into two task forces, one under British command covering the eastern area and the other under Rear (later Vice) Admiral Alan G. Kirk, USN, covering the western beaches.

There were several task groups under Admiral Kirk including forces commanded by Rear Admiral J. L. Hall, USN, and the late Rear Admiral D. P. Moon, USN. The force of more than two thousand naval craft of all types under Admiral Kirk's command formed but part of the total number of Allied vessels used in the invasion.

Since the assault areas were within range of American and British fighter aircraft based in southern England, there was little need for carrier planes. United States naval aviation, however, performed two vital functions throughout the entire campaign; one was protecting naval vessels operating between England and the invasion coast of France; the other was aerial spotting and observation for naval gunfire support ships. The first of these functions was the primary responsibility of the Navy's fleet air wing (Fairwing) 7.

Aside from air attack, the greatest potential menace to the naval forces in this campaign came from the still powerful German submarines operating from French bases on the Atlantic. It was of the utmost importance to the success of the invasion that U-boats be prevented entirely from operating in the English Channel or, failing that, "kept down" and thus rendered less effective.

Because submarines could operate under water for only limited periods at relatively slow speeds, their ability under such conditions to catch surface ships in a position favorable for torpedo attack was severely limited. Normally, they achieved position by relatively high-speed surface movement. By making the surface unsafe for them, coordinated aircraft and ship patrols neutralized U-boats by forcing them to remain submerged and unable to get into striking position.

The PB4Y-1's or Liberator patrol planes of Fairwing based at Dunkeswell, Devon, England, were important members of the Allied air-surface team which beat back attempted U-boat interference with the Normandy invasion. Operating under the direction of Royal Air Force Coastal Command, these navy planes helped maintain contin-uous air patrols over the Channel and its Atlantic approaches. Combined with those of surface ships, these patrols were so inten-sive that a submarine could not remain on the surface for more than a few minutes without being sighted and attacked.

Even the development of the "schnorkel" or breathing pipe which permitted U-boats to operate much more efficiently and for longer periods under water did not protect them from air patrols which covered every square mile of the Channel and its approaches every few minutes. Planes were able to spot the tips of the "schnor-kels" and thereby attack the submarines.

During the first few days of the invasion, U-boats surfaced only at night hoping for relative security from aircraft sighting. This move was continued by Royal Air Force planes equipped with searchlights for night spotting which were later joined by half a Liberator squad-ron similarly equipped. Air patrols made surfacing dangerous for U-boats at any hour of the day or night.

Except for a brief period in mid-1943, when Doenitz ordered his U-boats to remain on the surface and fight back, surfaced German subs were a rarity.

These combined patrols were so successful that on D-Day and for a period of three weeks afterward not one Allied ship was sunk by enemy submarine action. Then a small ship was sunk by possible, though not definitely confirmed, U-boat action. From that time until the close of the assault phase of the campaign, there were no further U-boat attacks in this area. Fairwing Liberators were credited with seventeen attacks on submarines during the first three weeks of the invasion, eight of which were made during a two-day period when the U-boats made an all-out effort to get through our patrols. Many of these attacks were assessed as damaging or as probable "kills."

Under the circumstances it is all the more remarkable that not one navy man or plane was lost during the operation, since on at least one occasion a wing plane engaged a surfaced submarine in a dramatic air-sea fight. Although damaged by the U-boat's antiaircraft guns, the Liberator's depth charges finally sent the enemy to the bottom.

Aerial spotting and observation for naval gunfire support was further developed in the Normandy campaign. The experience gained in the North Africa, Sicily, and Salerno landings was put to good use in that air spotting was done by shore-based, fast, single-seat fighter planes instead of regular ship-based float planes.

Spotting for all Allied fire support ships of both Eastern and Western task forces was performed by four squadrons of Seafires from the Royal Navy Fleet Air Arm, three squadrons of Mustangs and two squadrons of Spitfires of the Royal Air Force, and one squadron of Spitfires flown by seventeen United States Navy aviators from the six major United States naval units taking part in the operation — the battleships Texas, Arkansas, and Nevada, and the cruisers Augusta, Quincy, and Tuscaloosa.

All these squadrons were gathered at the Royal Navy air station at Lee-on-Solent, Hampshire, and were familiarized with the aircraft, methods, communications, and spotting procedure to be used in the operation, thus insuring uniformity of knowledge on every phase of spotting technique. Such training further enabled all pilots to spot interchangeably for British, American, or French ships.

The U. S. naval aviators formed Observation Fighter Squadron 7 and were trained in the Spitfires by the United States Army 67th Reconnaissance Group of the 9th Tactical Air Command. An air spot pool, varying at times between 95 and 160 planes, was created from all these squadrons at Lee-on-Solent from where they operated throughout the campaign.

Before D-Day, gun positions and other strong points from which the enemy might interfere with landing operations were listed by the Army and assigned as targets for air-spotted ships' fire. But in the first few days of the invasion, only three targets at a time were assigned to each pilot because even with their standardized training, it was impossible to expect every pilot to be an expert in spotting and reconnaissance.

With experience gained during the operation, these pilots increased their value by locating targets of opportunity and directing fire upon them. Prior to leaving Lee-on-Solent on spotting missions, pilots were briefed on prearranged targets or specific areas to be searched for targets.

In other cases, they were instructed to fly directly to the target area and report by radio to assigned fire support ships for target briefing. The ships often directed spotting aircraft to scout specific areas for intelligence information. When a pilot located and reported a target, the ship determined the advisability of engaging the target with naval gunfire. If the target was considered important, the spotting plane remained in the area directing ships' fire upon it.

As originally planned, a much larger share of the spotting was to have been done by Shore Fire Control parties but because these lost personnel and equipment in the first few assault waves, the major spotting tasks fell to aircraft. Four hundred and fifty-three missions were flown for the ships under Admiral Kirk during the first twenty days of the operation of which more than 100 were flown on D-Day.

Of this total, 81 missions were flown by the 17 United States Navy pilots. Approximately 10 percent of all spotting missions throughout the entire campaign were flown by these 17 pilots for both the Eastern and Western task forces.

Dependence on fighter planes for spotting was assured for some American naval units when their deck catapults were removed at Belfast before proceeding to the Channel thus preventing the use of their ship-based aircraft. However, faith in the fighter-spotter was not misplaced. Their success prompted several navy commanders to recommended continued development and more extensive training and use of this type of air spot.

The only objection to their use was in their range limitations. Coupled with the normally restricted range of a small, fast fighter plane, the two-way channel trip from the air spot pool to the target area and back, normally left a relatively short time during which each plane was available for spotting. Nevertheless, this problem was overcome by the establishment of an efficient shuttle service from Lee-on-Solent which kept spotting planes in the air whenever and wherever they were needed.

Most effective air spotting was done for the main 12- and 14-inch batteries of the battleships during the first few days of the assault. The Texas relied completely upon air spot on D-Day and was able to blast important enemy gun positions several miles inland even after its Shore Fire Control party was knocked out in landing.

The Nevada's log carries a report on June 8 of its main battery fire directed against a concentration of enemy tanks and other vehicles by a spotting pilot who radioed the ship with pleased excitement, "You really splattered them!" Every tank and truck was either damaged or destroyed and not one was able to make a getaway. A few days later the Nevada completely destroyed a battery of eight large enemy guns hidden in a hedge and located by air spot.

The Normandy campaign proved conclusively the value of the fighter plane for naval gunfire spotting. Furthermore, the spotting systems learned by our navy pilots at Lee-on-Solent were so simple and flexible that ships' gunnery officers had very little difficulty in adapting their operations to them in a few days' time.

Some commanders felt that the performance of this plane-ship, spotted-fire combination warranted its exclusive use over all other spotting methods in similar future operations. But development of the fast fighter for naval spotting and observation did not end here. The invasion of southern France a few weeks later brought even more improvements in the techniques of the navy air-surface combination.

The final major Allied amphibious invasion of the European continent in which United States naval forces participated was made on the southern coast of France east of Toulon in the area between Cap Cavalaire and Rade D'Agay. It was a combined American, British, and French operation in which United States naval aircraft played a vital role.

For the first time since the North African campaign American carriers were used, and air activity throughout the entire operation was spearheaded by carrier-based planes until the Army was able to move up its fighters from bases in Corsica to newly captured ones in the Rhone Valley. In original planning for the campaign, it was contemplated that carrier-based planes would be used primarily for spotting and observation.

Actually' they became a primary offensive striking force for the entire assault and consolidation phase of the invasion. In addition, regularly ship-based SOC's were used once again by some of the same gunfire support vessels which had participated in the Normandy campaign and whose deck catapults were reinstalled for this purpose at Oran, Algeria. Even navy blimps were used for spotting mines.

The assault area was divided into three sectors to each of which a naval attack force with its respective gunfire support group was attached. The aggregate complements of these support groups included 5 battleships, 3 heavy cruisers, 18 light cruisers, 3 destroyer leaders, and 28 destroyers from the Allied navies represented. Task Force 88, the aircraft carrier force, was under British command. This force was made up of 2 task groups, one of which, TG 88.1, was composed entirely of Royal Navy ships: 5 escort carriers, 2 antiaircraft cruisers and 7 destroyers.

The second task group, TG 88.2, under the command of Rear Admiral C. T. Durgin, USN, included 2 American escort carriers, the Tulagi (flagship) and the Kasaan Bay, 2 British escort carriers, 2

British antiaircraft cruisers and 6 American destroyers. The 24 Hell-cats (F6F-5's) of Observation Fighting Squadron 1 operated from the Tulagi and the 24 Hellcats of Fighting Squadron 74 from the Kasaan Bay. Other United States Navy aircraft were shore-based for a time.

A night fighter detachment of 7 Hellcats and a detachment of 5 Avengers operated from Solenzara, Corsica, to guard the carrier force against surprise attack. During the final phases of the operation, these planes were based on the two American escort carriers. In addition, Seafire squadrons flew from the British escort carriers.

Task Group 88.2 left Malta on 12 August 1944 as a part of Task Force 88 and arrived at its assigned station on the morning of 15 August (D-Day) in time to launch its first scheduled mission at 0550. The low clouds and haze that hung over the entire assault area during the morning hampered all air operations.

Poor visibility forced one pilot of Observation Fighter Squadron 1 so low to observe a target that he struck an LST's barrage balloon and landed in the water where he was picked up and returned to his carrier. Despite these difficulties, 10 missions were flown on D-Day from the Tulagi and 10 from the British carriers in the task group to spot gunfire for the Nevada, Texas, and Philadelphia and the French ships Montcalm and Georges Leygues; and 7 fighter bomber missions were flown from the two American carriers against gun positions. Near misses from these attacks set fire to two ammunition dumps and rockets and bombs accounted for numerous railroad cuts.

Observation Fighter Squadron 1 was the first United States Navy air unit to be trained and used specifically for the multiple duties of spotting and observing ground targets for naval gunfire support and assisting ground forces by armed reconnaissance and fighter bomber attacks on the enemy. This squadron's regular navy training was supplemented by a period of army training at Fort Sill, Oklahoma, in artillery ground spotting. The versatility of navy fliers reached a high point of effectiveness in this campaign when they successfully undertook missions normally required of army fighter pilots in addition to those expected of naval airmen.

While they flew strictly spotting or observation missions for the naval task forces, these aircraft were always under naval command, but flying support missions for the Army, they were controlled by the Commanding General, Tactical Air Command. The carrier force was committed to provide the air commander with seventy-two aircraft for beach cover, fighter bomber, rocket projectile and armed reconnaissance missions during daylight hours. This dependence upon navy fighters was especially important in the initial phases of the assault because of the distance from the beachheads and the nearest army air bases. Prior to the establishment of airfields in southern France, Corsica was the nearest point from which land-based fights could operate, but the rugged terrain of that island limited the size and number of fields constructed and, therefore, the number of planes operating from there.

Furthermore, the range of these planes was severely restricted by distances of from 100 to 330 nautical miles between Corsica and the target areas which left little flying time for effective attack operations. It was evident that navy planes based on carriers thirty miles offshore could make longer armed reconnaissance flights by skirting instead of flying through heavily defended flak areas, could make longer fighter bomber missions, and could remain over target areas for greater periods of time because their fuel was not consumed in long overwater round trips from distant bases.

Also, the proximity of the carriers to the beachheads enabled more prompt execution of these missions since planes were always available to the air commander, within a few minutes' notice for missions much further inland than those possible for Corsica-based planes. At night, the carriers retired sixty-five to seventy miles offshore, thereby reducing the possibility of surprise attack by moving out of ready range of enemy aircraft and providing ample area for evasive ship maneuvers should such an attack develop. Although the carrier task force was potentially strong, it was initially intended that its primary mission would be spotting naval gunfire, and its auxiliary mission, support of army ground forces.

As the invasion progressed, enemy air opposition did not materialize in the strength anticipated, and it became increasingly apparent that for long-range inland strikes, carrier planes were logical substitutes for army fighters until airfields could be established on the French mainland.

All these considerations were foreseen before the landings actually began, but the success of the carrier-based planes in carrying the initial brunt of air support for the entire campaign was considerably greater than anticipated. It was further expected that after the first six days of the operation, enemy action and mechanical failures would wear down the aggregate offensive capabilities of the aircraft to a point where their further retention in the assault area would be unwarranted. The carriers were to have been withdrawn at this point and spotting planes put ashore to continue their operations from land bases.

However, after moving out of the assault area on the night of D plus 6 for refueling, rearming, and resting of pilots, it was found that the carrier squadrons were in such excellent condition that the task group returned to the area on the morning of D plus 9 where the American carriers remained through D plus 14, their planes continuing effective offensive operations.

Throughout the campaign, the greatest problem facing all Allied aircraft was the intense and accurate flak put up by the enemy's antiaircraft defenses. This was particularly true of prepared coastal positions spotted by navy planes as well as of motor convoys which were prime targets for armed reconnaissance missions. The American carriers reported approximately sixty instances of flak damage to their returning planes exclusive of single hole or "nonvital" hits. A total of nine planes failed to return to their carriers, and it was assumed that flak accounted for most, if not all.

Hellcats were exceptionally versatile and served capably as combined spotters, fighters and fighter bombers, strafing with both rockets and .50 caliber machine guns on long-range tactical, armed or photo reconnaissance missions. They were frequently launched throughout this operation with bombs, belly tanks, rockets and machine-gun ammunition. Thus loaded, they were particularly destructive against ground targets such as gun emplacements, batteries, motor transports, troop concentrations, railroad trains, trestles, roadhouses, sheds, bridges, and barges.

Planes of the Tulagi and Kasaan Bay destroyed 825 trucks and other vehicles and damaged an additional 334. They destroyed or immobilized 48 locomotives and 353 railroad cars in addition to cutting railroads, highways, destroying fuel and ammunition dumps, and shooting down 9 enemy aircraft. Three enemy aircraft shot down on D plus 6 were German JU-52 transport planes thought to be flying to Marseille to evacuate key enemy personnel in line with previous German practice in similar situations. On this same day, United States Navy planes flew 58 and Royal Navy planes 60 missions, reaching the peak of their destructive effectiveness. German columns retreating by highway and rail were bombed and strafed, and at the end of the day more than 225 motor vehicles were destroyed or seriously damaged. This was but a part of the story of repeated successes of navy planes. From D-Day until D plus 6, their attacks mounted in fury and effectiveness as their value for new types of missions became increasingly apparent.

On D plus 1, most of their efforts were directed against rail targets and the commanding officer of Navy Fighter Squadron 74 successfully skipbombed the mouth of a tunnel between Fuveau and

Brignoles. The next day Hellcats destroyed, by bombs and rockets, a battery on Ile de Port Cros east of Toulon which was dominating the firing lanes for shore bombardment. The island surrendered a few hours later. On the same day, the Hellcats were called out on a special mission to attack a large concentration of enemy motor transport evacuating the Brignoles-Tourves-St. Maximin area. They destroyed at least eighty-five vehicles including troop and ammunition carriers and damaged many more. The Commanding General, XII Tactical Air Command, later stated that he had personally counted 202 vehicles destroyed in this area as a result of these and other attacks. Enemy air opposition did not appear until D plus 4 when the Hellcats shot down five planes in the air and destroyed one on the ground, but the next day heavy enemy flak shot down six of our pilots and damaged many planes.

At the conclusion of flight operations on D plus 6, Task Group 88.2 retired to Maddalena, Sardinia, as originally planned. It returned sixty hours later to relieve the British task group and continued operations in support of the invasion for six more days before retiring to Ajaccio, Corsica, on the first leg of its return to the United States. While American Hellcats were used most effectively in long-range sweeps deep inland, British Seafires performed identical tasks with equal creditability but closer to the shore line because of their less extensive range. By its constant and powerful striking ability, this Hellcat-Seafire team was instrumental in preventing the enemy from interfering with Allied consolidation of the beachheads, from making an orderly withdrawal from southern France, and from regrouping his forces farther north for a stand against the American Seventh and Allied Armies.

Enemy air opposition to this invasion was never expected on the scale encountered in previous European landings, especially since the Luftwaffe was occupied with more than it could handle in Normandy and on the Russian front. For this reason, it was considered feasible to use battleship and cruiser-based aviation once again for spotting naval gunfire and to insure adequate observation by pilots more experienced in heavy naval gunnery than the newly trained Hellcat pilots. Only those ships carrying aviation gasoline aft were permitted to retain their spotting planes aboard. All others were ordered to land their planes since they would be unable to refuel them easily during the operation, and the presence of high-octane fuel anywhere aboard these ships except as far aft as possible would introduce an additional and unnecessary hazard to ship's safety in the event of enemy artillery or air attack.

While this condition required a certain amount of reshuffling of pilots and planes among the cruisers and battleships, their effectiveness in spotting was in no way impaired. The SOC's proved their value many times during this campaign. So long as they remained clear of heavy flak areas, they were unmolested and were able to perform many vital missions. Only one of these planes was lost during the entire operation when the pilot ventured too near the flak areas and was shot down by antiaircraft fire. Some ships reported that launching and recovery of these planes during the early stages of the operation interfered with fire support when the ships had to maneuver specifically to recover these planes but their over-all utility far outweighed these temporary disadvantages.

Our control of the air during this invasion permitted the use of yet another branch of navy air power — lighter-than-air. Two navy blimps of Blimp Squadron 14 reported to Cuers airport near Toulon from Port Lyautey, North Africa, to assist two French PBY squadrons in mine spotting and minesweeping operations. The first blimps arrived on 17 September 1944. These blimps followed the faster patrol planes along the entire coast line of southern France from the Spanish to the Italian borders examining more thoroughly and accurately areas reported mined by the planes.

Their slow ground speed enabled them to spot individual mines to which surface minesweepers were directed. By coaching the minesweepers from the air with loudspeakers or radio, minesweeping operations along the entire French coast were greatly accelerated and many ships saved by timely warning from the blimps of the presence of mines. When our ships were finally able to anchor in Toulon Harbor, these blimps flew night patrols and provided valuable safeguards with their radar gear which warned against attacks by small enemy surface craft and submarines. From Sicily to southern France amphibious landings in the European theater with attendant combined supporting operations utilized in increasing degree practically every element of United States naval aviation.

While the exploits of no one military or naval unit can be singled out as most decisive to the success of such operations, the test of the value of each component group to the achievements of the whole lies in the skill and efficiency with which it carried out its assigned tasks. Throughout all these campaigns, our air navy made its conspicuous contribution to victory by performing with equal skill and efficiency missions far beyond the scope and importance of those originally expected of it.

11 – COAST GUARD AVIATION

NO ACCOUNT OF NAVAL AVIATION IN WORLD WAR II would be complete without some reference to the contribution of its wartime brother, Coast Guard aviation. Traditionally the Coast Guard is the "policeman of the sea," enforcing customs regulations and various laws of the ocean, and also is an important agent in protecting and saving life and property. Its air branch was established to assist in these same duties.

After World War II broke out in Europe, this traditional role was gradually thrust into the background at all the Coast Guard air stations except one, the exception being the San Diego station, which was selected by officials to maintain its peacetime functions. Supervision of neutrality rules, and, after the attack at Pearl Harbor, anti-submarine patrol and other defense duties became of prime importance.

The aviation arm of the U.S. Coast Guard during the war consisted of nine air stations and Patrol Bombing Squadron 6, which was attached to the Naval Greenland Fleet Air Group. The nine stations were at Salem, Massachusetts; Brooklyn, New York; Elizabeth City, North Carolina; Miami and St. Petersburg, Florida; Biloxi, Mississippi; San Diego and San Francisco, California; and Port Angeles, Washington.

In addition, a detachment was commissioned in April 1942, at Traverse City, Michigan, for patrol of the Great Lakes. A small parachute group operated in southeastern Alaska, a unit was maintained at headquarters, and individual planes were assigned from time to time for special duty with Coast Guard vessels in various districts.

On 1 November 1941 the President put the Coast Guard under the operational control of the Navy, and routine duties became subordinated to national defense. From our entrance into the war through 30 June 1943, Coast Guard aircraft made sixty-one bombing attacks on enemy U-boats, but their principal value in anti-submarine warfare was as "watchdogs" or harassing agents. During this same period the planes located over a thousand survivors and themselves rescued ninety-five persons.

The station which had the easiest transition to wartime status was the air station at San Francisco, commissioned in 1940 and the last Coast Guard aviation unit to be established. It was recognized from the beginning for its potential value in war duties in the San

Francisco Bay region, and anti-submarine patrols were begun at this station immediately after the Pearl Harbor attack. Assistance flights in 1943 numbered 98, and in 1944 totaled 99. During 3,688 other flights in 1943, 3,159 vessels or planes were identified, 169 persons and 7 medical cases were transported, and 57 other government agencies assisted. Flights during 1944 included 675 anti-submarine patrol and convoy coverage trips.

Farther north on the west coast the station at Port Angeles, Washington, experienced a rapid expansion when war came.

In 1935 it had started operations with three 75-foot patrol boats, four picket boats and one Douglas amphibian plane.

For the first three years of its existence, the station, lacking a landing field, was forced to operate only as a seaplane base, but in 1938 a hard surface runway constructed on the naval reservation was leased to the Coast Guard. When war came, old planes were armed and new ones assigned, and the station sandwiched rescue operations between anti-submarine patrols and convoy escort flights.

Protection of the Gulf region was entrusted to the Army, Navy and Coast Guard, working together. The joint operations plan specified that air patrols from the Coast Guard air station at Biloxi should cover the Mississippi Delta, the shipping lanes south and southeast of the station, and certain parts of the Gulf not under the jurisdiction of Pensacola and Corpus Christi. On 24 December 1941, it was made a task group unit in the new defense scheme, and in July 1942, an air detachment was established at Houma, Louisiana.

Planes operated by this station increased from 16 in January 1943, to 25 during the winter of the next year. Operations reached their peak during 1943, reports showing that the station at Biloxi had more flight hours during the year than any other Coast Guard air station in the eastern or Gulf areas.

In September 1943, one of the busiest months, the station conducted 109 submarine patrols, of which 43 were made by the Houma detachment. During that same month, 22 aerial convoy coverage flights were made by the two units, providing protection for 144 vessels. Daily patrols were continued until the summer of 1944, when the danger of enemy attack had subsided enough to permit resumption of normal rescue activities. Action reports record numerous attacks, which, if they did not result in sinkings, at least damaged enemy subs or drove them from the coastal area.

When the United States entered the war, the St. Petersburg station had nine planes in operation.

At that time the station was operating in air, on sea, and on land, but in October 1942, all duties related to aids to navigation, coastal lookout stations, and land and water patrols were removed from the station's jurisdiction, and from that time the station operated strictly as an air unit, retaining only closely related aviation activity. Ambulance and assistance flights become more frequent as the war progressed.

During this period the Coast Guard planes operated with the Navy, first under the Inshore Patrol and later as part of the Gulf Sea Frontier Command, and they executed patrol, convoy, and observation assignments, besides doing administrative and utility work. Seaplanes maintained regular anti-submarine and security patrol over the St. Petersburg area during 1942, 1943, and part of 1944. In June 1943, a detail at Port Saint Joe, Florida, was incorporated into the regular patrol.

Regular daily observation and security patrols were inaugurated at the station at Miami on 11 December 1941. As the Miami coast area was in the heart of the German submarine danger, the number of sunk or damaged vessels mounted alarmingly during 1942 and 1943. For several months, Coast Guard planes and boats were the only rescue agents available for the region, and the average number of monthly flights increased from 48 in 1941 to 349 in 1943.

Among the vessels torpedoed in this area was the tanker, Gulf-state, and survivors of the torpedoing owe their lives to an officer attached to the Miami station. While patrolling in an OS2U-3, the pilot was directed by radio to search for survivors of the Gulfstate. Upon sighting the wreck, he also spotted three groups of survivors.

After jettisoning his depth charges, he landed and picked up the three men in the first group, then taxied to the second group to give them a rubber raft and went on to take aboard a badly burned man of the third group. The pilot then stood by to protect the drifting survivors from sharks until other assistance arrived.

At the Elizabeth City station, the first station order from the Navy after our declaration of war was for an air patrol of "steamer lanes and offshore approaches to Chesapeake capes; on alert for enemy submarines."

This patrol, extending fifty miles out to sea and as far south as Cape Lookout, was maintained every day that weather conditions permitted. At the time thirteen pilots and ten planes were available for the anti-submarine patrols. At first, Coast Guard planes were unarmed, but on 22 January 1942 armed craft were assigned to the station.

Received at that time were two J2F-5's, equipped with machine guns and bomb racks, but best designed for observation and scouting. Planes more adequately equipped for the duty assigned were procured in December 1943.

The admirable location and size of its field made this station a nucleus for war expansion in the area. In May 1942, an air squadron of the 34th Coast Artillery Brigade was based at the station, and the Navy used the field for blimp operation.

Eventually the Navy completed its own station about a thousand yards from the Coast Guard site, but before that time the Coast Guard facilities and air station had been a navy operation base for squadrons flying between Bermuda and the United States.

During 1943, over 8,300 hours were flown, over 24,900 planes and vessels were identified, 29 others assisted. Altogether 69 assistance flights were made, in which 12 government departments and 76 persons were assisted and in addition 16 medical cases and 96 persons were given special transportation. During 1944 the 3,228 flights included 56 assistance and 1,692 anti-submarine or convoy coverage flights. Nine vessels or planes were assisted, and 21 rescues accomplished during the year.

From time to time pilots hazarded the dangers of crashing to effect landings. The first of such landings was made by an unarmed PH-No. 183 on 1 May 1942, when thirteen survivors were saved after drifting at sea for six days. Two men were flown to Norfolk and the remaining eleven were picked up by a Coast Guard cutter. The next day, another offshore landing saved the lives of two men adrift on a raft.

In July 1942, a plane picked up seven German survivors from the submarine Dergin which had been sunk by an army plane, and for this feat the pilot received his second Distinguished Flying Cross. These rescues represent only three of the assistance flights made during the period from 7 December 1941 to 1 July 1944, during which time 22,951.1 hours were flown, including 4,875 convoy and anti-sub flights and 2,550 training, test, and administrative flights.

In preparation for a "state of readiness," Coast Guard planes at the air station at Salem began regular patrols in 1940. By the summer of 1942 the station had five OS2U-3 seaplanes and three J2F-5 amphibians, all armed with .30 caliber machine guns and 325-pound depth charges. The remaining four planes, JRF's and a J4F, were unarmed. Four motor vehicles, one tractor, and a crash boat completed the equipment.

Daily inshore patrols protected Boston Harbor and the approaches to the sea. Offshore patrols covered areas north of Boston, off the coast of Maine from advance bases maintained at Bangor, Portland, Rockland, and Lewiston, and to the south around Cape Cod, Buzzards Bay, Nantucket and Vineyard Sound. Many submarines were contacted, some were attacked, others driven from the area, but no sinkings were reported. During the years of 1942, 1943, and 1944, one plane and two men were lost in line of duty.

To protect the New York area, planes at the Coast Guard air station at Brooklyn were eventually armed with depth charges. No proved sinkings of U-boats were recorded, but the planes no doubt scared off the enemy in a number of cases. As more aircraft were made available to the Navy, the station relinquished most of its purely military duties.

Reconnaissance patrols, as well as escort and convoy duties, however, continued. Tapering off of these duties came about after Floyd Bennett Field at the Brooklyn station was designated on 19 November 1943 as a helicopter training base, with three helicopters assigned to it by the Navy.

It was from this base that a helicopter rescue mission to Labrador was made which won commendations for six officers and men. A dismantled Coast Guard helicopter was picked up at Floyd Bennett Field by an Army C-54 transport, which then flew a thousand miles to Labrador, where the helicopter was reassembled and used to rescue eleven Canadian fliers who had been marooned for thirteen days in a frozen wilderness.

The helicopter made seven trips from Goose Bay to pick up the fliers, completing the entire mission in less than five days after the helicopter unit was notified.

An attempt to rescue them previously had been made by ski-equipped planes sent out by the Canadian Air Sea Rescue officials, and two of the group had been removed, but melting snows prevented further landings. To make matters worse, two more had been added to the party when one of the rescue planes crashed. Pontoon aircraft were of no more use than the ski-equipped planes because they could not land on the surrounding lakes until the spring thaw had progressed further.

British helicopters were added to the Brooklyn station's equipment when it was agreed that the Coast Guard should train some mechanics and pilots at the request of the British Admiralty. The Royal Navy Helicopter Training maintained its own personnel and planes except one HNS for joint use.

Other aircraft were used by both United States and British pilots. By October 1944, the school was flourishing and the number of helicopters in use had been increased to thirteen. Both a pilots' course and one for mechanics were given at the field.

By February 1945, when the sixth pilots' class finished its course, the school had trained 102 helicopter pilots — 72 Coast Guard, 6 Navy, 5 AAF, 12 British, and 4 CAAF aviators, as well as one NACA, one PU Corporation and one McDonald Aircraft aviator. Only qualified volunteer aviators were selected for training. Helicopter aerodynamics are more complex than those of fixed wing planes, and a long period of ground training was necessary.

Approximately thirty hours of flying time were required to complete the helicopter pilot training course. A deck-type landing platform was installed on the field to represent a rolling deck to give practice in landing the helicopter on a vessel during a storm. For more realistic operations the Coast Guard cutter Cobb was assigned in a training capacity.

Besides its helicopter training program, the Coast Guard has had a number of other training assignments. A preflight school was maintained at the Elizabeth City station from October 1941, to January 1942, for Coast Guard students, who went from this training to the Naval Air School at Pensacola, Florida.

Temporary schools for aviation machinists' mates and for cooks and bakers were conducted at Elizabeth City station in 1941 and 1942; and from 15 March to 15 July 1944 the station trained VP-6 personnel destined for Greenland duty. There was also a training program for new recruits and a cooks' and bakers' school at the station at Salem. During the early months of 1943 the St. Petersburg station assumed the unusual role of training Mexican pilots in anti-submarine warfare.

The most colorful of the Coast Guard aviation units was the special squadron assigned to the Greenland patrol, known as Patrol Squadron (later Patrol Bombing Squadron) 6, which was a special unit of the Atlantic Fleet. The squadron, commissioned on 5 October 1943 at Argentia, Newfoundland, was sent to Narsarssuak, Greenland, to relieve Bombing Squadron 126. It was the only naval squadron operating in the north Atlantic arctic region and the only squadron manned entirely by Coast Guard personnel.

Many months before the Jap attack on the island of Oahu, Coast Guard cutter-based planes were operating in Greenland, carrying out anti-submarine and coastal patrols and making heroic rescues.

From time to time during observation or mail delivery trips, bombers sighted stranded vessels in the stormy northern seas and sent out radio messages, which brought Coast Guard cutters speeding to the rescue.

Actually, the equivalent of a coordinated air-sea rescue mission was in operation before the Air-Sea Rescue service was instituted, and the changes effected were in organization rather than mission. It was not uncommon for ship-based or land-based planes to fly over several thousand square miles of the Greenland icecap under handicapping weather conditions in a single rescue search.

Perhaps the most daring Greenland rescue mission was that of the late Lieutenant John A. Pritchard, Jr., who lost his life attempting to save survivors of an American Flying Fortress stranded about forty miles from Comanche Bay. Pritchard and his radioman, Benjamin A. Bottoms, received the Distinguished Flying Cross posthumously for this feat. The two men took off from the cutter Northland in a Grumman Amphibian late in November 1942.

Despite the signal of the stranded men that they should not try to land, the landing was accomplished as planned, by retracting the wheels, going into a glide and setting the plane down on a long down slope where heavy snow covered the ice. To the cheers of the Northland's crew, the plane returned to the ship with two survivors. But on the next trip to rescue the last survivor, although the take-off was again successful, the return flight to the Northland was never completed.

The Greenland Patrol was inaugurated by Rear Admiral Edward H. Smith in October 1941, to operate as part of Task Force 24.8 of the U.S. Atlantic Fleet. It consisted of Coast Guard and navy vessels, manned by personnel from the former service, and its function was to transport men and supplies and to combat the German submarine menace.

Also, during the war, daily weather reports and ice observations from the Greenland area became necessary for transatlantic war operations. In the summer of 1943, the Coast Guard was directed to organize an air patrol squadron to be attached to the Greenland Patrol to provide air coverage for convoys, carry out anti-submarine patrols, deliver mail, undertake rescue missions and survey ice conditions.

The squadron's main base was at Narsarssuak, Greenland, and there were detachments at Argentia in Newfoundland, Reykjavik in Iceland, and in the Canadian Arctic. The patrol began operations with 6 PBY-5A planes, 15 aviators, 3 aviation maintenance officers, 1 radio electrician, 1 aerologist, and 131 enlisted personnel, of whom 7 were aviation pilots. In April 1944, there were 12 operational aircraft

(PBY-5A's) based at Narsarssuak, 5 officers, 24 aviators, 4 aviation pilots and a crew of 152 enlisted men, including the pilots.

Bad weather and difficult terrain limited aviation activities in Greenland. Fjords and harbors closed by pack ice during the winter, very high mountains, a great icecap covering about 85 percent of the interior and raging cross-winds did not make the pilot's life a care-free one. The complement was replaced annually because of these factors, and assignments to Argentia, where living conditions were better, were rotated.

The squadron's main base at Narsarssuak had a single concrete runway down a sheltered fjord. The three-inch incline to the east permitted landing uphill and taking off downhill, regardless of the wind. During the period from August 1943, to the end of November 1944, the squadron flew 6,234.6 hours, representing 638,998 miles cruised with an area of 3,213,605 square miles covered. The Army Air Force in Greenland maintained a special Arctic Search and Rescue Squad, locally nicknamed "The Find 'em and Feed 'em Boys," which assumed most of the rescue responsibilities, and there has been in operation since 1941 a special "Sled Patrol," made up mostly of Danes and Eskimos.

However, Patrol Bombing Squadron 6, during 1944, had its share of rescue missions, aiding 43 planes and vessels and rescuing or assisting 47 persons. One medical case and 87 other persons were transported during the year's operations, consisting of 71 assistance, 736 routine and 346 anti-submarine patrol and convoy coverage flights.

Because Coast Guard and navy officials agreed that some unit must carry on peacetime services, the San Diego station (which had grown out of a detachment to prevent smuggling across the Mexican border) continued with its regular duties and had few military assignments. The station had the same number of planes in 1943 as in 1938 and most of them were unarmed until early in 1944. The San Diego unit thus was the natural location for the initiation of air-sea rescue service, which took place when the enemy threat decreased.

The Air-Sea Rescue Agency was established in February 1944, but actually a unit had been organized at San Diego in December 1943, to rescue fliers forced down on land or at sea. This group came into being under the guidance of Commanders Max I. Black, USN and W. A. Burton, USCG as a result of the increasing number of airplane crashes in that region. Investigation showed that rescue equipment was inadequate, that aviation activity had increased along the southern Pacific coast, that available rescue agencies were not well coordinated, and that the system of disseminating information was inefficient.

The San Diego Air station was the first air-sea rescue unit to be organized and put into operation in the United States. All surface craft, blimps, and planes, together with any other rescue equipment used in rescue operations by army, navy, or marine agencies became a part of the new rescue organization. Information on accidents or emergency crashes was reported to the central rescue unit through the Naval Air Control Center.

By the end of February 1945, the personnel complement of the station was 42 officers, 341 enlisted men, and 73 students. During the first month of operation in 1944 there were 124 aircraft accidents in the San Diego area involving 201 persons, of whom 137 were saved. Considering that 59 were killed and no trace could be found of two others, the record is almost perfect.

Of those saved, 25 percent were rescued by Coast Guard planes or boats; the surface craft of the Navy rescued 27 airmen and fishing boats took aboard 37. Most rescues were executed in less than an hour, and some were made within six minutes after the crash.

Because of its past record, the Coast Guard, which had been organizing toward this end for some time, was the logical administrator for the new agency. As part of its beach patrol work, the Coast Guard had already established an efficient communications system which fitted ideally with the new setup, and air activities had been expanded to take over air-sea rescue duties.

Co-ordination of the army, navy, and Coast Guard rescue activities was achieved at the joint operations centers, while the actual rescues were the responsibility of each regional Air-Sea Rescue Task Unit, generally headed by the commanding officer of the Coast Guard Air station. The central organization consisted of an Air-Sea Rescue Agency, headed by the Coast Guard commandant and an advisory board of representatives from the Army, Army Air Forces, Marine Corps, Navy, and Coast Guard.

One by one during 1944 the various air stations took their places in the new organization. Actual operations were directed by sea frontier commanders and commanders of the various war theaters. The sea frontiers were subdivided into sectors with each sector comprising one or more task units. The Air-Sea Rescue task units were composed chiefly of Coast Guard vessels, planes and rescue facilities. Coast Guard control of ASR craft and equipment was purely logistical. District offices and lifeboat stations were independent of the organization except as they might aid in rescues or maintain equipment.

Besides rescue operations, Air-Sea Rescue service included development of equipment and training of personnel engaged in ASR duties. Under terms of the joint agreement of 17 August 1944, the Army Air Force was given the initial responsibility of coordinating searches and rescue procedures over land areas. Sea rescue operations were primarily conducted by the Naval Sea Frontier organization. The Coast Guard's part was to maintain aircraft units and facilities.

Unless it was necessary to attempt the rescue alone, the plane communicated with ASR stations and stood by to direct crash boats to the scene. Rescue boats, usually referred to as "crash boats," varied in size from less than 48 feet up to 104 feet and were provided with emergency equipment.

Experimental work indicated great possibilities in the future use of ASR blimps. Still more encouraging was the fact that an actual raft-to-blimp rescue had been successfully accomplished. The Coast Guard sent a small number of officers and men to blimp training in order to have personnel experienced in that type of rescue operations. A departure from the usual plane-and-boat or blimp-and-boat combination in rescue missions was the parachute unit, organized in the Ketchikan District early in 1944.

Its members, proficient woodsmen, were trained at the Forest Service Parachute Jumpers School at Missoula, Montana, and used a steerable chute which could be landed in a tree, from which the parachutist let himself down to the ground. A unique feature of the ASR training program at the Port Angeles station was the air-land rescue ski squad, trained and assisted by the Port Angeles chapter of the National Ski Patrol for service in snow-covered mountain areas. The largest aviation unit in the ASR organization was at Elizabeth City. One of the most popular rescue planes was the stripped-down PBY-5A, equipped with droppable life rafts and other lifesaving equipment.

Wartime activities had taken Coast Guard aviation far beyond realms dreamed of in the normal days of peace. In the performance of these new and exacting obligations, the Coast Guard exceeded even its own high standard of devotion to duty in the face of all obstacles.

PART IV: THE PACIFIC THEATRE

12 – READINESS IN THE PHILIPPINES

MOST PEOPLE THINK OF THE NEUTRALITY PATROL as a purely Atlantic Ocean venture. It was there that the submarine war was being waged, and it was through these danger-infested waters that the bulk of our shipping was passing. Our primary concern, therefore, was directed toward the protection of this Atlantic shipping.

On the other hand, there was always the potentially explosive situation in the Pacific to be considered. While Japan's overt action against the United States did not come until December 1941, tension in the Far East had existed long before this time.

The exposed situation of the Philippines made it inevitable that if war should come with Japan, these islands would be an early objective of the enemy. With these factors in mind, what might be termed a neutrality patrol was set up in Philippine waters in the fall of 1939, and in 1940 its size was increased.

From the outset, the role assigned naval aviation was one of patrol and limited defense. In view of the nature of our planes in this area and their small number, the Commander-in-Chief of the Asiatic Fleet, Admiral Thomas C. Hart, wisely insisted upon a policy of dispersal and concealment within his own command. In September 1939, a squadron of twelve PBY's had joined the Asiatic Fleet.

Twelve more Catalinas arrived in October 1940, and were followed by four more in the summer of 1941. To service these planes, the Asiatic Fleet had three seaplane tenders, Langley, Wm. B. Preston, and Heron.

Instead of making his operating plans in accordance with long-range projects for the defense of the area, Admiral Hart constantly revised his plans on an "as is" basis, never counting on what was expected or projected or promised until the forces or materials were actually at hand.

In connection with this policy, the Commander-in-Chief, Asiatic Fleet (CinCAF) favored the erection of dispersed temporary bases rather than a strategy equivalent to putting all his eggs in one basket such as the construction of a large base at Sangley Point, near Manila, that might not be completed in time for use.

Accordingly, an auxiliary air base was begun, in the fall of 1940, at Olongapo on Subic Bay, and was in full use many months before the war began.

Preparations were also made for operations from Los Banos, where the concealment of planes was practicable with extemporized facilities at hardly any expense. Each seaplane tender and each squadron made scheduled trips to advance bases in the southern Philippines for the purpose of learning the geography and weather of the area, so that wartime flights could be carried out under severe conditions of tropical storms with the possibility of basing at many places in the Philippine group. When war came, these preparations paid off in affording a large measure of security and dispersal for navy planes.

During 1941 there were two developments that lessened the role that naval aviation was to play in the early phases of the war in the Philippine area. By April it was decided that the United States Asiatic Fleet was not to be reinforced. The bulk of the burden of naval defense, on the other hand, was to rest on the British Far Eastern Fleet, which, it was expected, was to be reinforced by battleships and carriers. (As it turned out, the carriers did not arrive in the Far East.)

The second development was that the Army began to take more and more planes into the Philippines and assume the task of air defense of the area. Thus, it was that before war came, naval aviation's course was charted — patrol and spotting, with participation in delaying action as the Japanese actually made their sweep into the region.

Keeping these factors in mind, let us see what naval air squadrons, organized under Patrol Wing 10, accomplished in the months before the attack on Pearl Harbor. Late in October 1941, when a shift in Japanese high officials put the militarists on top, an offshore air patrol was established about a hundred miles off the west coast of Luzon.

On special patrols, planes investigated some of the islands in the South China Sea, bringing back photographs showing that considerable development had been taking place on Ita Aba and some on Spratly Island — straws in a foreboding wind. Meanwhile naval aviation stores were being scattered throughout the Philippines, and training exercises were carried out with the fleet.

Early in November, the Japanese Envoy, Saburo Kurusu, passed through Manila on his way to Washington and infamy. The usual courtesies were observed, and a reception was held in his honor. Admiral Hart held a brief conversion with the Envoy, at which time Kurusu remarked that his mission was "to keep your Asiatic Fleet idle."

By late November 1941, it was realized that the situation was becoming increasingly serious. In an effort to discover what the Japanese were doing, a considerable number of long-range reconnaissance flights were carried out to Formosa, Hainan, and Indo-China during the first week in December. The flights to the coast of Indo-China were not routine, but were directed on a day-to-day basis by CinCAF with orders to avoid being sighted from the coast and if possible, from Japanese shipping.

The PBY's, however, could not avoid being sighted by Japanese planes. These did not attack, but some of our pilots reported that on occasions Japanese aircraft did make practice runs on them. On these flights, two sizable Japanese convoys were sighted off the Indo-China coast.

As has been indicated, the Japanese were also conducting special patrols. On 5, 6, and 7 December, PBY's on patrol from Manila encountered Japanese planes in the vicinity of the Luzon coast. Each side had machine guns manned and warily avoided each other like stiff-legged dogs.

As war approached, naval aviation forces had been but slightly increased in size. Like Admiral Bellinger in the Hawaiian Islands, Admiral Hart had requested more planes, especially the fast and heavily armed, four-engined patrol bombers that could make sweeps to Formosa. For reasons already noted, these reinforcements were not forthcoming.

Consequently, the air elements of the Asiatic Fleet consisted of the following: Patrol Wing 10 with 2 squadrons of 14 PBY's each, and a utility unit with 5 OS2U's, 1 SOC, and 4 J2F's. The wing also had 3 seaplane tenders—two of them converted World War I destroyers, William B. Preston and Childs, and a converted minesweeper, Heron.

The ex-aircraft carrier Langley — nominally a seaplane tender — was transporting aviation supplies to the outpost bases. The three cruisers then in the Asiatic Fleet also had air units for scouting and anti-sub patrol duty.

CinCAF had well deployed his small air strength. The Heron, with four OS2U's, covered the western approaches to the Celebes Sea. The William B. Preston, with a detachment of three PBY's was in Davao Gulf. These patrols were linked in an informal arrangement with Dutch patrols long maintained on the boundary of the Netherlands East Indies. Late in November, the Langley arrived in Manila from the north to receive another cargo of aviation spares and equipment.

The Childs was also in Manila Bay. Except for their southern detachments, one of the patrol squadrons, VP-101, and the utility unit were based at Sangley Point. The other squadron, VP-102, was at Olongapo, and the extemporized base at Los Banos was ready for operations.

On 1 December, Admiral Tom Phillips, the new Commander-in-Chief of the British Far Eastern Fleet, arrived in Singapore with the Repulse and Prince of Wales. At noon on 5 December he reached Manila by plane.

A joint operating plan was worked out expeditiously on 6 December with General MacArthur present; yet even as the smooth copies of a joint dispatch to be sent to London and Washington were being made that evening, news came from British patrols that Japanese forces which had left Camranh Bay on 4 December were heading westward in the Gulf of Siam toward the neck of the Malay Peninsula.

Since PatWing 10 patrols had not gone in close to the Indo-China coast on 6 December, they had not seen these moves. Upon receipt of this news, Admiral Phillips left for Singapore by plane. The next day, 7 December in Philippines, but still 6 December in Hawaii, all was tense. Planes were bombed up; full crews and ammunition were aboard. Patrol Wing 10 was ready.

13 – TOWARD THE END OF TRAGEDY

FROM THE STANDPOINT OF MEN AND PLANES INVOLVED, the retreat of naval aviation forces from the Philippines through the Netherlands East Indies to Australia was but an episode, a campaign replete with the heroism of men and ships and planes struggling against great odds to hold off, to delay an enemy who appeared all-powerful. There seem, however, to be some good reasons for giving more than a cursory glance at this phase of the war.

In the first place, it provides a fruitful study in contrasts. Our final victory cannot be thoroughly appreciated unless we are aware of the problems and difficulties that were originally confronted. Our power was built at phenomenal speed, but no one should think that it was built overnight or that it was a simple task. In the early stages of the war, we were attempting to stave off the persistent advances of a formidable foe.

Every day that this foe could be stalled, every foot of ground or sea mile that could be contested meant just that much more time for our own war machine to get under way and that much less space to recover on the road back. Then, too, this early period should be noted more than casually and with considerable pride, for our men were never to fight more valiantly, even in the days to come when the results were more satisfying than they were in the first disheartening months of war in the Pacific.

Shortly after 0300 local time on 8 December, CinCAF received the news of Pearl Harbor, and at 0315 the Asiatic Fleet was notified, "Govern yourselves accordingly." Fifteen minutes later the order was received from Washington to put the War Plan against Japan into effect.

At 0710, just five minutes before a second dispatch from Washington confirmed the Pearl Harbor raid, formally declaring the existence of a state of war, the William B. Preston, lying at anchor far to the south at Malalag on Davao Gulf, reported by radio, "I am being attacked!"

Ten minutes later it reported that two of its PBY's had been sunk by Jap strafing, one officer (Ensign R. G. Tills) killed, and one enlisted man (A. E. Layton RM3c) wounded. On the other hand, the Preston's antiaircraft chalked up one Jap dive bomber as a "probable." The Navy's war in the Philippines had begun.

A little later four Jap destroyers entered the Gulf. The Preston at Malalag remained silent until they were well out of range, then slipped out. Had our planes been entirely land-based, where the Japs could have swooped down and obliterated all facilities, we would have lost control of the air far sooner than we did. As it was, the planes and tenders of PatWing 10 now began to lead a hunted existence, trying to keep one step ahead of a numerically superior enemy.

At midday the Japs made powerful and well-executed attacks on the army airfields in north and central Luzon, destroying the planes there and severely damaging ground installations. Fortunately, half the army bombers had been sent, two days earlier, to Mindanao in anticipation of an attack on the Manila area. The seaplanes of Pat-Wing 10 were shifted for dispersal and concealment before 0600 of 8 December, and dawn patrols had been sent to the northwest and northeast, the Army having taken over the two Formosa patrol sectors.

Two PBY's were at Sangley Point, 5 were based with the Childs in Manila Bay; 9 were stationed at Olongapo, and 9 were at Los Banos. The four J2F's, one OS2U, and one SOC based at Sangley Point continued their patrol of the area.

The cruiser air units were on their ships: the Houston was off Iloilo, Panay, awaiting Admiral Glassford (ComTaskForce 5) who left Manila by plane on the morning of 8 December; the Marblehead was at the Tarakan oil field; and the Boise, having brought an army convoy into Manila on the fourth and departed quickly southward, was ordered to join from Cebu.

The Langley and the two oilers, Trinity and Pecos, accompanied by two destroyers slipped from Manila Bay under cover of darkness to join the task force and head for Makassar Strait and fuel at the Dutch East Indies oil centers.

The mission of the naval forces in or near the Philippines remained as before, to support the Army's defense while damaging the enemy as much as possible. PatWing 10's planes were few and not expendable. First and foremost, they were to be scattered in lakes, swamps, coves, or any other place where dispersal facilities were available, taking advantage of all the hideout areas and gasoline supplies which had been established throughout the Philippine Islands. They were to search and attack from advance bases whenever possible, and also to search and scout for the Asiatic Fleet as it retired southward in accordance with the War Plan.

No information on enemy movements was obtained on 8 or 9 December, but from this time onward an extraordinary crop of mis-information flowed over the warning net. The intensive study of aircraft and ship recognition, which was to become fundamental in naval training, both on shipboard and in the air, had not yet begun. A normal jittery reaction in the Philippines sighted enemy ships when no ships were there, and when a vessel was really seen she was reported in one of two categories — irrespective of size, she was either a transport or a battleship!

On 9 December, however, an enemy freighter was bombed after a PBY had politely drawn within range before the ship identified it-self by firing and running up the Jap flag. On the other hand, the friendly Norwegian who did not respond when challenged had only himself to blame if he got shaken up by some near hits. Two planes were damaged on 9 December, one by Jap aircraft and one by a nervous Filipino ground battery. At noon that day a search plane was shot down with the loss of all hands.

At 0625 on 10 December, plane #17 of the northwest patrol sighted a float biplane headed toward Luzon and at 0645 reported sighting 2 battleships, 1 light cruiser, and 2 destroyers headed south at 10 knots. Thinking the force, a British one, the patrol continued on its way. Half an hour later the plane was ordered back to the con-tact, and at 0735 reported that it was under antiaircraft fire, having closed at 7,000 yards and cut across the stern of the formation at 15 feet off the water.

Meanwhile, at 0731, the attack group had been ordered out and the northeast patrol recalled loading with torpedoes. Five planes un-der Lieutenant Commander J. V. Peterson took off from Los Banos at 0910 and attacked the Jap ships at 1205. Five PBY's from Manila failed to join and so missed contact.

The attack group dropped twenty bombs, one salvo hitting the stern of a Kongo class battleship, certainly disabling its steering gear and possibly causing damage to its propellers. This was the last body of enemy combat ships found at sea by our air patrols before their withdrawal from the Philippines.

Meanwhile, the northeast scouts were being loaded with torpe-does at Olongapo. But by this time the Jap morning raid was on its way toward Cavite, and two of the PBY's which had been loaded with torpedoes were attacked on take-off and forced down, although one managed to shoot down a Jap fighter and the other damaged a sec-ond.

At 1255, presaged by the appearance of Jap fighters, Cavite Navy Yard and Cavite City were heavily bombed from above the range of the guns especially installed for their protection. The enemy attack was not interfered with at all by our fighters — the Japs had seen to the army fields earlier that day — and the bombing was very accurate.

Power facilities at Cavite were destroyed and ComAirAF, Captain (later Rear Admiral) F. D. Wagner, who was also ComPatWing 10, was forced to move aboard the Childs to re-establish communications. This attack made it entirely clear that, as far as the security of Manila Bay was concerned, the enemy had control of the air.

On 11 December, another PBY was damaged by a Jap bomber but managed to get in to Los Banos. More cheerful was the report that at the Cavite shop, the maintenance force had managed to get one of the two PBY's there in commission. The following morning (12 December) a half squadron took off from Olongapo for Lingayen Bay to attack a "Japanese fleet" on typically bad information.

They had returned, fueled, and were awaiting orders when Zeros came in low at the entrance of Subic Bay, climbed over a little hill and came in close to the water. At 1140 it was reported that seven PBY's had been burned at their moorings, two Jap planes being shot down in their attack.

Even so, the Navy turned its thoughts to attack as well as to defense. On 12 December, Captain Wagner proposed a joint Army-Navy attack on Palau, but on 14 December it was obvious that it stood little chance of fitting in with General MacArthur's plans.

On 13 December a bombing raid forced the evacuation of Olongapo except for an emergency ground crew. It was now obvious that PatWing 10, having less than one squadron of planes operable, would have to move south. By this time army planes were no longer keeping the air, except for one or two fighters at a time flying for reconnaissance purposes.

With Task Force 5 now well on its way toward the Netherlands East Indies, and all merchant shipping sent out of Manila on three successive nights, the mission of PatWing 10 was first to cover the fleet and, second, to transport key personnel.

Consequently, seven planes, with reserve crews, were ordered to proceed to Lake Lanao, Mindanao. They departed at dawn on 14 December, leaving Lieutenant Commander Guinn in charge at Los Banos. Since one plane had already been detached on 10 December to reinforce the southern detachment, this left 7 planes in the Luzon area — 5 flyable (4 of which were in fair shape) and 3 being worked on to get them in condition to fly out.

Of the planes at Manila and Los Banos, one was directed to report to CinCAF on 16 December. This plane flew the Chief of Staff (Vice Admiral Purnell) and a communication officer to Task Force 5, other key officers going by destroyer or submarine. On the same day another struck a reef near Los Banos while taxiing, and three flyable PBY's were now available.

On 19 December, just a few days after the commandant of the 16th Naval District (Rear Admiral Rockwell) had established his operational command post on Sangley Point, that area in turn received a high-altitude bombing attack which burned most of the gasoline that remained there in drums and also ruined radio installations.

Cavite and Sangley were thereafter virtually abandoned, and Com 16 moved his post to the prepared underground position on Corregidor. The Japanese landings on Luzon went on in spite of the efforts of submarines and army bombers against waves of amphibious forces.

On 23 December an army dispatch predicted an early retirement to Bataan and Corregidor, and on the following morning CinCAF was informed that the retirement had already been put into operation — that army forces and government headquarters would be moved to Corregidor during the day and that Manila was to be proclaimed an open city. Although foreseen, this eventually had never been officially or unofficially discussed since the beginning of the war.

After a full conference it was decided that the submarines should continue to operate from Manila Bay as long as possible, but that CinCAF should shift base to the Netherlands East Indies where he would be in closer touch with the fleet and with the British and Dutch naval commands.

Before dusk on 24 December, one of the flyable PBY's took off from the base at Sangley Point with eleven passengers aboard. At the last-minute half, the places were given to top-ranking army officers. At take-off the pilot spied a blacked-out "680" boat on patrol cutting across the course, fortunately just in time to avoid serious collision but not in time to avoid tearing off the port wing-tip float on the bow of the boat as the plane water-looped to starboard. The passengers were transported by land to Los Banos, where departure was safely made at about 2400.

The damaged plane, repaired that night, spent the next day hidden with the others among the mangroves at Los Banos. Orders were given for the two flyable planes to evacuate CinCAF and his staff from Cavite at 1800 on 25 December. Unfortunately, in the late afternoon, the Japs discovered them. One was burned, and the other, under repair, was sunk at its mooring. The third was shot full of holes but was patched up and managed to fly to Sangley Point that evening with one engine out.

However, since all of the Cavite area was to be evacuated or destroyed within the next few hours, orders were given to leave the plane on the beach and let the demolition squad destroy it. As for the two remaining planes, one was in the Cavite shop undergoing overhaul and was believed destroyed; the other not yet made entirely safe after its encounter of 10 December 1941, was flown southward at dusk. Admiral Hart and his staff left Manila Bay on the submarine Shark at 0200, 26 December.

The story of naval air action in the Philippines would not be complete if it did not tell of the valor of those naval air personnel who were left to fight it out on foot. These were mainly drawn from the ground operating crews from Sangley Point, Olongapo, and Los Banos, including the 65 men of the utility unit. All told, it was estimated that there were 8 officers and 175 men of PatWing 10 on Bataan.

On 10 December, when Cavite was first attacked, the utility planes which had been patrolling the Manila and Subic bays departed successfully, evaded Jap strafing and pursuit, and scattered to the dispersal area at Laguna de Bai. Two J2F utility planes departed for Lake Lanao on 14 December. This left five scout or utility planes in the Luzon area, which were flown on 25 December to the air base being constructed at Marivales Harbor at the tip of Bataan.

By 2000, 25 December, all PatWing 10 personnel were assembled at the Pan-American Airways base at Cavite. The greater part of them went by land route, but about fifty men under Lieutenant Commander (soon to become Commander) Francis J. Bridget went aboard the minesweeper Quail and the ferryboat San Felipe which had been loaded with gear and supplies. At about 0300, 26 December, they landed at Marivales, and the land caravan arrived at 0900.

Two hazardous night trips into the Manila area were later made for guns, ammunition and supplies. On the last of these Lieutenant McGowan — formerly of Admiral Hart's staff — and Lieutenant (jg) Pollock found that the road and bridges to Bataan had been blown ahead of the scheduled hour. They finally commandeered a barge and a badly frightened Filipino with a boat to tow it down the bay. That afternoon, as they left Manila, they saw the Jap motorcycle patrols entering.

Various missions for the five aircraft at Marivales were suggested during the first week of January but none materialized. Then, at daybreak on 10 January, a Zero raid sank all five in shallow water. From that time on PatWing 10 personnel fought on foot.

The same day, Commander Bridget was ordered to form Naval Battalion Marivales, to consist of all the miscellaneous naval units then on Bataan and two companies of the 4th Marines. While the history of their defense properly belongs to the history of that battalion, PatWing 10's record included at least two skirmishes with the enemy between 22 and 27 January when the battalion was relieved by a detachment of Filipino Scouts.

On one of these forays a naval pilot, meeting a band of Japs face to face at only a few paces, was badly wounded. On 29 January all but two of the naval aviators of noncommissioned aviation pilots were taken out by submarine. One of these, Commander Bridget, was later evacuated to Corregidor, but nearly 175 men of PatWing 10 were believed to have been on Bataan on 1 March.

There is one more heroic story of PatWing 10 activity on Luzon. Before Corregidor surrendered, two PBY's, piloted from Darwin, Australia, by Lieutenant Commander E. T. Neale and Lieutenant Deede, landed in the open sea off Corregidor at midnight of 29 March. At this time, Corregidor was being shelled from both Bataan and the Cavite side. The planes had been stripped to the bare essentials, carrying no bombs and no ammunition, but they were landed with medical supplies and the type of fuses and ammunition badly needed on Corregidor.

As soon as they landed, they were met by small boats from the island, unloaded their supplies and immediately thereafter loaded aboard some fifty persons: twenty army nurses, several army generals, one navy captain and other navy personnel. They refueled and hid at Lake Lanao the next day, and that night, with the Japs only ten miles away, one plane took off for Darwin and hence to Perth.

The other, on take-off, ripped a hole in its bottom on a rock that no one could see. To that plane had been assigned Commander Bridget and some of the army nurses; they went to the Army's outpost at Del Monte in the hope of getting another plane. Commander Bridget, for one, was captured there by the Japs.

Though it had at first looked as if it could not be saved, the PBY was patched up by the crew, with the assistance of the army and navy personnel there. They managed to get it aloft the next night and flew it to Darwin where it promptly sank on arrival, but was once again resurrected and eventually returned to Perth. The plane had left Lake Mindanao barely in time, for two days after the last PBY left, the Japs captured our Lake Lanao base.

Meanwhile, PatWing 10 had gathered its forces in the southern Philippine area, and sent them farther southward. Task Force 5, including the tender Langley, and all types of merchant shipping, were en route to the next line of defense, the Netherlands East Indies. This withdrawal, in accordance with the War Plan, brought 10 PBY's, 4 OS2U's and 2 J2F's together with the Childs and Heron at Balikpapan on 19 December.

As the Childs arrived at Menado in the Celebes, where the Dutch had a patrol base on Lake Tandano, a Dutch three-engined Dornier exchanged greetings, flew alongside, hauled off, came back, waved again, and then disappeared. At the tender moved slowly toward its anchorage, a petty officer casually remarked, "Here comes that four-engined Dutchman again."

"Four engines!" yelled Captain Wagner. "Commence firing!"

Fortunately, the Jap plane's bombs dropped some distance away and none detonated, and although recognition came too late for the Childs to deliver effective fire, it was believed that the Kawanisi suffered some damage. The lesson in recognition was effective; it was the last time that gunners on the Childs trusted an airplane.

Two days later the case of mistaken identity was reversed. On 21 December, when the wing moved to Soerabaja, Java, its planes were fired upon before the Dutch realized that visitors were American PBY's.

Thus PatWing 10 gathered with the Dutch and Australians for the air defense of the East Indies barrier. The Childs, William B. Preston, and Heron were at Soerabaja; the Langley was dispatched with the other auxiliaries to Darwin, Australia, to act as a receiving ship for aviation supplies and as a minor air station for seaplanes. A successful withdrawal from the Philippines, to unprepared positions, had been accomplished.

Judging correctly that the East Indies barrier might not remain secure, the Army and Navy chose Port Darwin in northwest Australia for development as a major military and naval base, but, of course, it was too far away to serve as a base for immediate operations north of the Soemba chain. The Commander-in-Chief of the Asiatic Fleet, therefore, arriving in the Shark on 1 January 1942, set up his post at Soerabaja in space already occupied by the Dutch and where Admiral Purnell, as Chief of Staff, was already exercising a de facto command.

One of the missions assigned PatWing 10 was to operate long-range searches from Ambon, in the Ceram group about eleven hundred miles to the northeast of Soerabaja. The Dutch already had a fairly good seaplane base at this spot, with considerable gasoline stores and a ramp for hauling out planes for necessary repairs. The task of the wing at this base was, in co-operation with Dutch and Australian groups, to patrol the surrounding area, especially the Straits of Molucca.

Until late in January, the wing ran almost continuous patrols over its assigned area, frequently meeting Japanese opposition. Both sides were equally unsuccessful in efforts to shoot down the other. Co-operation between the Americans and the Australians was excellent; a common radio frequency was used, and each made use of the other's codes and ciphers. Through this joint effort, Jap patrols that had been working out of Davao Gulf were scared off.

Though not yet famous as a fat and happy "Dumbo," the PBY, even this early in the war, proved itself to be adept at air-sea rescue. On 1 January, when a Hudson on fire made a forced landing, a PBY pilot heard the report, left his patrol track to land in a rough sea, picked up the surviving member of the Australian crew, administered first aid, and brought him back, more dead than alive, to the medico at Ambon.

PBY's also rescued the crews of several army B-17's that had landed in isolated areas. In addition, particularly in the later phases of the Japanese advance into Borneo, our patrol craft were told to pick up parties of Dutchmen who had been left behind as demolition squads to destroy important objectives before the Japs could land.

On two occasions, PBY's made rescues which the Dutch themselves had attempted but could not effect.

On 26 December 1941, the commandant at Ambon received what would have been a large order under any circumstances. He was told to send his PBY's in an attack on a Japanese force of 2 cruisers, 2 destroyers, and 13 transports at Jolo, one of the southernmost islands of the Philippines.

These were attractive, if difficult, targets, and it was also important to delay the Japanese consolidation of a position that would control the egress of our few remaining ships from the Sulu Sea and would be a steppingstone to the oil wells of Borneo. Since the Dutch did not have enough planes at Ambon and the Australians' Hudsons could not reach the target, six PBY's of PatWing 10 took off in two sections at about 2300 that night. Word had been received that there were no Jap fighters in the vicinity, but such was no truer at Jolo than at most other Jap landings.

The second section — getting in ahead of time — arrived over Jolo just before dawn on 27 December, and at 0625 started its approach. At 0645, when just past Dong Dong Island, enemy ships off Jolo Town opened fire. Having lost the element of surprise, the PBY's withdrew to gain altitude for a bombing run, but no sooner had they done this than Jap fighters appeared.

One of our planes, shot down about sixty miles to the south, was full of holes but stayed afloat; the other the managed to escape. The first section, after circling about thirty miles from the target at 0600 awaiting the others, went in to attack while dawn light was still favorable. As soon as they got over Jolo Island, but still too far away from the ships to start a bombing run, the three PBY's were in turn attacked by Japs.

A feature of the PBY which contributed as much as any other factor to its helplessness in combat was the pilot's inability to learn what was going on behind him. He was dependent entirely on the waist gunners for information on enemy tactics, but the waist gunners were exposed and were generally the first casualties. In any case they could not properly man both their guns and the voice telephone at the same time.

Such was the case at Jolo; the PBY pilots saw tracers going past their heads before they knew that the Japs were on their tail.

Besides the Jap fighters, the three PBY's braved intense antiaircraft fire from both ship and shore batteries. The PBY's which were all that the Asiatic Fleet had at the time, were entirely too slow for successful bombing attacks against any target which had antiaircraft protection. At Jolo the planes were hit by antiaircraft well before they were in a position to release their bombs. Nevertheless, amid almost continuous AA fire and fighter attack, and in violation of all the laws of probability, bomb they did.

From Filipino leaders who had been on Jolo resisting the landings it was later learned that one Japanese transport had been sunk and that a light cruiser was beached and burning. One plane had been shot down, with the loss of all hands, before it could reach its bombing point, but the other two hit their targets. Two Zeros were counted splashed.

Only two planes of the six returned to Ambon. It had been an expensive foray, both in planes and in men lost. Yet the wonder of it was that the crews of three of the downed planes got back to base alive. Most of that night at Ambon the base picked up radio signals, and the next morning a PBY patrol found one plane, a sieve of holes plugged with life preservers and mattresses, but still afloat. Its crew was taken aboard, the plugs removed, and the plane allowed to sink. The other two planes, however, had been burned. Aboard another plane the two waist gunners had been killed, but the others landed just off the village of Lapa on the opposite side of Jolo.

Of another PBY's crew only four survived after being in the water twenty-seven to thirty hours before reaching the beach of another island. How these men were cared for by friendly natives and managed to avoid others; how they were feasted by Chinese storekeepers, doctored and helped by local authorities and priests; and how they wangled their way from island to island down the long Sulu chain by native outrigger canoes and boats and a decrepit motor launch is an adventure story in itself. On 7 January they reached Tarakan, and on the tenth, fifteen days after having left Ambon, they reported to headquarters at Soerabaja.

The experience of the overage tenders in the first months of the war form sagas concerning which much could be written. Like the planes they serviced, they fought with continued courage an enemy that far outmatched them in size and equipment. Their activities consisted of supplying and repairing the planes and housing and feeding the crews. They had to be on the alert continually for Japanese scouts.

While the William B. Preston remained at Soerabaja for repairs, the Childs and Heron had proceeded to Ambon, arriving on 28 and 29 December respectively. The Childs moored alongside a coral ledge near the shore line, covered masts and stacks with palm trees and did her best to appear as part of the scenery. The Heron, however, was sent, on the afternoon of her arrival, to the rescue of the destroyer Peary, that had been bombed on its way down from Manila.

On the way back (31 December) the Heron was just south of the equator off Halmahera when it contacted two of our own patrols. At 0931 another plane was sighted coming in on approximately the same bearing as the patrol planes that had been seen. Inasmuch as that was the correct approach bearing, the plane was not identified immediately as Japanese, but fast action at the ship's guns forced it off its bombing run.

On two subsequent runs at 1015 and 1034 the bombs fell well clear, and Heron ran for the cover of an approaching rain squall. At 1120 the squall passed, but the Jap was found sitting on the water out of range and waiting. At 1520 reinforcements to the enemy appeared in the form of six Kawanisi-97's. It was a saying in the Asiatic Fleet then that any time more than one plane appeared they were sure to be enemy, so there was no question as to identity. By some fancy maneuvering during the next twenty minutes, Heron avoided the Kawanisis' bombs and damaged one plane.

At 1600, five twin-engined Jap land bombers appeared. These were well peppered on their runs by the Heron's .50 caliber machine guns, but this time it was impossible to escape all the bombs. One hit directly on the top of the mainmast and three others were near misses that damaged the port 3-inch guns and injured all the gun crews on that side.

At 1645, three more Kawanisis appeared and launched a torpedo attack — one on the starboard bow, one on the port bow, and one on the port quarter. It looked as if the little ex-minesweeper Heron could not escape. But this time the Japs' timing was bad, and the ship was turned to meet each torpedo as it was released. On the other hand, strafing by two planes did considerable damage, although one plane made a water landing and was immediately sunk by the 3-inch guns.

Two of the Heron's crew were dead and twenty-five, almost half its complement, had been wounded. That night the paint-locker fire was finally extinguished, the forward hold pumped out, and the ship made ready again to carry on. For the next three days, despite the terrific drubbing it had taken, the Heron engaged in its regular duties tending the seaplanes at Ambon, and on 4 January was sent to Darwin for repair.

Eager to get in some return shots at the enemy, it was a very unwilling Heron that left the scene of action for the quiet of Australia.

THE BOMBING OF MENADO AND TERNATE at the end of December portended the Japanese invasion of the Celebes and Halmahera. On 7 January it was reported that there were Japanese aircraft carriers off Zamboanga at the southwestern tip of Mindanao. On 10 January a PatWing 10 patrol in the Sangi Island sector sighted two large task forces in the Celebes Sea seventy miles south of Mindanao and attacked the large transports, but missed. These were the forces from Davao and Jolo which took Tarakan and Menado the next day.

Four PBY's attempted to attack the cruisers and transports effecting a landing at Kema, northwest of Menado, on 11 January but were driven off by a number of single-engined single-float fighters. One PBY was lost after making a water landing some sixty miles south of Menado. Drifting for five days in Molucca Passage, its crew of seven escaped detection by the foe — and unfortunately by our searchers also — by camouflaging their rubber boat with blue dungarees, which blended with the water very effectively.

On the night of the fifteenth they rode out two cyclonic storms, one of which they said raised waves of twenty feet or more, but on the seventeenth, they managed to paddle ashore on Mangoli, one of the Soela Islands south of the Molucca Sea.

Weak as they were, they contrived shoes out of life jackets and the next day climbed a five-thousand-foot range of mountains to reach the south shore — only ten miles away but twenty miles by native trail. On the nineteenth they crossed by sailboat to the Dutch settlement on Sanana Island.

A PBY was then sent and delivered the entire crew of the lost plane safely aboard the William B. Preston. Such fortitude as this saved more than one of PatWing 10's crews during their fight against odds of nature and enemy in the East Indies. As time went on, planes were lost — and sometimes crews with them — but not one of PatWing 10's men was ever counted out as expendable.

An 8-plane Australian raid on Menado on 12 January was made at a cost of four of the Hudsons, but little damage to the Japs. They continued to come on. After Menado was captured, our base at Ambon received almost a daily working over by Jap land-based planes. In the first raid, between 0330 and 0500 on 7 January, the Laha airfield on Ceram was bombed by five to seven seaplanes, without opposition. Although it was not until 15 January that the heavy raids developed, from then on Ambon became untenable as an Allied base.

Meanwhile, CinCAF had taken steps to use the dwindling resources of PatWing 10 to best advantage and to bring reinforcements. On 11 January ComPatWing 10's headquarters was moved back to Soerabaja for consolidation with the Allied command. On that same day the first reinforcements appeared when six PBY-5's of Patrol Squadron 22 arrived at Ambon from Pearl Harbor via Darwin.

More came on the fifteenth, but three were damaged, two of them beyond repair, as they landed just as the Japs were coming in for the first of their big raids on Ambon. In addition, five Dutch PBY's for which they had no pilots were also obtained to bolster the resources of PatWing 10 in the Java area.

In these first days — war had broken out barely a month earlier — the Navy had no reserve of aircraft or fleets of carriers nor a well-developed system of advanced bases and supply. The exploits of every plane and every person were individually important in holding the line.

Operating from Baoebaoe at the southeast of the Celebes, one PBY of VP-22 was lost when attacked by twelve twin-engined Japs in the Gulf of Tomini, but managed to make a water landing. Some of the crew had parachuted to safety, and all reached a near-by island and ultimately worked their way back to Java. Two days later another PBY-5, patrolling near the end to the Gulf of Tomini, was set upon by a lone Zero.

Highly skillful maneuvering got the Zero in the port waist gunner's sights and the PBY shot the Zero down in flames. The PBY, unblessed with self-sealing tanks or armor, was lucky to receive only half a dozen bullet holes.

But the loss of even one plane was critical — it limited the available eyes of our ships. By this time, even though the Dutch were desirous of standing fast at Kendari and at Ambon, it seemed to navy headquarters at Soerabaja that retirement was again the wise course. On 16 January, Ambon was ordered evacuated, but it was not until 28 January that all U. S. Navy personnel at Ambon were evacuated by plane to Darwin. A few days later Ambon was taken by the enemy, and the Allies lost a battalion of Australians and a crack NEI outfit.

From the time Ambon was abandoned as a base, PatWing 10 started a mobile operation basing two or three planes at a time with the Preston, Childs, and Heron at various places along the Netherlands East Indies chain of islands, and at Darwin in Australia. At the same time, stocks of fuel were laid down along the Soemba chain for the ferry flights of army P-40's from Australia via Timor — a plan which later had to be given up on account of the loss of too many of these light planes which were being pushed to the limit of their fuel capacity.

The tenders usually stayed in one area for two or three days, moving on as soon as a Jap scout was sighted. They fueled the PBY's at night and stood out to sea during the day so that they might have maneuvering room in case of attack. By this strenuous technique, PatWing 10 managed to keep alive as the Japs swept southward.

During this tenuous existence, the seaplane tender Childs had a narrow escape from destruction. In order to help the Dutch and British continue their fight to hold the Netherlands East Indies barrier, it was decided to use the tender as a tanker.

Loaded with 25,000 gallons of aviation gasoline and with an extra 5,000 gallons on deck, the vessel made a quick dash from Soerabaja to Kendari in the Celebes. Ready ground forces unloaded her, and at daybreak on 24 January 1943, the Childs was again under way for her return trip. Less than a half hour later she sighted a force of three Japanese transports and two destroyers entering Staring Bay.

The leading destroyer challenged, but the Childs was saved momentarily by a rain squall that obscured the area for twenty minutes. When visibility cleared, the vessel was running eastward, with the Jap force ten miles abeam, but with four more destroyers visible ahead. These vessels deployed in line and headed for the Childs, and then, to the relief of the Americans, they turned away. Possibly the large green awning that the Childs had draped across both stacks made identification sufficiently difficult that the Japanese preferred to take no chances in protecting their transports.

At any rate, a valuable seaplane tender had stuck its neck into a noose to deliver a sizable amount of gasoline to the Japanese, though it was hoped that the Dutch ground party might have had time to destroy it. The Childs subsequently fought off two air attacks while en route to base.

The diversion of a destroyer seaplane tender to make a run as a tanker was quite in accordance with a practice that expediency dictated not only in these first months of war, but throughout the next four years. It illustrates the inseparability of the naval aviation establishment as a whole with the problem of general logistics and support which is often unthought of among more spectacular reconnaissance and combat.

The Navy's air patrols now backed up to the wall in the long chain of islands, which stretch out eastward from Java, had the hottest sector of the whole area. The Preston operated from Alor, keeping patrols over the eastern area, while other PBY patrols were run from Childs out of Soerabaja up into Makassar Strait as far north as Samarinda, Borneo, over to the Celebes, down the coast to Makassar Town (Celebes) and back to the Java area.

As the Japanese forces gathered impetus for their continued pincers movement in the East Indies there was need for more than a patrol, and the Allies began to pull together their sea strength.

The navy pilots of PatWing 10 knew by experience how to fly through all kinds of weather, day and night; they were used to working with sea forces and consequently knew what they were looking for, and having found it they knew how to report it. In planes without protected gas tanks or armor, they went to the places where the enemy was bound to be. They reported his strength, disposition, course and speed, and then hung around whenever possible trying to guide the surface forces sent to attack.

This work could not have been accomplished without the seaplane tenders, which could move our bases quickly and, wherever they were, relay the radio reports from our planes to fleet headquarters in Java.

On 22 January, when the Japanese started to come down Makassar Strait, there were good submarine and air reports of enemy activity; PatWing 10 PBY's reported contact with ten Jap planes thought to be carrier-based.

At 0700 on 23 January, coastal lookouts at Samarinda, Borneo, reported a fleet of thirty-one ships offshore in Makassar Strait. Two of PatWing 10 PBY's had been stationed at Makassar at the southwest of Celebes, and every half hour from 1050 that morning they reported the location and strength of various parts of the Japanese force as it moved toward Balikpapan. That afternoon the Dutch bombers attacked and claimed 2 cruisers, 4 transports and a destroyer.

At the same time the 4 destroyers of DesDiv 50 speeded to the fray, supported by Marblehead coming up as fast as her damaged shaft would allow. Their action that night has gone down in naval history as an almost incredible feat against a superior enemy force: five or six ships were known to be sunk and others damaged.

The following morning a Dutch submarine torpedoed a cruiser and USAAF bombers also sank two transports and shot down 5 of the 12 Jap planes which had finally come up to provide air cover. For a short time, the Japanese were turned back, and that turning back was accomplished by old PBY's, old Martins, old Hudsons, and old destroyers.

The Makassar Strait action off Balikpapan was successful primarily because the enemy did not have air coverage to deny PatWing 10's PBY's the opportunity of providing reconnaissance for our surface and bomber commands. For some reason — perhaps the damage to a carrier by the submarine Sturgeon — normal Jap tactics had failed to materialize; it was the only time in this period of swift enemy advance that they did not depend heavily on building up both air and sea superiority at the point of their next landing.

Nevertheless, even though the Japs were temporarily checked at Balikpapan, the Allies did not have enough air or sea power to hold them as they kept driving the British back into Singapore, and sprawled into the Moluccas and New Guinea.

The Japanese expedition in Makassar Strait had been stalled for a short time, but the temporary loss of the Netherlands East Indies was inevitable. The opposition was simply too powerful. During the next month, PatWing 10 continued its operations under increasing difficulty but with continued vigor and bravery. PBY's continued to patrol the Makassar Strait.

It was now known as "Cold Turkey Lane," because there was virtually no protection for the patrols. Almost daily our planes were attacked, but by taking advantage of cloud cover, getting near the water, or using evasive tactics, the PBY's supplied the surface forces with daily, often hourly, reports on the southward movements of the Japanese naval vessels and transports.

Bombardment of our remaining bases, however, became increasingly heavy and toward the end of February our air forces at Soerabaja were compelled to remove to Australia. PBY's extended their patrols. PatWing 10 was still flying, but the going was getting tougher.

The Japanese realized the menace that a base at Darwin held for them and made serious efforts to reduce its potency, and their efforts were largely successful. A major air attack on 19 February gutted most of the port. The planes there were destroyed, and the seaplane tender William B. Preston was badly damaged.

Desperate as the situation was in Java under Japanese attack, it was felt that the island might be held if sufficient fighter planes could be obtained. It was decided to take a long chance and attempt to deliver fighter craft to the island. On 22 February, a convoy had sailed from Freemantle, Australia, bound for Ceylon. In the group were the tender Langley with thirty-two assembled P-40E's on deck and with pilots and flight personnel also on board, and the Seawitch, with twenty-seven crated P-40's in her hold.

On the day of departure, it was decided to divert these two vessels to Java. There was much discussion as to the port of entry of these two ships. Both Soerabaja and Batavia, where the planes could have been taken to an airfield, were being bombed daily. Consequently, Tjilatjap was decided upon, and preparations at that port were hurriedly made. The planes were to be unloaded on the dock, towed along the streets to a comparatively open field and flown out to various flying fields for operations against the enemy. Native laborers fled to the jungles, but streets had to be cleared, walls and shacks knocked down.

Somehow it was done, and hopes ran high. Both Langley and Seawitch were to unload, fuel, take aboard as many refugees from Java as possible — for which a list had been prepared — and depart forthwith.

As the Langley approached Tjilatjap, she was met on the afternoon of the twenty-sixth by a Dutch mine layer and two Dutch Catalina flying boats. A large Japanese invasion force was even then off the north coast of Java. Time was the important factor. Vice Admiral Helfrich said repeatedly that the planes must be got in and at the enemy, and every minute counted.

Rear Admiral Glassford shared the view that the risk of bringing Langley in ahead of the slower Seawitch without benefit of an approach to the coast under cover of darkness would simply have to be taken. Accordingly, on the morning of 27 February, Langley was met by the destroyers Whipple and Edsall and proceeded directly towards Tjilatjap.

It was a fair morning with only a few high, scattered clouds and a light northeast wind. The Langley was less than a hundred miles south of Tjilatjap when, at 0900, an unidentified plane was sighted. Realizing that the enemy had found the Langley, her captain sent a report to Admiral Glassford and requested a fighter escort.

There were not fifteen fighter planes in all Java, and none could be sent. When the Langley had been converted to a tender, she had lost half her original flight deck and the remainder was too short to launch the fighters she carried. At 1140 the Edsall gave the emergency signal "aircraft sighted." The Langley was zigzagging on a northerly course as nine twin-engined bombers approached at fifteen thousand feet. Probably they came from Bali.

As the planes approached the bomb release point, the rudder was put full right, and the bombs fell a hundred feet or more off the port bow. The ship shook violently and was sprayed with splinters and shrapnel, but sustained no serious damage. On the second run the planes dropped no bombs, perhaps studying the ship's evasion tactics.

As they made their third run, she made her turn just an instant too soon. The planes turned, too, before releasing their bombs.

The Langley shuddered under the impact of five direct hits and three near misses. One hit was forward, two hits were on the flight deck near the elevator, a fourth was on the port stack sponson, and a fifth bomb penetrated the flight deck aft, starting stubborn fires. After the bombs landed, six Japanese fighters which accompanied the bombers attempted to strafe the ship, but only one made a very determined attack.

Seldom has a ship been hit more severely by one salvo. Aircraft on deck were burning; there were fires below deck, fire mains were broken, and the ship was taking water and was listing 10 degrees to port. But she could still be steered, and her engines were still running in spite of the water rising in the engine room.

The ship was maneuvered to obtain a zero wind, and the fires were somehow put out. The shattered planes on the port side were pushed overboard and counter-flooding was carried out in an attempt to correct the list. It was useless; water continued to rise in the engine room, and the list was increasing.

At 1332 the order was given to abandon ship. The Edsall and Whipple maneuvered skillfully to pick up survivors with the gratifying result that out of the entire crew there were only six killed and five missing. Whipple fired nine four-inch shells and two torpedoes into the Langley to insure her sinking. The position was about seventy-four miles south of Tjilatjap. Whipple and Edsall with the survivors cleared the area at high speed, going off to the west.

The next day weather permitted transfer of Langley's crew to the tanker Pecos in the vicinity of Christmas Island, two hundred miles south of Java. Within four hours after this transfer, the Pecos was attacked by three carrier-based planes and sunk. Again, the Whipple, which picked up the call, rescued the survivors, but by this time a number of the crew of the gallant Langley, the largest of the Asiatic Fleet seaplane tenders, had perished.

With the sinking of the Langley perished also whatever hopes the Dutch had of saving Java. The twenty-seven crated P-40's did arrive at Tjilatjap on 28 February, and were destroyed, still in their crates, to prevent their falling into the Japanese hands.

The bombing of Darwin cut off the direct route of supply from the Pacific which until this time had been in the hands of the Allies. Landings had been made by the enemy on Bali and Sumatra, and new Japanese forces were known to be forming to the north. Thus, even before the loss of the Langley, it had been apparent that there was no way of overcoming Japanese superiority.

On 25 February, General Wavell departed for Ceylon, and the defense of Java rested on the Dutch, with both the British and American fleet units remaining under the direction of Vice Admiral Helfrich until the Allied command was dissolved.

The story from this point is mainly one of surface engagements, but the handful of PatWing 10 planes still remaining in the area continued their work.

The best tribute to their effort is found in the statement of Vice Admiral Glassford who said of their patrol that no information that reached the forces "was as valuable, complete and reliable as that from the PBY's. At the end they became practically the exclusive source of systematic combat information.

Planes flew at night in moonlight after the daylight phase of the final battle in the Java Sea, (27-28 February), for the purpose of shadowing the enemy, especially the true objective — the transports — in order that (Admiral) Doorman's force might attack. No praises could be too high for the work of PatWing 10."

In the Battle of the Java Sea, the enemy successfully prevented the Allied striking force from reaching the convoys. The Jap invasion was successfully accomplished, and since no port in Java was tenable any longer, it was obvious that the time had come to withdraw. On the morning of 1 March 1942, the Joint Allied Command in Java was dissolved, and the American ships were ordered to Australia. American planes and personnel, such as remained, likewise withdrew from the area.

From the Philippines to the Java Sea, naval aviation had scored a courageous record, in the deeds of PatWing 10, against overwhelming odds. Two light cruisers and 2 large transports had been damaged, 1 large transport definitely sunk, and 8, possibly 12, Japanese aircraft had been shot down. Something in the neighborhood of three hundred bombing, reconnaissance, rescue, and utility missions had been flown, to say nothing of the many flights of all types of planes for administrative and supply purposes.

Yet during all this time there were only three operational, as distinguished from combat, losses — a remarkable record. The losses in PBY's were 15 shot down and 17 destroyed or damaged beyond repair while they were based at various places in the Philippines and in the East Indies.

Five officer pilots were known to have been killed, and twelve more were listed as missing on patrol. From the perspective of the war as a whole, these first bitter months may be viewed as the beginning of the end of tragedy. Farther over the Pacific, American task forces were already successfully testing their strength, giving a portent of the power that was to sweep back across lost territory to the shores of Japan itself.

14 – THE BEGINNING OF GLORY

THE JAPANESE CARRIER ATTACK ON PEARL HARBOR, a bold and successful stratagem whatever its moral aspects, set the pattern of the role that naval aircraft would play in the Pacific war. It is one of the ironies of that war that the Japanese Navy set the example which the United States was to exploit in a series of offensive raids across the breadth of the central Pacific Ocean that culminated in the great blows against the Philippines, the Ryukyus, and the empire, and broke the will of the enemy to resist.

The first of these raids, conducted with our prewar carriers, were unpretentious in scale but important in effect; they discovered the weaknesses of the enemy, provided the hard core of experience needed by commanders who would soon take the responsibility of new fleets, and stimulated a jaded morale with a fillip of success. They were experiments in carrier warfare, conceived in improvisation and born of the necessity of defending a shattered fleet and of learning the methods to be employed by the fast carrier task force of the future.

Vice Admiral (later Fleet Admiral) Halsey, Commander, Aircraft Battle Force, led the first offensive strike of the Pacific war against Japanese installations in the Marshall Islands on 1 February 1942. Such strange names as Eniwetok and Kwajalein had not yet become bywords, and lack of information concerning these atolls made the raid an adventure in tactics.

Two striking forces participated: Task Force 8, led by Halsey in the Enterprise, attacked the northernmost islands; Task Force 17, commanded by Rear Admiral [later Vice Admiral] Fletcher in the Yorktown, struck the southern atolls and Makin Island in the Gilbert group.

The perspective of time has shown that this operation was more productive of lessons in aerial warfare than of damage done to the enemy. Not only were the flying squadrons tempered in battle, but the Enterprise, under vicious attack by enemy planes during the withdrawal of Task Force 8, discovered the weaknesses of carrier defense.

One Japanese plane which deck-crashed the "Big E" portended the day of the Kamikaze suicide plane. Admiral Nimitz remarked that the results obtained by the raid were "noteworthy." If the material damage inflicted was more temporary than permanent, the psychological damage was not. This first indication of an American counteroffensive was, to the Japanese, a prognostication of things to come.

They were not long in coming. Three weeks later, on 24 February, Halsey led his force of 1 carrier (the Enterprise), 2 cruisers, and 6 destroyers against Wake Island, which we had lost in the first sweep of Japanese conquest. Opposition was feeble, and the raid was little more than a training exercise. As one officer of Air Group 6 remarked, "It was just a matter of going in and unloading your bombs. We found no surface ships at all and no airplanes except three 4-engined big boats, and one of the Japanese destroyers which was probably damaged in their attack on Wake and which they had beached."

But lack of opposition did not invalidate the experience. From the lessons gained in the actions against Wake and the Marshalls, Admiral Halsey recommended an increase in fighter strength on carriers, the installation of leakproof tanks in all aircraft, an increased use of incendiary ammunition, and a revision of methods for the identification of friendly planes. There was much to learn, but it was being learned quickly and, fortunately, at little cost.

Such raids, of course, had another value. The Japanese offensive in the South and southwest Pacific areas were unchecked, opposed only by handfuls of ships, planes, and men. Diversions seemed necessary while we were building and training our forces at home, and organizing in the southwest Pacific; the central Pacific offered ideal diversionary targets.

In one effort to hamper the efforts of the enemy, a raid was planned on Rabaul for 21 February, three days prior to the attack on Wake. Although Rabaul was not even reached by the Lexington, which was to carry out the raid, the venture cannot be put down as a failure. As the carrier headed toward the objective, she was attacked by Japanese bombers. Effective antiaircraft fire and outstanding performances by our protecting planes prevented damage being done to the Lexington.

The most astonishing feat of the day was the exploit of Lieutenant (later Lieutenant Commander) Edward "Butch" O'Hare, who in his fighter craft was individually responsible for the destruction of five enemy bombers. The attack on the task force made it evident that the element of surprise had been lost.

Consequently, after steering a course directly for Rabaul during daylight hours to create as much alarm as possible among the enemy, Vice Admiral Wilson Brown ordered a withdrawal.

In another effort to irritate and confuse the enemy, Admiral Nimitz ordered a carrier feint at Marcus Island, a triangular air base within 1,200 miles of Japan. Vice Admiral Halsey made the raid on 4 March 1942, with three ships, the Enterprise, the Northampton, and Salt Lake City, a force which illustrates not only the daring of the operation but also the weakness of our offensive abilities at this time. The attack achieved surprise and some destruction of enemy installations. Its particular value, however, was to re-emphasize the need for reserve pilots on carriers.

This long thrust into dangerous waters sapped the strength of airmen who flew an attack mission and then returned to the routine tasks of search and patrol. The success of the raid as a diversion can only be conjectured. Undoubtedly it caused a flurry of excitement in Tokyo.

Another raid was carried out in an effort to check the Japanese advance southward in the New Guinea-New Britain area, which had gained considerable headway by the end of February 1942. A rather sizable task force, built around the Lexington and the Yorktown, undertook the assignment. Because of the desirability of surprise, a moonlight air attack on Rabaul and Gasmata was first planned.

This project was abandoned, however, when it was learned that a majority of pilots on the Yorktown were not qualified in night launchings and landings and were relatively inexperienced in night bombing. A dawn air attack, to be followed by a bombardment of shipping and shore installations by cruisers and destroyers, was next fixed upon.

On the seventh of March, however, information was received from the ANZAC forces that changed the situation and necessitated a drastic revision of plans. This was the information that a Japanese convoy consisting of a cruiser and several destroyers, with transports, had been sighted off Buna, New Guinea. The following day information was received that the enemy was landing troops on Salamaua, and later word came that both Salamaua and Lae had been taken by the enemy. It was evident that the Japanese were moving in force on New Guinea.

In order to protect our own forces at Port Moresby and elsewhere, certain surface vessels were detached from the task force. The carrier task force then moved toward New Guinea. The singular feature of the raid that followed was the fact that it entailed a flight from carriers across a high mountain range to attack the enemy on the other side of New Guinea.

It was common knowledge that the Owen-Stanley Range, over which the planes had to fly, had peaks up to 15,000 feet high, and that the interior, in addition to being rugged, was largely unexplored jungle. To secure more accurate information both as to territory to be flown over and weather to be encountered, two planes were dispatched to Townsville, Australia, and to Port Moresby for data that proved to be of great value.

It was learned that the best pass over the mountains, the highest point of which was at an elevation of about 7,500 feet, happened to be on a direct line between a point in the Gulf of Papua (from which it had been hoped that the carriers could launch their planes) and Salamaua, the objective. Great care had to be exercised to avoid arriving at the pass when it would be obscured by clouds.

Since weather was to be such a factor, Commander Ault flew a scout bomber to a position about midway across the mountains and over the highest point of the pass, to act as a combination weather bureau and guidepost.

Responsibility for carrying out or abandoning the attack was placed in his hands. He remained on station between Mount Chapman and Mount Lawson, broadcasting weather and operational information to both the planes and the surface units until all the attacking planes had returned.

In this attack 104 planes left the Lexington and the Yorktown. Our only loss consisted of one scout bomber shot down by shore-based antiaircraft fire, and eleven other planes slightly damaged.

That surprise was achieved was thoroughly demonstrated by the fact that no enemy fighter opposition was encountered at any time during the raid. It was difficult under the circumstances to determine accurately the extent of the damage inflicted upon the enemy. Our planes had seen more than the bear did when he went over the mountain.

In the last of the preliminary raids in the central Pacific — the Tokyo attack of 18 April 1942 — the naval air arm played a supporting role in a drama which starred Lieutenant Colonel [later Lieutenant General] Doolittle's B-25's. The circumstances of that attack, which took the war to Japanese homeland, are well known.

Task Force 16 was the strongest force to appear up to this time in the central Pacific area and with 2 carriers, Enterprise and Hornet, 4 cruisers, and 8 destroyers, Vice Admiral Halsey had a potent weapon under his command. Although the operation provided little experience for naval airmen, it proved that our fast carriers could steam at will through the enemy's ocean, and it foreshadowed the great carrier strikes of 1945 against the empire islands.

These first raids in the central Pacific were necessary preliminaries. They were the means by which we felt out the enemy, blooded our carrier air groups, and proved the potentialities of shipborne air strikes. When the great tests came, in the battles of the Coral Sea and Midway, naval airmen were experienced and confident. The training period was over.

TO THE NORTH AND EAST OF AUSTRALIA lies a strange assortment of islands, varying in size from the second largest island in the world, New Guinea, to tiny coral atolls, scarcely able to rear their heads above the emerald sea. In prewar America these islands probably existed only for readers of the National Geographic Magazine, who settled back in their easy chairs and read of kangaroos that climbed trees, of birds that did not fly, of fuzzy-headed black men who had bizarre methods of passing away the time. With the coming of the war, however, these lands emerged from their obscurity to land in newspaper headlines.

Names never before heard became a part of the language of today, first as symbols of the fiery expansion of the Japanese war machine and then as lasting monuments to the heroism and the uphill victory of American and Allied fighting men.

North of Australia lies New Guinea, shaped like a huge dragon facing west. At one point it is only a hundred miles from Australia, and scientists have estimated that if the waters of the strait dropped only sixty feet, a person could walk dry-shod from one land mass to the other. Some miles to the northeast of the back of the "dragon" lies a chain of islands called the Bismarck Archipelago. Important in this group are two large islands, New Britain and New Ireland, and a smaller collection of islands, called the Admiralty Islands, of which Manus Island is the most important.

South and east of this group is a double chain of islands known as the Solomons. San Cristobal is the most southern of the lot; next in the chain is an island whose name has been engraved on American hearts — Guadalcanal. Due east of San Cristobal lie the Santa Cruz Islands and directly south of them are the New Hebrides. Southwest of New Hebrides, and about a thousand miles northeast of Sydney, Australia, is 8 New Caledonia.

That the Japanese had been quick to recognize the strategic value of these islands was indicated by the events of early 1942. In January they landed on New Britain and began fortifying Rabaul into a mighty base; they were establishing other bastions on Buka and at Kieta on Bougainville, the largest and most northern of the Solomons.

In early March they sized Salamaua and Lae on New Guinea, and despite air attacks by the Lexington and Yorktown on these ports on 10 March 1942, the onward surge of Japanese conquest continued. By April the enemy had purloined the Shortlands and Faisi Island in the Solomons.

The strategic importance of these areas to the Japanese was twofold. They might well be landmarks on the road to a conquest of Australia and New Zealand. Failing this, they could be used as vantage points from which to cut the life lines of supplies that soon were to pour from our west coast to these outposts of the British Commonwealth of Nations.

These islands likewise possessed great strategic significance for us. With bases established in the Solomons and the neighboring islands we could vitiate Japan's effort to halt the flow of supplies to Australia, and when sufficient supplies and manpower were on hand, we could use the islands as a springboard for our advance into Japanese-held territory.

We could move on to enemy bases in the Gilberts, Marshalls, and the Carolines, the latter formidably protected by that seemingly impregnable base, Truk. Operating in another direction, we could close in on the Bismarck Archipelago, and then central Pacific bases, and ultimately retake the Philippines.

All this, of course, lay in the future. In the early days the picture was black, and it looked as if Japanese strategic plans were the more important. Japan was in the ascendancy. Bataan was lost, and even as Corregidor fell, the Japs were occupying Hollandia on the north coast of New Guinea.

The remainder of the Philippines was submerged. Stilwell retreated from Burma, having taken, as he put it, "a hell of a beating," and Singapore, looking stiff-necked out to sea, was taken by wriggly fighting men sliding down the Malay Peninsula.

Meanwhile, despite the appalling blows they had received, the Allied nations were not inactive. As we have seen, our fledgling carrier task forces were testing their wings in harassing raids in the central Pacific. In addition, while the Japanese were moving into the southwest Pacific and the Solomons area, the Allies had been preparing bases which it was hoped would be out of range.

These spots were in the Ellice, Fiji, and the Samoan islands, and were essential if the onrush of the enemy was to be stopped and the tide of battle turned. Marine and navy forces were reorganized to prepare for future action. Major General F. B. Price assumed command of military forces in the Samoan area, and Rear (later Vice) Admiral John Sidney McCain was ordered detached as Commander, Patrol Wings, Pacific Fleet, to command all aircraft in the South Pacific area.

Admirals ordinarily meet their appointments, and when one is a day late it usually means that something is up. Rear Admiral Fletcher had a date with Rear Admiral Fitch on 4 May 1942. It was an important engagement, for Fletcher commanded Task Force 17 and he was to rendezvous at a prearranged spot with Fitch, CO of Task Force 11, in conjunction with units of an ANZAC (Australian and New Zealand) squadron. The purpose of the meeting was to take steps to prevent the further advance of the Japs into the islands north and east of Australia.

But the admiral was late, and therein lies a tale. On 3 May the admiral received intelligence reports from General MacArthur that the occupation of Florida Island north of Guadalcanal had begun and that Japs were going ashore in transports in Tulagi Harbor. Admiral Fletcher had been waiting for this news for two months. The rendezvous could wait a day. The admiral and his Task Force 17 headed for the Solomons, even though units of his forces were fueling and not prepared to join him in battle.

By 0700 on 4 May, Admiral Fletcher's force was within a hundred miles of Guadalcanal, facing a weather front so heavy that it blanketed Guadalcanal and the area south of it for seventy miles. Notwithstanding poor visibility, the cruisers had launched an inner air patrol, and the Yorktown a combat air patrol of six Wildcats, followed immediately by the attack group. When the planes reached Tulagi and Gavutu harbors they found a considerable number of ships and five seaplanes moored off Makambo Island.

The strike was satisfying. In the three attacks made by the Yorktown's planes, 2 destroyers, one cargo boat, 4 gunboats, and various small craft were sunk, while a seaplane tender, a cruiser, and cargo ships were damaged, and the 5 seaplanes destroyed. The cost was, in war terms, slight; the Yorktown lost 2 fighter planes, but the pilots were rescued. A torpedo plane and pilot were lost, and 2 torpedo planes, 3 bombers, and 3 scouts were damaged by bullets.

As a result of the Tulagi interim, Admiral Fletcher was a day late at the rendezvous, but no one, except the Japs, minded very much.

The meeting was held on the morning of 5 May somewhere in the Coral Sea, south of the Louisiades Islands. In accordance with the flexible custom of task force organization, the whole unit became Task Force 17. All the ships of the combined task force were refueled from the Neosho during 5 and 6 May, and the operation order was placed in effect. Ominous reports were coming in — the enemy was on the move from Rabaul and headed for the occupation of bases in the Louisiades and the possible seizure of Port Moresby, on the southern shore of New Guinea.

At the same time, it was reported that the enemy's Carrier Division 5, with the Shokaku and the Zuikaku, was in the Bougainville area. Admiral Fletcher's objective was to make night and day attacks on all enemy surface units and furnish a screen of cruisers and destroyers to protect the Lexington and Yorktown, whose air groups were to check the further advance of the enemy into the New Guinea-Solomons area.

When Admiral Fletcher became aware of the impending movement of the Japs, he proceeded to the northwest in order to be able to strike by daylight of the seventh. The oiler Neosho and its escort Sims were detached to the south in accordance with fueling directives, and both were overtaken and sunk by enemy bombing the next day.

Admiral Grace's support group comprising 2 Australian and 1 American cruiser and 2 American destroyers, was sent ahead to attack enemy transports and light cruisers, which had been sighted headed for Port Moresby through Jomard Passage and the Coral Sea. In the early afternoon, when Grace's group was about 110 miles southeast of South Cape, New Guinea, 10 to 12 single-engined Japanese monoplanes passed within 6,000 yards of the Farragut, and although ships' guns opened fire no hits were made.

An hour later 10 to 14 planes, probably Mitsubishi 97-type heavy bombers, were seen bearing down from dead ahead. The leading ships opened fire at about 6,500 yards, and 2 planes were shot down almost immediately. The Japs then dispersed from the tight V formation and came over in small groups on port and starboard of the Perkins.

Our ships, by radical maneuvers and heavy fire, avoided all torpedoes. After the Japs had dropped the torpedoes, they strafed the ships they passed in retiring, and seven men topside the Chicago were wounded; two of them died later. The attack had lasted only ten minutes, but from 4 to 6 Japs had been splashed. An unsuccessful attack on the Australia followed some time later.

Task Force 17 was headed north on the morning of 7 May, and extensive searches were being made for the Japanese force, especially carriers. It was certain that three enemy carriers were within striking range of Fletcher's force, but there had been no additional information since the afternoon of the sixth.

Ten scout planes from the Yorktown covered the area of Deboyne Island, and one of them reported two enemy carriers and four heavy cruisers north of Misima Island. Two other scouts encountered and shot down two two-float monoplanes, one near Misima Island; there were no other contacts. To the east, however, blotted out by a heavy front, were the Shokaku and Zuikaku of Japanese Carrier Division 5 which a Yorktown scout had missed because he had to relinquish his search as a result of bad weather.

The Lexington and Yorktown both launched planes almost simultaneously to attack the carriers reported by the Yorktown scout at 0845. At 1100, however, when the combined attack group was well under way, other Yorktown scouts returned to the carrier and reported that what the pilot had actually seen were two heavy and two light cruisers.

The attack groups were not recalled, since later information supplied by shore-based reconnaissance planes of the Australian command reported one carrier, sixteen sundry warships and ten transports near the latitude and longitude from which the Yorktown scout had erroneously reported two carriers. Word was then given the attack groups, which changed course, and at 1130 the Lexington group made contact with the enemy north of Misima Island.

The battle was on, and for one Jap carrier, at least, the day of reckoning was at hand. The first of the Lexington squadrons to go in was Scouting Squadron 2, followed a few minutes later by the bombing and torpedo squadrons. Despite hits by all three, the enemy carrier prepared to launch planes and only one small fire was visible from the air when the Yorktown planes arrived shortly afterwards.

The same type of attack was repeated by scouting and bombing squadrons together with Torpedo Squadron 5, with the added advantage that the enemy ship had turned into the wind to launch its aircraft and could not easily take evasive action. Within three minutes after the last torpedo had struck, the waves closed over the Japanese carrier, originally thought to be the Ryukaku but later identified as the Shoho.

The Lexington group accounted for five Japanese planes shot down at a cost of two for us. The pilot of one of our planes was Lieutenant Edward Allen who had won the Navy Cross in the defense of the Lexington against Jap bombers near Rabaul the February before. The pilot and crew of the other plane were forced down on Rossel Island and later rescued by the Australians. Four Japanese planes were shot down by the Yorktown fighters, but one of her dive bombers was lost with its pilot.

While the combat patrols were being landed on the Yorktown at dusk on the seventh, three enemy planes flew by on the starboard side burning running lights which indicated they mistook the Yorktown for a Japanese carrier.

Our own planes were circling for landings and although antiaircraft brought down one of the enemy planes, the others went undamaged as a result of our fear of shooting down our own planes. The radar screen of the Lexington showed that thirty miles to the east, planes were circling apparently to land on a Japanese carrier, confirming the conjecture that Carrier Division 5, until now unaccounted for, was in the area.

Admiral Fletcher decided to proceed southwest so as to be in a position to attack the enemy carriers in the morning. A Lexington scout on dawn search patrol, at 0820 on the eighth, sighted an enemy formation of 2 carriers identified as the Shokaku and Zuikaku, 4 heavy cruisers and 3 destroyers 170 miles northwest.

An attack group composed of 35 dive bombers, 15 fighters, 11 scouts and 21 torpedo bombers was launched at 0900 at about the same time that it became evident the Japs had located our force. This meant that an enemy attack could be expected about 1100, and the task force prepared to receive it.

Admiral Fletcher turned over tactical command of the force to Commander, Air, Admiral Fitch. History was about to record its first carrier-against-carrier battle, for while our planes were winging toward the Jap force, its planes were headed for Task Force 17.

The day was clear with unlimited visibility and ceiling.

At 1100, radar reported many aircraft approaching from the south, about seventy-five miles from our position which was southeast of Rossel Island. The first planes, torpedo bombers, were sighted thirteen minutes later at 6,000 to 7,000 feet, coming in on both bows.

As soon as the enemy planes were sighted, the ships of Task Force 17 advanced speed to 30 knots. Our fighters had gone out at 10,000 feet but when the Japanese torpedo planes were sighted 20 to 30 miles distant from the task force, the enemy had an altitude advantage of 7,000 feet, and our fighters could not gain altitude soon enough to intercept him before the pushover point. Our Wildcats did engage the 18 protecting Japanese fighters and shot down or damaged 6.

The position of the anti-torpedo patrol, however, was 6,000 yards from the ship at 2,000 feet and enabled the enemy torpedo planes to come in over this patrol at high speed. However, our SBD's on the port side of the Lexington shot down eight planes, four of which were carrying torpedoes. A bomber and two fighters were also destroyed by the SBD's, while we lost one dive bomber to a Jap fighter.

At 1120 the first torpedo struck the Lexington and exploded in front of the port forward gun gallery. Only a minute later another hit farther aft, opposite the bridge. Meanwhile, Jap dive bombers were attacking at 70-degree dives from 17,000 feet.

A 1,000-pounder hit the after end of the port gun gallery in the ready-ammunition locker just outside Admiral Fitch's cabin. Two near misses hit close aboard aft on the port side and at first it was thought the ship had been struck by torpedoes. Another 500-pound bomb hit the gig boat pocket on the port side, and a 100-pound bomb, realizing one Jap pilot's dream, struck the stacks and exploded inside, killing and injuring a number of men in the stack machine guns, sky aft, and in the after-signal station. Altogether there were seven hits on the Lexington.

Fire started in the main deck near the admiral's country, and the ship was listing six degrees to port. Damage Control, however, shifted oil to trim the ship, and fire parties were fighting numerous fires. In spite of the blasting she had taken, all units of the Lexington were in commission, although three of her fire rooms were partially flooded. Pumps, however, were controlling that damage. Her steering gear was intact, and she was making twenty-five knots, under good control. Both elevators, however, were out, jammed in the up position.

By 1300 the ship was on even keel, three fires had been doused, and the one in the admiral's country was under control. She was still landing and reservicing planes.

Simultaneously with the attack on the Lexington, the Japanese launched an attack on the Yorktown. Torpedo planes came in over and through our protecting screen of fighters and separated into small groups as soon as antiaircraft batteries opened fire. Three torpedoes were observed approaching on the port quarter and four more on the port beam. Skillful maneuvering enabled the carrier to evade them only to have others come in on the starboard quarter.

As the torpedo planes completed their attack the dive bombers came in. Six near misses on the starboard side and two each on the port and starboard quarters shook the ship so that the screws were lifted from the water and bomb fragments pierced the hull above the water line. Only one direct hit was made, but it pierced the flight deck and penetrated into the ship, exploding above the fourth deck and killing thirty-seven men and injuring many others.

Damage was not sufficient to impair the Yorktown's battle efficiency, however, and she continued operations.

While this attack was in progress, the Japanese carriers were having a similar experience. Our strikes, which had been launched at 0900, arrived over the Jap force at 1030. They found 2 carriers, escorted by a battleship, 3 heavy cruisers, and 4 destroyers. The dive bombers, which had arrived first, waited for the slower torpedo planes to come up and then launched a coordinated attack on a carrier which had started to send off planes. In spite of fighter opposition, Yorktown dive bombers pressed home their attacks and as soon as they pulled out, were followed by the torpedo planes.

Although the bombing squadron from the Lexington failed to locate the Jap force, the other planes from that ship participated in the attack. Results were indeterminate, but it is known that at least two bomb hits were scored.

The planes returned at 1400 to find the Lexington in bad shape but still capable of landing planes. All damage had been checked or cleared, when from somewhere deep amidships a heavy explosion shook the ship.

All telephones but one were out, and another fire started. All pressure in the forward fire main was gone and the gyro compass system was inoperable although the after gyro and repeaters were functioning. The explosion had doomed the ship.

Ironically, Japanese planes had failed to do what gasoline vapors caused by small leaks had done. As the fire spread, all communications were lost, water pressure was nil and minor explosions kept recurring, intensifying fires.

By 1600 the last main control telephone was so weak that Captain Sherman, fearing he would lose contact with the engineering personnel, ordered them topside. The safeties were opened, the ship halted, and preparations to abandon ship were made as all water pressure was gone and firefighting was now impossible.

Admiral Fitch ordered destroyers with fire hoses to stand by to receive excess personnel. The Morris stood alongside with hoses while personnel descended by lines to her decks and safety. The fires were now beyond control, and explosions were occurring more frequently.

Captain Sherman, meanwhile, fearing the ship would blow up, had had the sick and wounded transferred to whaleboats. Admiral

Fitch directed him to abandon ship. Captain Sherman passed the word, and orderly disembarkation began.

By 1800 the admiral and captain prepared to leave the ship, but not until Captain Sherman had made an inspection and found, on the starboard side in an after-gun gallery, some men having difficulty disembarking. He ordered them to shift aft and disembark from that point. The executive officer, Commander Mortimer Seligman, his final inspection finished, reported all hands were off the ship. At that moment another explosion shook the Lexington, and the two officers had to duck to avoid debris showering them near the elevator. The skipper ordered Seligman off, and when he saw that his executive officer had swum to a motor sailer, Captain Sherman himself went hand-over-hand down a line, dropped into the water, and was picked up by a boat from the Minneapolis aboard which he reported to Admiral Fitch.

It was a magnificent but sad spectacle made by the doomed and burning ship. She had performed so gloriously only to perish in her hour of triumph. Gallant to the end, she remained afloat seven hours and finally was torpedoed by the Phelps. Three out of five torpedoes hit her, and she sank on an even keel. Even as the Pacific received her, a tremendous explosion rocked ships for miles around.

The battle of the carriers was over. The enemy had incurred such losses that he would have to take pause before he could continue his progress south. The balance was heavily weighted on our side. From 0700 on 4 May to dusk of the eighth, in addition to losses suffered at Tulagi, the Japanese had lost the Shoho and a light cruiser and had seventy planes shot down, as well as those which sank with the carrier. Both the Shokaku and Zuikaku and their planes were badly damaged, and in the case of the Shoho, the Japanese probably lost the entire complement of the carrier. For the time being the Japanese invasion of Port Moresby was halted. Our own losses, while serious, were not nearly so grave as those of the Japs.

We had lost the Lexington but 2,735 of a complement of 2,951 were saved, and eighteen of her planes were safely landed on the Yorktown. In addition, the Neosho and Sims had been sunk on the seventh.

Twenty-seven of our planes were lost to enemy action. The Yorktown was also damaged but not so badly that she could not turn up a month later at the Battle of Midway.

The Battle of the Coral Sea indicated that despite the weather advantage the Japs had enjoyed, and the superiority they had both in number of planes and in their performance ability, our personnel were superior in quality and skill. Our tactics were better, as was our antiaircraft fire.

This was the first time in history that a decisive naval engagement had been fought without surface ships taking combatant roles, and in that engagement the nectar of victory that the Japs had been drinking for five long months was for the first time flavored with the gall of defeat. Furthermore, had the Japanese won the Battle of the Coral Sea, they might have been able to take Port Moresby from the sea. In their attempt to take it by land they were unsuccessful; failing in that, their campaign in the southwest Pacific ultimately collapsed.

As we have seen, the Battle of the Coral Sea, 4-8 May 1942, was the first decisive engagement in the history of the world in which aircraft carriers played the principal part and in which opposing surface ships did not fire a shot at or get within sight of each other. The Battle of Midway, coming about a month later, was the after event which definitely established the pattern of future naval warfare. As such, it stemmed the Japanese offensive at its flood tide, saved our outer defenses from occupation, and put the enemy forever after on the defensive. It was the turning point in the battle for the Pacific.

After their South Pacific offensive had been stopped in the Coral Sea, the Japanese mustered ships and men for a grand assault on our Aleutian and Hawaiian bases. American naval strategists knew that the situation was extremely serious. The Lexington had been sunk in the Coral Sea and the Yorktown damaged, and the air groups of both carriers had been weakened by four months at sea. The Japanese still had carriers and ships in home waters that could be used against Hawaii and the Aleutians, and there was reasonable doubt that we could bring our forces back from the South Pacific in time to do any good. If we guessed wrong, especially if the enemy renewed his attacks in the South Pacific, the situation might become extremely serious.

The forces at our disposal were distributed as seemed best, and we awaited the next enemy move. On the west coast were the battleships with a light destroyer screen. Only five cruisers and four destroyers could be spared for the Alaskan area in case of a thrust in that direction. Two task forces, composed of three carriers and supporting ships, and land-based air squadrons were available to protect the Hawaii-Midway area.

Task Force 16, under Rear Admiral (later Admiral) Raymond Spruance, and Task Force 17, commanded by Rear Admiral (later Vice Admiral) Frank Jack Fletcher, rendezvoused northeast of Midway on 2 June 1942. In the former force were the carriers Enterprise and Hornet; in the latter, the Yorktown, restored to fighting condition in three days at Pearl Harbor. The defensive capabilities of Midway Island itself had been strengthened by the addition of army, navy, and marine air units. Nineteen submarines were assigned to cover the approaches to the Hawaiian area. All were on station by 3 June.

The Japanese began the battle on that date with a feint at the Aleutian Islands. Moving in under cover of a weather front, an enemy carrier force attacked the Unalaska base at Dutch Harbor and Fort Glenn on Unimak Island. The raid, as is shown elsewhere, was useful both as a cover for landing operations on Attu and Kiska and as a diversion for the main effort against Midway. The diversionary tactic failed, however, because we refused to divert our naval strength from Hawaiian waters; distance and time factors made such a shift impossible. Part of the price of the victory at Midway, therefore, had to be the minor Japanese success in the Aleutians.

On 3 June the usual searches were being conducted in the Midway area, coverage being excellent except for the sector to the north-northwest where visibility was extremely poor. At 0904 a plane of Patrol Squadron 44 made the first contact of the battle: "Two Japanese cargo vessels sighted bearing 247° from Midway, distance 470 miles. Fired upon by antiaircraft."

At 0925 another plane reported: "Main body bearing 261°, distance 700 miles, 6 large ships in column." During the next few hours other units of the Japanese Fleet were sighted by our patrols, and it became evident that a force of powerful proportions was converging for an attack.

At about 1230, nine B-17's, each carrying bomb-bay gasoline tanks and four 600-pound bombs, took off from Midway for the first attack on the enemy. At 1623 this flight found two or three heavy cruisers escorting approximately 30 auxiliaries, 570 miles from base. An attack was made by three flights of three planes each at altitudes of 8,000, 10,000, and 12,000 feet, respectively. Results were unobserved.

Before the B-17's returned, a volunteer flight of 4 PBY-5A's, each carrying a torpedo, took off from Midway at 2115 on a mission of historic interest, for this was the first night torpedo attack by our patrol planes on surface ships. Station keeping was difficult in the darkness, and one plane lost the formation and did not participate in the attack. At 0115 on 4 June a radar sighting was made on a group of ten ships off the port beam of the PBY's. The planes approached with engines throttled back. The result of the torpedo drops was difficult to ascertain, and the returning pilots could only report that one or two ships were "possibly damaged."

In the meantime, there was cause for anxiety. The Japanese carrier force had not been located, and the imminence of a raid on Midway was real. In the absence of sightings, the enemy carriers were believed to be approaching from the northwest under cover of the weather front. Preparations were made accordingly; all planes were put in the air or kept in a state of take-off readiness.

At 0545 on 4 June, anxious American airmen received the most important contact report of the battle, a brief message in plain English: "Many planes heading Midway, Bearing 320°, distance 150." Five minutes later the Midway radar picked up the planes at a distance of 93 miles, altitude 10,000 feet. By 0600 every plane able to leave the ground had taken off. Pilots of Marine Squadron 221, flying Buffalos and Wildcats, intercepted a Japanese formation of 60 to 80 navy bombers escorted by about 50 Zero-type fighters at 12,000 feet altitude, distance 30 miles.

Despite this interception, however, the enemy planes proceeded in ragged formation to their objective and at 0630 dropped the first bomb on Midway. By 0175 the raid was over, and the marine fighters were ordered to land and refuel. Of the 27 planes which attacked the enemy flight, 15 were missing and 7 severely damaged.

They had, however, inflicted punishing damage to the enemy; known Japanese losses amounted to 43 aircraft. Damage to Midway installations was severe, but fortunately the attacking planes had spared the runways, apparently for their own anticipated use.

Shortly after 0615, four torpedo-carrying B-26's left to attack the enemy carriers. At 0705 they sighted the target — three large carriers, a battleship, several cruisers, and about six destroyers — and leveled off for the approach. The Japanese shot down two of the planes and prevented the other two from making an accurate assessment of the results obtained.

It was believed, however, that one carrier was damaged by two torpedoes. At the same time the Midway-based detachment of Torpedo Squadron 8, consisting of six Avengers, made a run on the same group of enemy ships. Of this flight only one badly shot-up plane returned to make a landing.

At 0755, Marine Scout Bombing Squadron 241 moved in to the attack. This squadron was divided into two groups, the larger group of 18 Dauntless aircraft under Major L. R. Henderson being the first to reach the target. Enemy fighter defense was effectively organized. Of the 18 SBD's, 9 returned to base, and of these only 6 remained fit for service.

Major Henderson and his gunner were shot down before the final attack was underway, and it was only by determined devotion to duty that the squadron obtained three direct hits and several near misses on a large Kaga-class carrier.

Following a high-level attack by army B-17's, that reported three hits on two carriers, the second group of marine bombers under Major B. W. Norris arrived on the scene. For a loss of two planes his twelve Vindicators obtained two hits and two near misses on the Japanese battleship. With their return to Midway the first phase of the assault on the enemy fleet was accomplished.

Admiral Nimitz described the situation as of that moment in his report of the action:

"The Midway forces had struck with full strength, but the Japanese were not as yet checked. About 10 ships had been damaged, of which 1 or 2 AP or AK may have been sunk. But this was hardly an impression on the great force of about 80 ships converging on Midway. Most of midway's fighters, torpedo planes, and dive bombers — the only types capable of making a high percentage of hits on ships — were gone, and 3 of the Japanese carriers were still undamaged or insufficiently so to hamper operations. This was the situation when the carrier attack began."

During the night of 3-4 June, Task Forces 16 and 17 moved to a position about two hundred miles north of Midway. At 0700 the Hornet and Enterprise began sending off planes, Admiral Fletcher having by this time received through radio interception the first contact of the Japanese carrier force. Inasmuch as only two enemy carriers had been reported, the Yorktown delayed launching until 0840.

Meanwhile the Japanese had apparently sighted our own carriers and had turned northward. This maneuver took us unawares, and the air strike that had already been launched consequently failed to reach the target, except for the Hornet's Torpedo Squadron 8, which had become separated from the rest of the air group, had turned north and discovered the enemy ships Akagi, Kaga and Soryu close together, the last damaged and smoking, and the Hiryu standing off to the north.

In a gallant but futile attack made at 0920, without support of any kind, the fifteen planes of the squadron were shot into the sea by overwhelming fighter opposition. One pilot, Ensign G. H. Gay, survived to witness the subsequent action, from the insecure vantage point of an uninflated life raft wallowing in the water.

Less than an hour later Torpedo Squadrons 3 and 6 from the Yorktown and Enterprise arrived at the scene. In a wild melee of flying planes and flying shrapnel, hits were obtained on two carriers and probably a third. But the price was heavy. Of the 41 torpedo planes in the three squadrons only 6 returned. It was a sacrifice, however, which forced the enemy carriers into such radical maneuvering that they could not launch their own planes, and it left few of the Zero fighters in a position to intercept our dive bombers, which now attacked with devastating effect.

At 1022 the bombers struck. Antiaircraft fire was light and there was no fighter opposition until after the planes had pulled out of their dives. The Enterprise planes broke up the decks of the Kaga and Soryu, and the Yorktown bombers hit a carrier believed to be the Akagi as well as a battleship and a cruiser. The action cost 18 SBD's but the results were spectacular. Burning furiously, the three stricken carriers marked the high tide of the Japanese advance. Later in the day the submarine Nautilus delivered the death blow to the Soryu. Only the Hiryu escaped undamaged to the north.

The temporary escape of the Hiryu cost us the carrier Yorktown. At 1159, radar contact was made with a Hiryu attack group of thirty to forty aircraft which necessitated the waving off of the returning Yorktown strike. Defensive fighters were vectored out to meet the enemy planes about twenty miles from Task Force 17. Only eight bombers evaded the Wildcats, but before they were shot down by ship's gunfire, they scored three hits on the Yorktown, which, however, reduced her combat efficiency only temporarily. By 1215 the action had been completed; by 1350 the wounded carrier began refueling planes.

At 1427 the second enemy air strike approached. Although interception was again made at some distance from the Yorktown, the Japanese put two torpedoes in the carrier, causing her to smoke heavily and finally stop with heavy list to port. At 1445, with all power gone and danger of capsizing imminent, the ship was abandoned.

The epilogue requires no elaboration. On 5 June a salvage party boarded the listing vessel and it was determined to take her in tow. The operation proceeded successfully until 1335 on 6 June when she was struck by two submarine torpedoes. At 0501 the next morning she turned over on her port side and sank in about 3,000 fathoms of water with all battle flags flying.

Revenge for the Yorktown was not delayed. Scout bombers located the Hiryu at 1430 on 4 June, shortly after her air group had completed the attack on the Yorktown. Enterprise planes reached her first and left her burning so fiercely that the Hornet attack was diverted to a battleship and a cruiser. With the destruction of the Hiryu, control of the air had been won, and the tide of battle turned. Pursuit of a disorganized and fleeting enemy now occupied our air force.

During this critical afternoon, Army B-17's from Midway had found and attacked damaged units of the Japanese armada. The island garrison, meanwhile, feverishly prepared for the possibility of a landing attempt which seemed close at hand when an enemy submarine shelled installations at 0130 on the morning of the fifth.

Shortly before dawn on 5 June search planes took off, followed by a mixed force of Fortresses, Dauntlesses, and Vindicators — the remnants of Midway air power. This force located and bombed cruiser targets and claimed some hits. Later in the day the B-17's made other attacks. The results of the day's operations were reported as one hit each on two heavy cruisers and three hits on a large cruiser.

An attempt by Task Force 16 to close with the enemy on 5 June produced no successful results. On the morning of 6 June, however, Enterprise search planes found two Japanese dispositions within fifty miles of one another. The task force chose the northernmost group as its target, leaving the southern group for the Midway Fortresses. The Fortresses failed to make contact but at 0950 the Hornet strike hit a battleship and cruiser and sank a destroyer. Enterprise planes followed this up, sinking the heavy cruiser Mikuma and severely damaging the Mogami.

At 1645 a second Hornet attack group found two cruisers and two destroyers and obtained hits on three ships. This was the last strike of the day. Fuel shortage, reports of Japanese submarines in the vicinity, and the danger of coming into range of Wake-based enemy planes forced the task force to retire. The Battle of Midway was over.

For the loss of the Yorktown, the destroyer Hammann (sunk by submarine torpedo while assisting in the attempted salvage of the Yorktown), and 150 aircraft, our airmen had sunk 4 large Japanese carriers, with an estimated 275 planes, 2 heavy cruisers, and 3 or 4 destroyers, had damaged numerous other units of the enemy battle fleet, and by doing this had relieved the danger to Midway, Hawaii, and the west coast.

The Battle of Midway will be an object of curiosity for succeeding generations of historians. It was a contest primarily to retain naval air forces, and it established the fact that naval air power was a principal factor in the control of the sea. It is a monument to America's naval heroes.

JAMES H. DOOLITTLE (1896-1993)

15 – WINGS OVER GUADALCANAL

THE SANGUINARY BATTLE FOR GUADALCANAL is primarily the story of the hard-won victory of the foot Marine and soldier against incredible odds. It is no reflection on their achievement to note that the campaign was given considerable assistance by our naval and marine air forces. As in the case of the land forces, the way of the air units was hard. We had our reverses, and we learned costly lessons.

Through it all, however, we were testing and hardening the techniques of war in the air. To alter slightly a famous utterance, there were times when too much was required of too few. We found that the refueling of fighting forces during a battle, furnishing and distributing supplies in advanced areas formed major problems of modern warfare.

We also found that skilled American fighters in American-made planes could lick considerably more than their own weight, and that they were not discouraged by what appeared to be, and often was, overwhelming opposition. We learned that if need be carrier planes could fight from land, and that navy, marine, and army pilots could weld themselves into a fighting unit against the enemy, and when the job was completed return to their special tasks.

As we have seen, the Japanese, in their first rush of wartime offensive, had swept down the Pacific through the Philippines and the Netherlands East Indies and were literally knocking at the door of Australia. Their continued plans had been seriously checked by the battles of Coral Sea and Midway. If we were to keep the advantage won by these engagements, however, and if we were to prevent the severance of our supply lines, we had to take the offensive at once.

By 25 June 1942, therefore, Vice Admiral Robert L. Ghormley, the commander of the South Pacific force and the South Pacific area, had received orders to attack as soon as practicable. Establishing headquarters in the USS Rigel at Auckland, New Zealand, Vice Admiral Ghormley met with Major General (later General) Alexander A. Vandegrift, USMC, who two weeks previously had arrived in New Zealand.

The plan as evolved was to use the 1st Marine Division, reinforced by the 2d Marine, the 1st Raider, and the 3rd Defense battalions in a projected assault on the Tulagi-Guadalcanal area. The landing place was to be Red Beach, on the north shore of

Guadalcanal, some 6,000 yards east of the airfield that the Japs were assiduously constructing. D-Day was tentatively set for 1 August.

While we were planning our offensive, the enemy was landing a large force of laborers and soldiers on the island soon to be so hotly contested. The laborers were building, not far from Lunga Point, the airfield soon to be known as Henderson Field.

Consequently, it became more imperative than ever to attack, in order to prevent the enemy from using the new field as a launching point for air blows at our bases in the New Hebrides-New Caledonia area. The urgency of striking at once while the foe was still weak from the staggering blows inflicted at Coral Sea and Midway overbalanced the exceedingly short hiatus between these battles and the opening of the Solomons campaign.

In addition to the scarcity of time, there was a shortage of civilian manpower for unloading and reloading transports, with the result that Marines, weary from a long voyage in crowded vessels, had to turn stevedores for the time being. The weather, furthermore, was not on our side, for an Antarctic rain blew down continuously, soaking the workers and saturating cardboard containers until they spilled their contents.

Despite these obstacles, combat troops were loaded with the barest minimum of personal equipment, and on 22 July the first transport group, carrying the 1st Division, left Wellington with a naval escort bound for Koro Island in the Fijis where rehearsals for the imminent invasion took place from 28 to 31 July 1942.

Meanwhile, our naval air forces were preparing to give assistance. On 22 July, Admiral McCain, Commander, Aircraft, South Pacific area, advised Admiral Ghormley that his planes would begin a search two days before the scheduled attack to forestall enemy counteraction. Aircraft of the southwest Pacific command were also to carry on this preliminary activity.

The task force commanded by Admiral McCain consisted of seven groups, and illustrated the effective blending of the various air forces to complete a specific task. Two groups were USAAF bombardment groups, three were naval aviation units, and the remaining two were marine units. The shore-based planes within Admiral McCain's command aggregated 291 aircraft. Mainly, these conducted patrol sweeps. The Assault Force was under Vice Admiral Fletcher, with the Air Support Force composed of the carriers Saratoga, Enterprise, and Wasp and their screening ships under Rear Admiral Leigh

Noyes. The function of this force was to supply air offense and defense for Admiral R. K. Turner's amphibious force.

The landing force was under the command of General Vandegrift, and was composed of the 1st Marine Division (less the 5th Battalion, 11th Marines, 1st Tank Battalion less two companies and detachments), 2d Marine (reinforced), the 1st Raider Battalion and the 3rd Defense Battalion. The rehearsal, simulating combat conditions as nearly as possible, was over on 31 July, and that afternoon the fleet left the Fijis for the Solomons. The stage was set, and the characters were ready for the opening act of one of the great dramas in World War II.

On 3 August the fleet passed through the southern New Hebrides and by the fifth reached the 159th meridian and headed north for the Tulagi-Guadalcanal area. Even before the force sighted the islands destined for invasion, Admiral McCain's planes had been executing pre-invasion missions. By the sixth the weather afforded cover for a stealthy approach to the Solomons. Clouds and a mist that limited even surface visibility made enemy air reconnaissance impracticable.

By 1600 on the sixth, the armada disposed itself for approach; the squadron destined for the Tulagi landing followed six miles astern by the force covering the Guadalcanal landings. By 0300 of D-Day, which had been changed to 7 August, they were off the northwest tip of Guadalcanal. There they separated, the Tulagi squadron passing to the north of Savo Island, now discernible in the waning moonlight, the Guadalcanal group heading south for the north shore of Guadalcanal. There was no exchange from the enemy, and the completeness of the surprise was confirmed. Meanwhile, the carriers were disposing themselves seventy-five miles from Tulagi, and before sunrise they launched their first strikes.

As the Chicago, leading 15 transports deployed in two columns, headed a ghostly procession along the dim shore of Guadalcanal, our carrier-based bombers appeared and concentrated their attacks on gun batteries, vehicles, and supply dumps, while fighters from the Wasp destroyed 15 patrol planes and 7 seaplane fighters in the water at Tulagi. Nine hundred yards off Red Beach the Marines were awaiting the signal to land. At 0650 came the signal to "land the landing forces," and the amphibious operation was under way.

There is unfortunately no place in this account for the heroic struggle of the land forces on Guadalcanal. The landings were accomplished easily and, as the commanding general of the 1st Marine Division said, "proceeded with the precision of a peacetime drill."

There was little opposition at the outset, and the defending forces fled precipitately to the hills.

This initial response of the Japanese, regrettably, was not a harbinger of things to come, as the ensuing months were to show. In contrast to Guadalcanal, the landings on Tulagi, and two other spots, Gavutu and Tanambogo, were bitterly contested, and gave carrier aircraft their first opportunity of World War II to demonstrate their effective close support of an amphibious operation.

The first phase of the Guadalcanal campaign might be termed the digging-in process. During this procedure, the question of air support soon became of vital significance. The sizable air resistance that developed immediately after the initial landings, placed a strain on the resources of the carriers. As a result, the flattops had to retire from their positions south of Guadalcanal because their fuel was running low. In addition, they had lost 21 of their 99 fighters. Enemy air strength, meanwhile, was mounting.

As a consequence of these factors, a conference was held, and it was decided to retire the carriers as soon as possible so that they could be refueled and again be ready for action. With their departure on the ninth and the ensuing, virtually unopposed air raids by the enemy, transports, now dangerously vulnerable to air attack, could unload only a portion of their supplies on the Guadalcanal shores. By sunset of the ninth, the transports were forced to leave.

The next days were difficult ones, but despite obstacles, the Marines went ahead with their digging-in process. Important from the standpoint of aviation was the work carried on for the completion of an airfield. A strip 2600 feet long and 160 feet wide had been left by the Japanese in their first hasty flight from the area.

Handicapped by the lack of power shovels, and working with hand shovels, trucks and dump carts, the Engineering Battalion moved 100,000 cubic feet of fill and within six days had a runway nearly 4,000 feet in length, despite continuous enemy bombings.

These bombings continued, but holes were filled, and on 12 August 1942 the first plane landed on what five days later was christened Henderson Field.

This first plane was a PBY-5A which landed and evacuated two wounded men when only 2,800 feet of the runway were operable. On 20 August, marine planes arrived, having been preceded by marine aviators who had helped prepare the field for combat operations. These marine planes were divided into two squadrons, one fighter and the other bomber, equipped with Wildcats and SBD's. The members of these squadrons were to give a brilliant account of themselves before the termination of the Guadalcanal campaign.

Due credit should also be given to Army Air Force fighters, who lent distinguished service in P-400's. The total of this first force was small; 19 Wildcats, 12 SBD-3's, and 5 P-400's. Intended for British high altitude use and equipped with oxygen gear, the P-400 (a stripped-down version of the Airacobra), had to be employed as a low-altitude fighter since there were no oxygen bottles available on Guadalcanal. Opposed to this small force were the groups of eighteen to twenty-four enemy bombers, escorted by twenty or thirty Zeros, that came down every day from Bougainville or Buka fields. It was the quality of pilots, rather than the quantity of planes that depleted the Japanese air force in the Solomons.

The air squadrons had arrived just in time to be of assistance in the bitter struggle — called the Battle of the Tenaru — that followed. The Japanese had thrown reinforcements into the island that were of a much different caliber from the laborers who had fled to the hills from Red Beach.

With the limited air strength at our disposal, there had to be a division of effort. The army P-400's did excellent service as interceptors and as ground support, just as in 1944 and 1945 the marine air forces were to assist army ground forces in the Philippines. The Wildcats served as bomber escorts, and the SBD's bombed enemy vessels, though on occasion they supported ground forces.

At noon on the twenty-first, 4 Wildcats of Major John L. Smith's squadron engaged 6 Zeros at 14,000 feet between Lunga Point and Savo Island. In that engagement Major Smith shot down the first Japanese plane to be destroyed in the air by a marine pilot in the Solomons. Before the Guadalcanal campaign was over, Major Smith was to account for 19 enemy planes.

Between 24 August and 27 September, the air force was augmented by 11 SBD's from the Enterprise. Launched on the morning of 24 August, this unit, known as Flight 300, had been sent in search of a Japanese carrier. Failing to make contact, the force landed after dark at Henderson Field, and for more than a month gave valuable service to the American forces on Guadalcanal.

During this period the American carriers had refueled and were ready to enter the picture once again. At the same time the enemy was pulling its sea and air forces together in the area. The result was an engagement known as the Battle of the Eastern Solomons. From 23 to 25 August the enemy had three forces spread out in the Malaita area in an arc sixty to eighty miles wide. One was composed of the carrier, Ryujo, a heavy cruiser, and three destroyers. The second consisted of two large carriers, believed to be the Shokaku and the Zuikaku, both of which had participated and had been damaged in the Battle of the Coral Sea.

These carriers were supported by 4 heavy and 6 light cruisers and 8 destroyers. The third force contained 3 heavy cruisers and from 3 to 5 destroyers. In the engagement we inflicted more damage to both enemy ships and planes than we received, but failure of communications prevented our striking the two large carriers, which would have been more strategic objectives, and which were closer to us. Despite this fact, the Battle of the Eastern Solomons was extremely important, since it permitted continued consolidation of our positions in the Solomons.

Japanese air raids on our establishments on Guadalcanal, meanwhile, continued with daily precision. Our land-based air forces conducted themselves with great bravery and ability. Instances of teamwork were supplemented by numerous cases of individual heroism, and time and time again the will to survive was illustrated. A case in point was that of Second Lieutenant Richard R. Amerine, USMCR. Forced to bail out of his plane, he landed in the Coral Sea some four miles from Guadalcanal. Swimming to shore, he stumbled upon a sleeping Japanese officer. Braining the officer with a stone, the Marine Corps lieutenant helped himself to a .44 caliber Smith and Wesson pistol, holster, ammunition belt, and the dead man's shoes.

Thus equipped, he started a thirty-mile trek through dense, enemy-controlled jungle to his base. After a harrowing experience, which included beating two more Japanese to death with the butt of his pistol, Lieutenant Amerine finally made his way back to safety and the Silver Star medal.

Enemy activity, despite the setback received in the Battle of the Eastern Solomons, was greatly accelerated the first two weeks of September. Even the heroic efforts of our greatly outnumbered aviation force could not circumvent the enemy's reinforcement of his ground troops nor prevent his landing supplies in preparation for a major and decisive contest for the possession of Guadalcanal. Our own supply transports, on the other hand, arrived irregularly and hardly ever completed unloading because of air, surface, and submarine attacks. On the afternoon of the thirtieth, for example, in spite of our shooting down eighteen enemy planes in the noon attack, the enemy dropped bombs on the USS Colhoun, a transport, and sank her in less than five minutes, although ninety of her crew were saved.

Our air forces were being slowly augmented by planes from carriers and outlying American bases. As a result, air assistance was possible in some of the raids on outlying Jap positions, as at Tasimboko, where a minor defeat was inflicted upon the enemy. By September, we had 74 F4F's, 37 SBD's, 3 P-400's, and 12 TBF's on Guadalcanal, ready to render aid in the fierce battle of Bloody Ridge that took place on the thirteenth and fourteenth of September.

During this period our growing air force made continuous strikes at enemy shipping and land positions. Even General Geiger slipped away from the watchful eye of his staff, climbed into an SBD and dropped a 1,000-pound bomb on enemy installations in the Visale area. During this period the ratio of Jap planes shot down was five to one in our favor.

During these months, also, the squat little "Duck," or J2F-5, and the antiquated cruiser biplanes proved more than once their ability as rescue planes, and were used both for food dropping and for the actual rescue of downed airmen. For the last half of the month, the 1st Marine Air Wing not only carried on intensive tactical operations but also operated a transport service. Thirty-eight R4D's were utilized to evacuate the wounded and to deliver equipment and reinforcements to Guadalcanal.

It was also during the period between the arrival of the planes on Guadalcanal and the end of September that Admiral Nimitz awarded three Navy Crosses and eleven Distinguished Flying Crosses to navy, marine, and army pilots for outstanding service in the campaign. Before the campaign was over, there were not only many more of these medals awarded, but eight marine pilots received the Congressional Medal of Honor.

On 15 September, naval aviation suffered an important loss at sea. The carrier Wasp, while covering a movement to land reinforcements and supplies, was struck by three torpedoes from a submarine and was so badly damaged that she had to be sunk.

During October the Japanese made major efforts to break through our defenses by means of land attacks. Heavy reinforcements were made by means of the so-called "Tokyo Express," a shipping effort that continued despite our attacks on it.

By 9 October the enemy had landed ten thousand additional troops. The discovery of the Japanese plans on the body of a dead Jap officer plus the fact that we also received reinforcement in the shape of the 164th Army Infantry Regiment, made possible frustration of their scheme.

One attempt to halt the Tokyo Express, though it did not stop the run, was a successful venture for us. On the night of 11-12

October, a task force, commanded by Admiral Norman Scott, surprised the Tokyo Express between Cape Esperance, on the northwest coast of Guadalcanal, and Savo Island. In a thirty-minute engagement named for the cape, we sank a heavy cruiser, four destroyers, and a transport, in addition to putting a cruiser out of action and badly damaging a destroyer, all in exchange for the loss of one vessel, the USS Duncan. Intensified air raids had preceded the affair, and our planes performed yeoman work afterwards in directing rescue operations for personnel of the USS Duncan. Later that same day air attack groups encountered another Japanese force and inflicted serious if not fatal damage to two destroyers. It was on the following day that Captain (later Major) Joseph Foss shot down the first of the twenty-six planes that he was to destroy before the close of the Guadalcanal campaign.

Despite these losses, the enemy continued to pour reinforcements into Guadalcanal. On the fourteenth a heavily escorted convoy of seven transports was seen headed for the island. A lone American bomber managed to sink one of the transports on the fifteenth, but, though additional bombers were sent over from Espiritu Santo, many Japanese troops managed to get ashore.

The enemy continued his activity; shore bombardment and daily air attacks harassed our land forces and depleted our air forces. Nearly three months of sanguinary battle under atrocious conditions were beginning to take their toll on bodies and nerves, yet the determination to hold on was not diminished.

After the middle of October, the situation began to change. Some of the old veterans were given relief as new forces came in. Both sides were feverishly preparing for the showdown. For a time, it looked as if Japanese endeavors were achieving greater success, and it was during these days that the outcome was most in doubt. Beginning the night of 23 October, the enemy made four successive desperate night attacks in the effort to dislodge our forces from Henderson Field. Heavy rains reduced the ability of the American planes to assist in the defense, as at times the field was a bog from which no planes could take off. On the ground a bitter hand-to-hand struggle finally repulsed the invaders.

While these desperate struggles were taking place on Guadalcanal, three Japanese task forces were headed for the area. One of these included three carriers — the Shokaku, Zuikaku and Zuiho — and another included the carrier Hayataka. Our own naval forces in the area were extremely weak in comparison. The Enterprise was under repair at Pearl Harbor, the Wasp had been sunk on 15

September, and the Saratoga was out of action as a result of a torpedo hit on 31 August. We had, in the South Pacific, a task force composed of the Hornet, supported by the heavy cruisers Northampton and Pensacola, the antiaircraft light cruisers Juneau and San Diego, and a number of destroyers. The only battleship in the area was the Washington, which had supported an army convoy into Guadalcanal on 13 October.

This was not a very imposing opposition to confront the mighty three-pronged fleet the Japanese had concentrated for the attack. Consequently, repairs were hurriedly completed on the Enterprise, and she departed at flank speed for the Solomons with the new battleship South Dakota, to make rendezvous with the Hornet group, and to operate under the command of Admiral William F. Halsey, Jr., new Commander, South Pacific. About midnight of 25 October our ships, reinforced by the remaining units of Admiral Scott's task force that had fought the Battle of Cape Esperance two weeks before, rounded Savo Island and retired southward before daylight, observed intermittently by Jap planes. Admiral Kinkaid was in command of the Enterprise and Hornet task forces.

The Battle of Santa Cruz that ensued on the morning of 26 October was of a pattern similar to that of the battles of Coral Sea and Midway in that most of the damage was wrought by carrier-based planes and no surface vessels opposed each other. In the battle we lost the Hornet and suffered damage to the Enterprise, the South Dakota, San Juan and the destroyer Smith. The Porter was so badly damaged by a submarine torpedo that she had to be sunk. We also lost 74 carrier planes, 20 of which were shot down in combat. On the other side of the picture, we inflicted damage on 2 enemy carriers, a battleship, 3 heavy cruisers and a light cruiser or destroyer, but none was sunk. In addition, we destroyed 66 enemy aircraft.

On the basis of these figures, it is difficult to avoid the conclusion that the battle was more costly to us than to the enemy, because of the reduction in our carrier strength. The only advantage we gained from the engagement was the fact that the plane losses and carrier damage we inflicted gave the enemy a shortage of aircraft resulting in insufficient air coverage for the enemy invasion fleet in the decisive battle that was looming three weeks ahead.

THE BATTLE OF GUADALCANAL grew out of a desperate attempt on the part of the Japanese to strengthen their forces on the island, and an equally desperate effort on the part of the Allied forces to prevent this reinforcement from taking place. For some time, the Japanese had tried to take in men and supplies by cruisers, destroyers, and landing boats from nearby islands. They were not making much headway, however, against the combined depredations of our planes from Henderson Field and of our PT boats.

Since they could not appreciably improve their situation by these piecemeal means, the Japanese undertook an operation of major proportions. By 12 November, 2 carriers, 4 battleships, 5 heavy cruisers and about 30 destroyers and transports were concentrated at Rabaul, including 4 probable battleships, 6 cruisers and 33 destroyers in the Buin-Faisi-Tonolei anchorages.

We had nothing adequate to ward off the all-out attack the Japanese were about to launch. The Enterprise was at Noumea but was damaged and not expected to be ready for action before 21 November; the Saratoga was almost ready to rejoin the fleet but was still at Pearl Harbor. Even our considerably augmented land-based aviation was inferior in numbers to the force that the enemy was mustering.

A preliminary phase of the battle was our own attempt to reinforce the island by an extensive supply operation. The work on the Enterprise, meanwhile, was speeded up. When our transports began unloading at Guadalcanal, they were subjected to severe attacks by enemy planes.

Marine pilots took to the air, and casualties were fairly heavy on both sides, but the Marines demonstrated great ability in first knocking down the escorting Zeros and then turning their attention to the bombers.

On 12 November one of the first suicide attacks of the war was made, as a plane crashed into the San Francisco and then plunged into the sea, taking a toll of thirty lives and causing fire damage.

The Battle of Guadalcanal may be broken conveniently into three phases. The first was a cruiser engagement on the night of 12-13 November, the second centered about aviation activity on the thirteenth and fourteenth, and the last was a battleship engagement on the nights of 14-15 November 1942. It is clear that for the most part, the first and last phases lie outside the scope of this account. The cruiser action of 12-13 November was a short but exceedingly bitter struggle in which we were outmatched in numbers of ships engaged. Virtually the only thing that saved our forces was the fact that the enemy had brought the wrong ammunition with them.

The Japanese had come to bombard our shore establishments, not to engage armor-protected vessels. The bulk of their ammunition, therefore, was bombardment ammunition. The San Francisco, for example, took fifteen major caliber hits and still stayed afloat. In the engagement we lost a light antiaircraft cruiser, Atlanta and 4 destroyers, with a number of ships badly damaged, including the cruiser Juneau which was sunk the following day by a Japanese submarine. The Japanese lost 1 light and 1 heavy cruiser and 4 destroyers, with a number damaged.

In addition, in this so-called cruiser phase, the Japanese had one battleship, the Hiyei. This vessel was badly damaged during the night and was spotted by American planes the following day. Nine torpedo planes and 6 fighters from the Enterprise, on their way to Henderson Field to act as a relief force, joined marine aircraft in attacking the injured battleship. Throughout the day she was hit again and again by planes from Guadalcanal, but at night she was still afloat, despite the fact that she had taken an amazing total of eleven direct hits from torpedoes, four 1000-pound bombs, and several smaller bombs.

The explanation of this invulnerability apparently lies in the fact that the torpedoes were set to explode at a depth of ten feet. This was considered the proper depth for operations around Guadalcanal, but because the Hiyei was very low in the water, it is believed that several torpedoes exploded against armor instead of against more vulnerable parts of the ship. No ship could stand up indefinitely under such a beating, however, and late that evening some of the crew were observed leaving the vessel, and it was assumed that she was being abandoned. On the following day the Hiyei had disappeared and an oil slick two to three miles off Savo Island was evidence that the first enemy battleship of the war had been sunk.

On the fourteenth a strong force of enemy cruisers and destroyers was sighted. This was the group that, possibly with the help of the Hiyei, had bombarded Henderson Field the night before. Marine and navy fliers attacked this force and succeeded in sinking at least two cruisers. At the same time, search planes had sighted twelve Japanese transports in the "slot" moving north of New Georgia toward Guadalcanal. Other search planes located a strong cruiser and destroyer force in another area.

Our work was now clearly cut out; it was up to the air and surface forces to prevent the transports from getting through. All day long on the fourteenth, planes from Henderson and the Enterprise attacked the transports, gradually cutting them down and engaging in running fights with the Japanese air support. Naval and marine fliers were joined at times by two flights of army B-17's. At the end of the day, 4 enemy transports had been sunk and 4 were aflame and dead in the water, but 4 were still doggedly headed for Guadalcanal.

Meanwhile our surface strength had been considerably increased, especially by the addition of two battleships, the Washington and South Dakota, which by noon of the fourteenth were only fifty miles from Guadalcanal. The presence of this force made possible the third phase of the battle, the battleship engagement of the night of 14-15 November. Scouting planes played a part in the early stages, but essentially it was a struggle of surface forces. In the engagement our naval forces sank 5 enemy ships and damaged 6, and lost 3 destroyers. Our victory was due in a considerable degree to superior ship-borne radar.

Four Japanese transports, in the meantime, had reached Guadalcanal and were attempting to unload. As soon as daylight came, our air forces, aided by marine land units and the destroyer Meade, began attacking the enemy. By dusk of the fifteenth the beached ships had been completely destroyed. The all-out effort of the Japanese had failed. While it was to take three hard months to force the final evacuation of the enemy from the island, Guadalcanal had in reality been won.

The enemy, in the Battle of Guadalcanal, tried neutralization by naval bombardment rather than by carrier-based planes probably because of the enormous losses in carrier planes in the Battle of Santa Cruz. Except for fighter coverage, the Japanese Air Force took no part in the battle. One of the most decisive factors in the Battle of Guadalcanal was the successful use of our carrier groups as temporary land-based groups — the Enterprise flew the planes off and then retired to a safe distance, while the planes operated with devastating effectiveness from Henderson Field.

Another important feature of our operations at Guadalcanal was the co-operation of all our fighting forces, Army, Navy, Marines, Australians, New Zealanders, and even the natives. With the taking of Guadalcanal the Allies were definitely on the road back to victory.

ADM. FRANK JACK FLETCHER (1885-1973)

16 – THE NORTHERN DAGGER

IN THE DAYS BEFORE THE WAR, the summer tourist bound for the Orient and a few days out of a west coast port would put on an extra sweater and go on deck. Gazing north, if the bone-chilling fog condescended to lift a trifle, he could dimly discern a forbidding shore. He would be seeing the Aleutians, and the sight would bring home forcibly to him the fact that ordinary travel maps are misleading and that the shortest distance between two points is sometimes not what it seems to be.

In those days of summer vacations few people in the United States thought of the Aleutians or of the Great Circle route. They were not aware of the fact that unless lured by the sound of soft music, the swish of grass skirts, or more tangible reasons, to the Hawaiian Islands, ships bound for the Far East followed the contour of the earth that took them far to the north.

With the advent of war, however, all this was changed. Professional and tablecloth strategists began to see the potentialities of the Aleutians and Alaska. The long chain of islands that bears the name, Aleutian, was seen as a dagger. Who should grasp this weapon? This was a vital question, since, held by the United States, it might be thrust deep into the heart of Japan. Held by the Nipponese, however, it might be utilized in a plunge into our northwest areas that would make the long-talked-of Japanese invasion a reality.

As it turned out in the course of the war, the Aleutians developed into a dagger that was little used, and one that, from a view of the war as a whole, inflicted not gashes but pinpricks. This observation in no way minimizes the heroism of the men involved, but demonstrates a point that they would be the first to stress. This was the fact that in the Aleutians, the primary enemy was not the Japanese, troublesome though they were, but the weather. Paradoxically, the weather, cursed and reviled by those who contended with it, was in the last analysis on our side. Had the days been normally fair on Attu, and in view of the early striking power of the Japanese Empire, a powerful thrust might well have been made in this area before our war machine could get under way.

Having thus theorized, let us see what actually did happen — the days not being normally fair on Attu. The Aleutians form a chain of some 120 islands that stretches from the Alaskan peninsula to within 90 miles of Kamchatka, on the shores of Siberia. Of these 120 islands, there are a few that stand out in view of wartime operations.

These are Unalaska (on which Dutch Harbor is located), Unimak, Atka, Adak, Kiska and Attu. The last named is nearly a thousand miles from the mainland of Alaska and is 750 miles from a chain of Japanese-controlled islands, the Kuriles, which lie north of the Japanese main island empire. The Aleutian Islands are volcanic in origin and are unify rocky and barren, making up in precipitous mountains what they lack in vegetation.

Coupled with this inhospitable terrain, is the weather that makes the region consistently hostile even to normal ways of life and more so to life during war. The reason for this foul weather lies in the fact that the Aleutians are at the meeting point of two major air masses. On the one side are the warm ocean currents and accompanying warm air masses flowing up from Japan. On the other side are cold currents and air from the Bering Sea and eastern Siberia. The meeting of these two masses results in an almost constant weather front over the Aleutians which produces heavy fogs and low visibility as the normal weather in this area.

As if this were not hazard enough to flight operations, the natural terrain of the islands is such that airfields are of necessity tucked between snow-covered mountains, and subject to sudden winds, known as "williwaws," that sweep through the valleys with high intensity and are frequently accompanied by heavy snow. In the face of all this, only perseverance and a continued will to overcome all obstacles made it possible for our air forces to operate at all.

The Aleutian campaign was conducted largely by land-based aircraft, both army and navy. It was carried out, as will be seen, with planes not specially adapted to meet local weather conditions and operating from inadequately equipped land bases and seaplane tenders. From the beginning it was possible to assign squadrons and men to this area only to the extent allowed by the strategic situation in other areas. Since, primarily in view of adverse weather conditions, this area appeared unlikely to develop immediately as of great strategic importance, and since conditions were more pressing in the South Pacific, equipment was assigned only in small lots as it could be spared.

At the outbreak of war, the Alaskan area was treated and administered as an outpost. The direction of naval air activity came for Patrol Wing 4, located at Seattle, Washington. There were three naval air stations in operation, one at Sitka, one at Dutch Harbor, and a third at Kodiak.

These were manned in December 1941, by a single patrol squadron, with the navy designation, VP-41. This squadron's six planes, PBY's, were serviced by the seaplane tender Gillis. Since weather was the worst enemy, even in peacetime some sort of relief was necessary for this squadron. Another squadron, VP-42, was accordingly based in Seattle, as a future relief for the unit in Alaska.

Aside from a few odds and ends of scouting and utility planes, the main weapon of naval aviation in the Aleutians during the first months of the war was the two-engined PBY or Consolidated Catalina. This lumbering craft, derided as a "cold turkey" in the heart-rending first weeks of the war in the Java Sea, was to emerge as one of the real work horses of naval aviation, courageously attacking any task set before it, even substituting, in the Aleutians, as a dive bomber.

In the Alaskan area, the Catalina possessed a number of important assets, and one vital weakness. Among the factors that brought about its use in the north was the fact that it was a seaplane and in case of failure to reach an established landing place could be set down in almost any small inlet or harbor until weather conditions improved.

Another asset was its lack of speed. Numerous mountain peaks, obscured by fog or snow, presented a terrific obstacle to a speedy plane. When the PBY's were equipped with radar, these hazards were cut down appreciably by the fact that this craft could make use of its slow speed and the radar to avoid accident. The Catalina, furthermore, had a long cruising range which enabled it to make long patrol sweeps. In February 1942, PBY-5A's were taken into the Alaskan area. These were the amphibious model of the PBY, and proved even more versatile than the original PBY.

As a result of these factors, the Catalina made a very satisfactory patrol plane — up to a certain point. This point was the entrance of the enemy on the scene. The slow speed which enabled the plane to avoid mountains made it vulnerable to faster planes, and blind spots not protected by its own guns increased this vulnerability. Consequently, later in the campaign fast land planes, such as the PV (Ventura) were added for certain uses.

During the first months of the war, the primary task of naval aviation was defensive. As soon as the news of Pearl Harbor reached the commander of the Alaskan sector, maximum daylight patrols were made. Early in February, VP-41 was relieved by VP-42, with twelve PBY-5A's. These planes were sent back to Seattle, two at a time, to be equipped with radar gear. The squadron had three main tasks, to protect shipping in the region, to defend, if necessary, the air stations, and to obtain all information possible concerning enemy movements into the area.

By the spring of 1942, as already recounted in the story of Midway, a Japanese diversionary operation in the North Pacific appeared probable. Interview with enemy officers after the conclusion of hostilities revealed that, as a preliminary to action, the Japanese had launched a seaplane from a submarine off the west coast to discover the possible concentration of men-of-war at Seattle and that similar reconnaissance was made of Dutch Harbor. Anticipating such a diversion CinCPac (Commander-in-Chief, Pacific Fleet), on 22 May, transferred a force of two heavy and three light cruisers with destroyers and submarines from the central Pacific to the north Pacific area under the command of Rear Admiral R. A. Theobald.

Patrol Wing 4, composed of Squadrons 41 and 42 moved its headquarters north from Seattle. It was immediately designated the Air Search Group and assigned the responsibility of making long-range patrols in search of the expected enemy force. Four seaplane tenders, Casco, Williamson, Gillis, and Hulbert were detailed to work the Patrol Wing 4.

Meanwhile the air base facilities were still in the formative stage. At Kodiak there was both an airfield and a seaplane base; Dutch Harbor had facilities for handling a limited number of seaplanes; and the Eleventh AAF had a newly constructed field at Unimak. These were the main bases, and, at each, construction was still in progress. Secondary bases, with seaplane anchorages, were located at Sand Point and Cold Bay.

At 0545 on the morning of 3 June 1942, Dutch Harbor was attacked by 15 carrier-type fighters, launched from 2 Japanese carriers 100 miles offshore, which were followed immediately by three flights of 4 bombers. The attack was detected some distance out by the tender Gillis, and the alerted garrison opened with heavy antiaircraft fire.

Despite this warning, the damage was considerable. The bombs destroyed two barracks and two Quonset huts and damaged several buildings in the Fort Mears area. Approximately twenty-five men were killed and an equal number wounded.

All but one of the Patrol Wing 4 planes based at Dutch Harbor were out on patrol. The remaining plane was destroyed by the enemy. On the other side of the ledger, the Gillis claimed that two Japanese planes were shot down. On the following day the enemy struck again, this time causing heavy damage to fuel tanks and the oil supply. Army pursuit planes downed two of the attacking planes before the enemy retired from the scene. Casualties for the two days' attack totaled about one hundred.

The next phase was an attempt to locate the Japanese main force. The first contact with the enemy surface force was made by an army pilot, who was not heard from again after reporting that he had seen a large carrier with a large bomber on deck.

Seven hours later Lieutenant (Junior Grade) Lucius Campbell sent a message that he had sighted the enemy, and gave the estimated position. Despite antiaircraft fire and the presence of enemy fighters, he tracked the force so long that he ran short of fuel on his return to base. Fortunately, he and his crew were rescued the next day by a Coast Guard craft.

Despite the terrific odds against them, two naval aviators made contact with the enemy, reporting the presence of a carrier and two destroyers. One made an unsuccessful bombing attack during which he lost an engine by antiaircraft fire, but managed to reach his base in safety. Lieutenant Commander Perkins was ferrying a torpedo to Unimak when he heard Ensign Freerks make his report.

He turned toward the enemy, located him and tracked him for two hours. Then, before returning to base, he made an unsuccessful torpedo run through antiaircraft fire.

The patrol wing's task, as has been seen, was to search for the enemy and locate him sufficiently accurately for the army striking force to attack. The full attack of the army striking force, however, could not be brought to bear on the enemy because of adverse weather conditions. The navy patrol planes continued their work, losing both aircraft and lives in the effort, but gradually the Japanese slipped away out of range.

If it was believed at the time that Japanese force had left the area, this belief was soon shattered. The attack on Dutch Harbor was seen to have had a triple purpose. First, there was the attempt to destroy American installations. In the second place, as we have seen, the Japanese were making a feint to cover a strong blow at Midway. Thirdly, they were creating a diversion to cover the occupation of two islands of the Aleutian chain. On 10 June, two patrol pilots discovered the presence of ships and installations at Kiska and Attu.

Upon receipt of this information, CinCPac ordered Patrol Wing 4 to bomb the installations. Patrol Wing 4, therefore, became more than a mere patrol unit. Its size was increased rapidly to four squadrons, and it set forth on its new task. The seaplane tender Gillis was moved to Nazan Bay on Atka Island, and during the next four days Catalinas carried out an amazing exhibition of continuous aerial attack. Pilots rearmed and refueled their planes upon returning from one attack and took off for another as soon as possible. In four days, the supply of gasoline and bombs carried by the Gillis was completely exhausted.

The strain on the pilots was terrific, one pilot flying 19½ hours during one 24-hour period. PBY's were called upon to perform work for which they had neither been planned nor built. Creaking at every joint, they dropped through the clouds as dive bombers at 250 knots as compared to their normal speed of 85 to 90 knots. Their bombardiers dropped their bombs by "seaman's eye," and the pilots pulled back into the overcast for concealment after the bombing run. In the face of the handicaps of construction, only one plane was lost, though many were badly shot up and damaged.

Despite this heroic effort of the patrol squadrons, and although damage was done to their installations, the Japanese remained. On 15 June Patrol Wing 4 resumed its normal function, and that for which it was prepared: long-range patrol. Its operations, however, had demonstrated more than courage.

They had shown, in addition, the great mobility of naval aviation. Four days prior to this attack, two of the squadrons had been in Seattle. Within this short period, these squadrons had moved a distance of more than two thousand miles and had gone immediately into action against the enemy.

The remainder of the first year of the war saw the beginning of a program in the Aleutians that was ultimately to result in the expulsion of the Japanese from this area. In general, the plan was a dual one. In the first place, efforts were directed to confine the enemy to his original positions on Attu and Kiska, and, by making things as uncomfortable for him as possible, to prevent the development of these bases. In the second place, our own bases were gradually expanded westward to increase the potency of raids on Japanese-held positions and to pave the way for the final assault.

These were the aims of the entire military force in the Alaskan area. Shore-based aviation had certain functions in connection with these general plans. They were, in brief, to conduct long-range patrols, to detect further Japanese moves into the area, to prevent reinforcement of existing garrisons, and to attack enemy shipping and installations. In addition, aviation was to protect our own shipping and surface forces, and to aid in the rescue of personnel downed at sea.

In order to carry out these sizable duties, aviation was reorganized. The Eleventh Army Air Force was designated the Air Striking Unit and Patrol Wing 4, the Search Unit. Brigadier General Butler, Commanding General of the Eleventh AAF, was placed in command of these air groups under Rear Admiral R. A. Theobald, commander of the North Pacific Force.

Under this new organization a new and more effective search plan was put into effect for patrols.

Toward the end of summer, steps were taken toward the westward expansion of bases. On 30 August 1942, Adak was occupied by our forces. The only mishap was the torpedoing of the seaplane tender Casco in the bay. The ship was beached, repaired, and returned to service. Work was begun immediately on a runway, and it was ready for operations by the twelfth of September.

Thus, it was that before the end of the first year of war, the United States forces had started west on their gradual approach toward the Japanese-held islands of the Aleutian chain. In many ways the year had been a discouraging one. The Japanese had gained a foothold and were hanging on tenaciously. We had yet to inflict a defeat upon them. On the other hand, the Alaskan area had been reduced for the time being to a minor theater of war, and the heavy fighting was to be done away from our own shores.

17 – STARTING THE ROAD BACK

THE FIRST WAR IN THE PACIFIC SPELLED THE DOOM of the Japanese on Guadalcanal. But although they were down, they were not yet out, and it took some time to enforce eviction proceedings. As during the earlier phases of the struggle for the island, the Army, the Navy, and the Marines took part in assaults on the enemy, but the mopping-up period will always be most clearly remembered as a time when the Marines, both on land and in the air, established a high point of courage and stamina.

Closely tied to the campaign on Guadalcanal were our attempts to dislodge the growing enemy forces on the near-by island of New Georgia. As early as November 1942, Japanese craft were seeking anchorages in the inlets and bays of this island, and just as early, pilots from Henderson Field were selecting these anchorages and additional land sites as targets. Despite constant surveillance, however, the enemy had completed 90 percent of an airfield at Munda on New Georgia, before our reconnaissance planes discovered it on December 1942.

During the rest of the month the field was continually blasted, but despite this fact, by the end of the year the airfield was not only paved and servicing fighting planes but was also sending out medium bombers. As the war moved into its second year, the Japanese also improved a base at Kolombangara that likewise seemed more or less impervious to air attack.

It was decided, therefore, that the only solution was to attack and take the fields, since neither naval nor aerial bombardment of the weight available could do more than temporary damage. Meanwhile, however, there was the problem of Guadalcanal. The Tokyo Express was still making its run, though the cause was lost.

Although our reinforcements continued arriving and our air power was increasing, and despite the constant "derailing" of the Tokyo Express, it was believed that the enemy would make a last desperate attempt to reinforce 1 Guadalcanal. There were certain signs that pointed to this. In January there was an increase in Japanese air strength, indicating, it was believed, that enemy carriers might be in the offing, and there was a resurgence of naval operations in the Buin and Rabaul areas. To forestall any last, desperate effort at reinforcement, the most powerful U. S. naval force yet to appear in the South Pacific was concentrated in the Solomons and the

Fiji areas on 27 January. It included 6 major task forces, composed of 5 carriers, 7 battleships, 12 light and heavy cruisers, and more than 30 destroyers.

One of these task forces went to the support of a convoy of four transports and four destroyers bound for Guadalcanal with supplies. This force became the victim of the first night torpedo attack that the Japanese had yet attempted. Despite the effective work of our planes in destroying the attacking planes, the enemy succeeded in launching enough torpedoes to cause serious damage. The chief casualty was the Chicago, which was lost after a hectic air battle. The loss was heavy, but the objective of the task force had been accomplished; the convoy had safely deposited its cargo at Guadalcanal.

Then, early in February, it appeared that the Japanese were at last ready to make their attempt at reinforcement. On 4 February, twenty Japanese destroyers left Buin for Guadalcanal, and on the following day many 30-foot landing barges were seen drifting off Cape Esperance. Shortly afterwards, our patrols located a large force headed from Rabaul. Included in this group were 2 enemy carriers, 4 battleships, and 6 heavy cruisers. It looked as if the push might be on. Our forces in the area were alerted, and Air Group 10 was ordered from Enterprise to Henderson Field to reinforce the marine and army planes there.

Then came an anticlimax rather than a climax. The activity of the Japanese from Rabaul and elsewhere proved to have been a feint to cover the real move, which was not reinforcement, but evacuation of troops from Guadalcanal. Exactly six months from the date of the marine landings on the island, the Japanese completed the removal of their forces from Guadalcanal except for occasional patrols.

Thus, by early February when our main body joined forces near Esperance with the enveloping detachment from Verahui, the Guadalcanal campaign officially ended. We had taken a firm step on the road back, and the enemy's forward thrust had been definitely stopped.

Radio Tokyo, of course, expressed it somewhat differently, reporting that the Japanese Army had been "transferred" from Guadalcanal after "its mission had been fulfilled."

The Japanese did not indicate whether or not a part of this mission had been the annihilation of 6,066 of their own men in the last three weeks of the campaign, or the estimated total loss of Japanese personnel of between 30,000 and 50,000. Nevertheless, had we suspected the real purpose of their accelerated activity of early February we would have turned their successful withdrawal into a disastrous rout and added to their already staggering casualties.

Although it was to take another year to complete the conquest of the Solomon Islands, this was only twice as long as it had taken to wrest one island in the chain, Guadalcanal, from the enemy. The way was still hard, and the Japanese were both vicious and tenacious, but the growing might of American production was becoming evident on the field of battle.

Outstanding evidence of this growing superiority was the introduction of a new plane in the Solomons. This was the F4U, or Corsair, that was to blaze a glorious name for itself all the way across the Pacific. Prior to this time the Marines had been forced to rely on the F4F (Wildcat), which was a sturdy plane but outperformed by the Japanese Zero with which it had to contest in the Solomons. The Zero, on the other hand, was no match for the Corsair, which made its combat debut on 13 February 1943, and from the outset proved a nemesis to the Japanese Air Force.

It should be noted also that the marine squadron that first introduced the Corsair to the area was also the first marine squadron to go into combat against the enemy from a carrier.

Not only new planes but also new equipment was being rapidly supplied the growing air force, and the development of airborne radar made possible night searches and shipping attacks by patrol planes. Known as Black Cats, from the fact that the familiar Catalina was painted in somber hues for this type of work, the aircraft of Patrol Squadrons 12 and 54 ranged far and wide though the Solomons spying out the enemy's nocturnal activities, spotting the fire of our surface ships in night bombardments, and themselves going in to attack whenever possible.

The first step ahead from the Solomons came on 20 February 1943. This was an unopposed landing on the Russells, a small group of islands in the great Solomons chain lying between Guadalcanal and New Georgia. Out of the ooze and muck of the jungle, with little equipment, and with constant dogfights overhead, the Marines managed to construct an airstrip 2,000 feet long that was ready for operation in June.

Meanwhile, there were sparring engagements on the water and in the air. Some were bitterly contested as, for example, the day that Henderson fighters went up to meet 160 enemy planes and knocked down 38, in return for 7 aircraft and 1 pilot lost.

The occupation of the Russells was merely a preliminary move toward an invasion of New Georgia. Our facilities were being constantly improved. In the early days, flights had been canceled because of insufficient gasoline; there was now storage space for 45,000 barrels of aviation gasoline, with a pipe line that took the gas direct from the tankers at Koli Point to the tanks near the fields.

Admiral Halsey issued the operation plan for the invasion of New Georgia and the capture of Munda airfield on 3 June. The thirtieth was set as D-Day with one of the objectives Rendova, a small island off the coast of New Georgia. The air support group was commanded by Vice Admiral Aubrey W. Fitch, and was responsible for reconnaissance and striking missions and for direct air support during the landings. Rear Admiral (later Admiral) Marc A. Mitscher had tactical command of all land-based aircraft operating from Henderson and the other Solomon air bases.

For a month before the landings on New Georgia, army, navy, marine, and New Zealand pilots harassed enemy fields and bases in the northern Solomons, and struck at Japanese shipping in the Bougainville area. One of the most spectacular air battles took place on the seventh of June. An enemy force of 40 to 50 Zeros was intercepted by our planes near the Russells, and in the ensuing struggle the enemy lost 23 Zeros, while we lost 7 planes and recovered 3 of the pilots.

The case of one of these, Lieutenant Samuel S. Logan, USMCR, is one that Marines will not soon forget nor forgive. This officer, forced to bail out of his plane, found that the Japs were deliberately making firing runs at him. Desperately trying to save his life by spilling air from his parachute, Lieutenant Logan barely avoided a collision in the air with a Zero, but had half his right foot severed by the propeller of the strafing plane.

A New Zealand flier drove the Zero away before he could complete his deadly work. The injured pilot, meanwhile, managed to apply a tourniquet, inject morphine, and take sulfa tablets while descending. Landing in the water, he had enough strength to inflate his life raft, and place dye markers in the water, thereby enabling a J2F (Duck) to come to his rescue.

Even more unbelievable than this harrowing experience, was that of Lieutenant Gilbert Percy, USMCR. This officer was pursued by five Zeros, one of which he shot down, but in the fray his plane's oil, gas, and hydraulic systems were wrecked, a wing tip disintegrated, and an aileron was shot away. At three thousand feet the engine sputtered, and Percy jumped — but his parachute failed to open.

Putting his arms straight down at his side and keeping his feet together, he hit the water. Instead of being killed, Percy was knocked out, recovered sufficiently to make his way to shore and, with the aid of natives who went for assistance, was taken to safety. His injuries which included a broken pelvis and broken ankles, kept him out of active duty for a year, but he was alive and held a record that few men would care to attempt to take from him.

As D-Day for the Rendova landing in New Georgia approached, air activity on both sides increased. There were raids and counter-raids, with the percentage of planes downed heavily in our favor. On 30 June 1943, the Rendova landings were made as planned. The landings were affected before serious air opposition developed. As the destroyers stood out through Blanche Channel for the return to Guadalcanal, however, twenty-four Mitsubishi-type torpedo bombers with a large number of Zeros came in for the attack.

The raid lasted only eight minutes, and at the end of this period all the enemy planes but two had been splashed, and the two remaining planes were later downed. In return, we had lost one 7,712-ton transport, the flagship USS McCawley. In subsequent waves of attacks our planes downed a total of 101 planes out of 130 that were engaged. Marine fliers topped the lists with 58 planes. We lost 17 planes, but recovered 7 pilots, and once again some of the recoveries were in the realm of the fantastic.

Other landings followed, some successful, some filled with discouragement and heartbreak. Sea engagements, also, took place in this campaign that was raging on land, sea, and in the air. The Helena was our worst loss, going down in the Battle of Kula Gulf on 5 July as a result of torpedo hits. Our forces, however, moved ahead. It took five weeks to capture Munda airfield, and throughout that period air warfare reached its zenith for the campaign. Perhaps the highlight of the period came in a series of three raids our planes made on Japanese shipping and harbor installations at Kahili.

More than two hundred of our planes, army, navy, and marine, took part in the raids and were attacked on the morning of 17 July by Japanese interceptors. Forty-nine enemy planes were shot down and considerable damage was done to the main targets, with half a dozen vessels sunk.

While the Japanese were receiving this rough treatment in the Bougainville area, they were retaliating with punishment of our forces in the vicinity of Banks Island. The USS Chincoteague, a small seaplane tender, had been badly hit by enemy planes.

After three grim days of caring for wounded, trying to check fires, and preparing a destroyer tow, it appeared as if all work had been in vain, for out of the horizon at sunset of the seventeenth came three enemy bombers, ready for the kill. The tender, however, was saved to serve another day, because four marine Corsairs appeared at the appropriate moment and knocked down all three bombers.

By the twentieth, the enemy garrison at Munda was in a precarious state, since sieges and blockades had prevented the arrival of reinforcements. The land engagement went ahead; with strong air support, and finally, on 5 August, the airfield fell to our advancing forces.

The capture of the Munda airfield completed the first phase of the advance into the New Georgia group. The other major Japanese base, Kolombangara, was for a time considered to be the next logical objective. By July, however, it was decided to by-pass this stronghold and attack a weaker Japanese possession, Vella Lavella, in the hope of building up a new base that would render Kolombangara impotent.

Aviation's part in this invasion was along the customary pattern. Army, navy, marine, and New Zealand planes dive-bombed important gun positions on both Kolombangara and Vella Lavella, and army bombers dropped forty-nine tons of bombs on the field a Kahili as a part of the softening-up process. When the actual landings were made, marine Corsairs and Kitty Hawks flown by New Zealand pilots provided air coverage.

Although we had by-passed Kolombangara, the enemy continued his operations at that base. Despite the fact that the Japanese could not rely on the Tokyo Express getting through, they sent large fleets of barges, and when these failed, supplied the base by means of submarines and float planes, and eventually by parachuting supplies to the more remote spots. Meanwhile, we attempted to starve out the enemy by constant air patrol and by laying mines in the sea approaches. Marine pilots, especially, were successful in downing enemy planes and sinking barges.

In addition to making these harassing attacks, American forces began to build an airfield on Vella Lavella. With the construction of this field at Barakoma, toward the end of September, the position of the Japanese on Kolombangara became untenable, as they were now located between this new field and Munda. While the Japanese evacuation was taking place, our air forces divided their attention between heckling this retirement and striking at a new objective. This was the Japanese stronghold of Bougainville, the most northern and the largest island in the Solomons.

We needed it as a base for further strikes against Rabaul and other Bismarck bases. The task was not an easy one, as the enemy had been fortifying Bougainville for about two years, and by the time our invasion was to start, he had two airfields in the north, two on the southern coast, one in the Shortland area, and another on the east coast.

As a result of the success of by-passing Kolombangara for Vella Lavella, it was decided once again to strike the enemy at a weak point rather than at his strongest holdings. Aviation softened up the area and the supporting airfields. Employing as many as a hundred planes, Marines averaged four attacks a day in this work.

As the time approached for the actual landings, planes from the Saratoga rendered assistance by bombing Japanese airfields, while the Marines gave direct support to their own troops which comprised the landing forces on the beaches of Bougainville. A Japanese cruiser force that attempted to prevent the landings was intercepted and routed. The landings were made successfully, and although we were not to conquer the entire island during the course of the war, we had succeeded in our objective of neutralizing much of the importance of Bougainville for the Japanese. As the campaign on the island progressed, the Marines introduced a new form of air fighting. The first night fighting squadron went into action against the Japanese, in an effort to check the night bombing raids that were conducted against our fields. As a result, the enemy was reduced to less effective daytime attacks.

During this period also, one of the most famous marine squadrons, the "Black Sheep," led by Major (later Lieutenant Colonel) Boyington, came into prominence as a result of startling performances against the enemy. The struggle in this area was to continue throughout the war, with aviation carrying its full share of the load.

The work of naval aviation in western Australia may perhaps best be described as insurance against the possibility of a sneak attack by the Japanese or against submarine warfare in the area. It may be likened somewhat to the operations in Alaska, where little actually happened during much of the war, but where a potential threat long existed.

As was the case with Alaska, patrol operations were carried on under severe handicaps, though of a different nature. Except at Perth, which is on a river, there could be no permanent seaplane bases. Elsewhere there were 22-to 28-foot tides making correspondingly high tidal currents of from 3 to 4 knots at a minimum and 10 to 12 knots in some places.

Both Sharks Bay and Exmouth Gulf, being large areas, were subject to heavy swells. In fact, several patrol planes were lost because they broke their backs between waves. The nearest harbor was Darwin, sixteen hundred miles away.

Another handicap was the lack of adequate maps of the west coast of Australia. It was necessary, therefore, to make new maps of much of the coast line, and a permanent reminder of the presence of our seaplane tenders in the region was left in new place names, such as Preston Point, Childs Cape, and Heron Haven.

The problem of supplying the outlying bases that were set up was extremely difficult, and PatWing 10 was forced to uneconomical practices because there was no land transportation at all north of Geraldton. In this part of Australia, a plane could fly for six hundred miles without seeing a house, road, or a single person.

The sole trans-Australian railroad had been purchased, one might think, at fire sales, since there were five different gauges of track, and every piece of freight had to be transferred at least four times en route. Then to get supplies, food, and gasoline to the seaplane operating bases, there was no way but to use the tenders themselves.

Since it took three days for a tender to go from Perth to Exmouth Gulf and about four and a half to Heron Haven, the fuel problem became critical because by the time the tender reached the base it did not have enough fuel to remain on station until the planes used up the store of gasoline.

To help solve the problem, lighters and tanks were built, and a torpedoed Dutch tanker was utilized for storage. All food had to be brought in the same way, since no planes were available for transport duty. The result was that much of the time the outposts were on hard rations.

Fortunately, PatWing 10 had received a shipment of advanced base gear originally destined for it in the Philippines. In other respects, however, the problem of maintenance of equipment was of considerable magnitude. The wing operated planes consistently with too many hours between engine overhauls. At first, engines had to be sent to Wagga Wagga, across the continent near Sydney, and it took four months or more to get an engine back. In 1943, the RAAF set up an engine overhaul shop at Kalgoorlie, four hundred miles inland from Perth, and the engines were sent to this shop by truck; but it still took a couple of months.

Engine accessories were simply not to be had, and the only way out was by "cannibalizing" planes. Western Australia was at the end of the supply line, either east or west. Consequently, when materials did finally begin to arrive, they frequently lacked important component parts scrounged by other aviation units along the line which were also in need of maintenance equipment.

One exception to this gloomy picture was the then highly secret radar. When this equipment was first installed, a special crew of experts was sent from NAS, Pearl Harbor, to instruct the plane crews in proper use and maintenance. This mode of operation was greatly appreciated by PatWing 10 and it contributed to efficient use of the equipment.

The United States naval air defense of western Australia was entrusted to a refurbished PatWing 10, later renamed Fleet Air Wing (Fairwing) 10, the same organization that had fought so heroically in the early months of the war in the Philippines and in the Netherlands East Indies. Until fall of 1943, the wing's work consisted primarily of long patrols and convoy coverages that were productive of little contact with the enemy. In spite of this, it was coverage well spent — the fact that one has insurance does not mean that he wishes his house to be burned.

On the other hand, the law of diminishing returns began to apply. Over a period of two years, patrol planes in the area had sighted not more than six or eight enemy submarines and no Allied ships had been sunk in these waters. The western Australia area came to be used more and more as a training and respite region, and at length, in April 1944, General MacArthur finally permitted the withdrawal of one of the two squadrons remaining there.

Meanwhile, a decision had been reached to shift the principal operations to eastern Australia and then to New Guinea. PBY's began to operate on 7 August 1943 near Samarai, on Milne Bay, with one other squadron nearby at Port Moresby. On 15 September, Pat-Wing 17 was created in the southwest Pacific to take over operations in eastern New Guinea and along the Rabaul-Buka shipping lane, in support of General MacArthur's growing offensive in that area.

To prevent the Japanese from slips supplies into the area under cover of darkness, extensive use of "Black Cats" was made. In view of the vulnerability of the PBY in daytime combat, it was decided in September 1943, to experiment with the use of this craft for night raids on enemy shipping and installations. The Catalinas of VP-11 and VP-52 were painted black for the purpose; only one nickname could possibly result — "Black Cat."

By the end of 1943, squadrons VP-11, VP-52 and VP-101 of Fleet Air Wing 17 had greatly improved the technique of "Black Cat" operations, performing feats that their predecessors even as late as the summer of 1943 would have considered risky. Pilots were never told that they had to attack from low altitudes; they were merely told what the likelihood of success was from various altitudes, and to the "Black Cat" pilots goes the credit for proving that low altitude night flights can be made in comparative safety with effective results. Even up-moon attacks on moonlit nights were found to be practicable.

Generally speaking, the "Black Cats" were able to home on the target sufficiently by radar to guide them to a point at which they could see enough in the darkness to make corrections for radar error. The approach consisted of a glide from 1,000 feet or more to from 75 to 50 feet at the dropping point. The bomb load usually was composed of two 500- and two 1,000-pound bombs armed with four-or five-second delay fuses, to protect the relatively slow PBY from the blast effect of its own bombs. The unknown factor at the outset of this venture was whether or not planes could carry out this type of attack frequently without being shot down.

Experience showed that they could. Since the approach was made in a glide, the Japs were often unaware that a run was being made, and more times than not they failed to commence firing until the drop had been completed. By that time, if they were still among the living, the Japanese could do little but fire blindly in the direction of the disappearing plane. Enemy planes which sometimes covered the Jap convoys were not at this period equipped with radar, and as a result rarely attacked the PBY's. Furthermore, being land-based for the most part, they did not dare follow our "Black Cat" evasive maneuvers close to the water at night.

This attack technique, coupled with the long range of the PBY, made them a potent menace to Japanese shipping. The PBY's had a normal range of fourteen hundred miles round trip, with allowances for gas consumed in attacks and in "snooping." To add to their range and safety of operation, emergency bases with extra supplies of aviation gasoline were set up.

Weather conditions were a mixed blessing. While they made for rugged flying, they also gave protection against interception. Squadron aerologists, affectionately called "Thunderbirds," were in an unenviable position and took a continual ribbing from the pilots. It seemed that the inevitable forecast included a couple of bad weather fronts and scattered thunderheads in the area to be covered, yet the planes took off regardless.

The "happy hunting ground" of the "Black Cats" came to be St. George's Channel, the front door to Rabaul. Even though the Japs placed patrol boats and planes across the mouth of the channel at the time of night that "Black Cat" marauders were expected to pass through, they blocked it off with only partial success. It became a relatively simple task to use other routes, and the effectiveness of our patrols was not diminished. Other Japanese countermeasures consisted of routing ships dangerously near the shore of New Ireland so that radar blips would be hard to isolate, or, on hearing the sound of an unidentified plane, of stopping the forward progress of ships to avoid leaving a telltale wake. By the end of December, the Japanese even diverted a good deal of their shipping from night to daytime dashes.

In the push that was taking place from the Solomons back through New Guinea toward the Philippines, the indebtedness of our armed forces to the work of all the "Black Cats" was well expressed in a Presidential Unit Citation presented to these PBY squadrons:

For outstanding performance above the normal call of duty while engaged in search missions and anti-shipping attacks in the enemy Japanese-controlled area of the Bismarck Sea. Rendering pioneer service in changing the passive defensive search into a bold and powerful offensive . . . utilizing the full potentialities of the PBY seaplane and its equipment, locating enemy task force units and striking dangerously by night in devastating masthead glide-bombing attacks to insure vital hits on the target.

Dauntless and aggressive, in the fulfillment of each assignment, the gallant pilots . . . conducted daring lone patrols regardless of the weather in a continuous cover of this area, intercepting and attacking so effectively as to inflict substantial damage on hostile combat and other shipping, to deny the enemy the sea route between New Ireland and New Britain Islands and thus prevent the reinforcing of important Japanese bases. The splendid record of this combat group is a tribute to the courageous fighting spirit of its officers and men and reflects the highest credit upon the United States Naval Service.

In the latter part of 1943, the Catalinas in New Guinea once again demonstrated that they were versatile workhorses of naval aviation. In addition to usual convoy coverages from Port Moresby, the Catalina squadron there was assigned to air-sea rescue, to the transportation of men and supplies to inaccessible points, and to miscellaneous other missions. During this period, there were thirty-five air-sea rescue missions flown.

Some of the most important flights of the Catalinas were made in support of land scouting forces far in the interior jungles of New Guinea. Thirty-six of these flights, for example, were made to the junction of the Yellow and Sepik Rivers in central New Guinea, approximately a hundred miles west of Wewak and behind enemy lines.

Ninety-six thousand pounds of supplies were transported during the period, in addition to a number of personnel. The normal take-off was before dawn, and the destination was reached, weather permitting, through passes over the Owen-Stanley Mountains, which in that area rise to 12,000 feet.

North of the mountains, since ground fogs or low-lying clouds almost invariably prevailed, it became the practice to drop below the clouds when the Sepik River was reached and follow the river at treetop height to the landing area. Unloading was accomplished in minimum time by natives. Increasingly bad weather during the day often forced the planes to near maximum altitude of 19,000 feet on the return trip.

The round trip was approximately a thousand miles, and the fastest round trip was made in less than eight hours. When the position became untenable late in December 1943, 219 Australian officers and men and 25,000 pounds of gear were evacuated by the navy planes. These operations were in support of a forward movement by army and Australian forces which had been augmented with landings at Salamaua and at Lae. While these were entirely army operations, the Navy's planes of Fairwing 17 at Port Moresby were utilized by the V Air Force Command.

On 3 September, ten PBY's (which may have included Australian planes) were sent to bomb Rabaul. This was the first raid made on this enemy stronghold since early July, and twenty-five tons were dropped on the town and airfields, causing fires and explosions.

On 22 September sea-borne invasion forces were landed near Finschafen at the eastern end of New Guinea, and this base fell on 2 October, although fighting continued in the interior. During the following week large concentrations of shipping and enemy planes were seen at Rabaul, and on 12 October a heavy army raid, supported by navy rescue Catalinas, was sent from the newly acquired bases on New Guinea. Throughout November, the PBY's regularly attacked shipping between Kavieng and Rabaul, which was also the object of two raids by Task Force 58 on 5 and 11 November, and which, in addition, was constantly bombed by shore-based army Liberators.

18 – THE TASK FORCE SHOWS ITS STRENGTH

AFTER THE VICTORY AT MIDWAY, as we have seen, the air war shifted to the South Pacific theater where the bloody battles of the Solomon Islands were fought and won. In the meantime, new carriers were commissioned and readied for combat — Essex, Yorktown, Lexington, Independence, Princeton, Belleau Wood, and Cowpens.

These ships assembled at Pearl Harbor during the summer and autumn of 1943 and began the training exercises which were the necessary preliminaries to an offensive sweep across the central Pacific. On 31 August the first of the dress rehearsals took place at Marcus Island where a task force built around the Essex, Yorktown, and Independence struck that Japanese outpost for the second time during the war. On 1 September, in a concurrent operation, the Princeton and Belleau Wood covered the unopposed landing on Baker Island; and eighteen days later, with the new Lexington, they raided Tarawa in the Gilbert Islands. Finally, on 5-6 October, the largest carrier task force yet assembled, composed of the seven new carriers then available, attacked Wake Island in a test of revised air and surface tactics. The curtain was now ready to be raised on the first amphibious operation of the central Pacific offensive.

Our seizure of the Gilbert Islands was a logical development in the war against Japan. This group of coral atolls was the first step in the shortest and most direct route to the empire. A successful invasion would shorten our supply lines to the southwest Pacific, force the enemy to divide his strength in order to meet a new and ominous threat, and transfer the action to an area ideally suited for the operation of our new, fast carrier task force.

Task Force 50, commanded by Rear Admiral (later Vice Admiral) Pownall, was the first edition of the famed air-surface striking force, later known alternately as Task Force 58 and Task Force 38, which swept the sea from the Gilberts to the Sea of Japan. All the experience of the South Pacific campaign, the early raids across the central Pacific, and the recent rehearsal strikes had gone into its development. The early hesitancy to operate more than one carrier in a single formation had been gradually overcome; by groping toward and eventually achieving the technique of maneuvering carriers together, navy men had forged a mobile air force of overwhelming strength.

Of no less importance than the appearance of this force was the introduction of escort carriers into the Pacific amphibious operations. These frankly makeshift, yet efficient ships had hitherto been employed only in transport and convoy activity, except for a single combat assignment in support of the North African landings. Their participation in the Gilberts campaign established their importance in support operations. Thereafter they became an integral part of the amphibious forces which neutralized and stormed the beaches of the central Pacific to the gates of the empire.

The combined carrier force, comprising 6 large carriers, 5 light carriers, 8 escort carriers, and about 900 planes, had an ambitious program: In addition to establishing and maintaining air superiority in the area, this force was to neutralize enemy defenses, support the assault, conduct medium searches ahead of the assault forces, provide fighter protection, maintain anti-submarine patrol, provide gunfire spotting, and maintain continuous observations and reports over the objectives, Tarawa, Makin, and Apamama. These tasks not only reflect the importance of naval aircraft in amphibious operations; they suggest that air units were the principal factor in the control of sea areas.

In the effort to establish air and sea control of the Gilbert Islands the land-based aircraft of Task Force 57, flying from the Ellice, Phoenix, and Samoan islands, were directed to conduct photographic reconnaissance missions, attacks against enemy bases within range to the westward, and long-range searches.

Commencing on 13 November this air force began nightly raids on Japanese installations in either the Gilbert Islands (Tarawa and Makin), the Marshall Islands (Maloelap, Mili, Jaluit, Kwajalein, and Wotje), or Nauru Island. Although there was always the risk that this activity might alert the enemy, nevertheless it did provide information for planning officers, tested enemy defenses, and inflicted considerable damage.

D-Day for the initial landings in the Gilberts was 20 November 1943. During the week before, the four groups of Task Force 50 began converging on the islands, two groups on a track from Pearl Harbor and two from Espiritu Santo. The interceptor group (Task Group 50.1), which was built around the Yorktown, Lexington, and Cowpens and commanded by Rear Admiral Pownall, operated in an area between the Marshalls and Gilberts in order to intercept attacks launched from the northern atolls against our assault troops. In the period between 19 and 26 November this group inflicted severe damage on the enemy, principally at Mili and Jaluit.

For a loss of five pilots and two crew members, aircraft of the three carriers shot up Japanese airfields, sank a small cargo ship, and burned fifty enemy planes on the ground and in the air. Their efforts reduced enemy plane availability in the Marshalls and prepared the way for the amphibious forces to enter the Gilberts area unmolested by air attack.

Task Group 50.2, the northern group, was commanded by Rear Admiral (left Vice Admiral) A. W. Radford and included the Enterprise, Belleau Wood, and Monterey. This group was assigned the mission of gaining and maintaining control of the air at Makin and providing direct support to the Northern Assault Force, whose task was the conquering of that atoll. Operations were unspectacular until the night of 25 November when the task group came under a determined air attack which was repulsed by gunfire. The next night the Japanese snooped the group and Admiral Radford ordered night fighters launched against approaching torpedo bombers.

Two Betty's [Mitsubishi G4M bombers] were shot down but Lieutenant Commander E. H. "Butch" O'Hare [Lieutenant Commander Edward Henry "Butch" O'Hare, 1914-1943], one of the Navy's first air heroes and an experimenter in night tactics, failed to return to the Enterprise. Both the Interceptor Group and the Northern Group had sortied from Pearl Harbor; the carriers of the Southern Group (Task Group 50.3) and the Relief Group (Task Group 50.4) had recently participated in the early November raids on Rabaul and consequently approached from the south. The Southern Group was commanded by Rear Admiral A. E. Montgomery and included the carriers Essex, Bunker Hill, and Independence.

Formed on 15 November near the Ellice Islands it proceeded north to Tarawa where it supported the Marines in their epic struggle on Betio Island. Under sporadic attack throughout its operating period, the task group met a flight of sixteen to eighteen Betty's at dusk on D-Day which succeeded in putting one torpedo in the Independence. The planes had come in low over the water and were not picked up by radar after they had been seen by an alert spotter. Excellent damage control enabled the Independence to steam, under its own power, to Funafuti for temporary repairs.

On 19 November the Saratoga and Princeton launched strikes against 8 Nauru Island and, with air opposition negligible, neutralized the airfield and removed any threat of its planes to the success of the Gilberts operation. Following this action, the group escorted the Makin and Tarawa garrison forces to the combat area, thereafter, operating to the southeast of Tarawa in its capacity as a relief carrier group, subject to call in support of ground troops.

By 26 November the air battles had been won, the strategic islands stormed and taken, and the success of the operation assured.

The Commander-in-Chief, Pacific Fleet, therefore, effected a reorganization of the fast carrier force in anticipation of interim operations before the invasion of the Marshall Islands. A striking force composed of the Yorktown, Lexington, Essex, Enterprise, Cowpens, Belleau Wood, and screening ships under the command of Rear Admiral Pownall, hit Wotje and Kwajalein on 4 December, destroying an estimated seventy-eight planes and sinking and damaging several ships.

After recovery of the air groups the task force commenced retirement. The Japanese made two small, sneak torpedo attacks in the early afternoon and sent continuous but hesitant attacks from dusk until after midnight. During the night the Lexington was discovered by flares and struck by one torpedo, but withdrew successfully without further injury.

Four days later a force of six fast battleships under the tactical command of Rear Admiral (later Vice Admiral) W. A. Lee, Jr., Commander, Battleships, Pacific Fleet, bombarded Nauru Island. Task Group 50.4, led by Rear Admiral F. C. Sherman in the Bunker Hill with the light carrier Monterey, provided air support. A total of 150 sorties was sent against that shattered outpost and the target areas were quickly enveloped in smoke, flames, and dust. Retirement was made toward New Hebrides bases where the force reported to the Commander, South Pacific Force, for duty.

The story of the escort carriers during the Gilbert Islands campaign is one of experiment mixed with tragedy. On D-Day, Rear Admiral H. M. Mullinix and his group of three carriers, the Liscombe Bay, Coral Sea, and Corregidor, began operations in support of the Makin landings as part of Rear Admiral R. M. Griffin's Air and Surface Support Group (Task Group 52.13).

The success of the landings can be attributed, in part, to the bombing and strafing missions of the CVE planes and to the air coverage which they provided for the amphibious units in the Makin area; but aircraft losses were high, and the experimental nature of the operation was evident. On the morning of 24 November, while sixteen miles off Makin Island, the Liscombe Bay was struck by a submarine torpedo.

The ship exploded amidships, burst into flames, and showered a destroyer, five thousand yards away, with sparks and burning debris. An observer reported that "a few seconds after the first explosion, a second explosion which appeared to come from inside the Liscombe Bay burst upwards, hurling fragments and clearly-discernible planes two hundred feet or more into the air. The entire ship seemed to explode and almost at the same instant the interior of the ships . . . glowed with flame like a furnace." At 0535, twenty minutes after being hit, the carrier sank by the stern with the loss of Rear Admiral Mullinix and seven hundred officers and men.

To the south, Rear Admiral V. H. Ragsdale led his division of three carriers, the Sangamon, Suwannee, and Chenango, plus the Barnes and Nassau, in support operations against Tarawa. This group (Task Group 53.6) maintained combat, intermediate and anti-submarine patrols, made searches, conducted hunter-killer operations, and provided bombing-strafing support for forces ashore. Upon the annihilation of the Japanese garrison at Tarawa, the group moved eastward and provided routine air coverage for our units at Apamama. The work of the escort carriers at Makin, Tarawa, and Apamama can best be judged by what was learned and later applied in support operations across the Pacific to Okinawa. The Gilberts campaign was the medium of their formative experience. In it they learned the technique of support for ground troops and of tactical maneuvering with amphibious forces. The loss of the Liscombe Bay emphasized the vulnerability and limitations of the type but did not change the previous concept of its use. After their experience in the Gilbert Islands the CVE's became efficient and necessary components of each succeeding invasion.

The campaign was more than a proving ground for the escort carriers. The four-group carrier task force appeared for the first time and learned the lessons of amphibious support. Experiments with night fighters uncovered their defects but established their future in combat operations. And new planes, the Hellcat and the Helldiver, revealed a superiority over Japanese types. Here, then, was a re-hearsal on a ground scale for the events of the future.

The role of naval air power in the Gilbert Islands campaign demonstrated the ancient but fundamental principle of warfare that overwhelming force is the key to victory. From the beginning the carrier planes of Task Force 50 were able to deny the adjacent seas to the enemy fleet; in a matter of hours they reduced Japanese air opposition to scattered nuisance raids from the Marshalls. The defenders of Tarawa, fighting grimly from their battered forts, called in vain for help which might have changed the fortunes of war. In a real sense, therefore, our carrier planes insured the victory. For the fate of the enemy was sealed even before the Marines struggled ashore to go about the bloody work of conquest.

19 – THE END OF THE ALEUTIAN CAMPAIGN

THE BEGINNING OF THE SECOND YEAR OF WAR saw a continuance of the American dual plan of operations in the Aleutians. It will be recalled that our military forces sought, in the first place, to harass the enemy in his installations on Kiska and Attu in order to prevent development or expansion; and, second, to extend our own bases westward preliminary to an assault on the Japanese stronghold.

Adak had been occupied by our forces in August 1942, placing them within 250 miles of Kiska. On 12 January 1943, Amchitka was occupied without any opposition. The Navy's Catalinas provided anti-submarine patrols as a cover for the invading forces. By May regular patrols were being run from Amchitka. As bases were moved westward, it was possible to lessen the number of patrols to the east. In fact, searches were cut out entirely from Sand Point and Cold Bay, releasing planes for coverage far to the west. After the occupation of Amchitka, patrol plane search could be and was carried out as far as the northern Kuriles.

As a result of these patrols the Japanese supply route to its Aleutian bases was continually menaced, and the way was paved for an attack on these bases. In preparation for this campaign, Patrol Wing 4 was considerably strengthened. It moved its headquarters from Kodiak to Adak in March 1943. The two original squadrons, which had been sent back to the states, were brought back as bombing squadrons, with a new plane, the [Lockheed] PV-1, or Ventura [Lockheed B-34 Lexington]. This, in comparison to the Catalina, was a land plane and was capable of greater speeds. In addition, there were four Catalina squadrons.

These squadrons were to provide anti-submarine patrol prior to and during the attack which was to be made first on the island of Attu on 11 May 1943. When the actual attack came, they were to furnish air cover for the surface vessels, while a sizable army air force was to cover the landings. To support the landings, Vice Admiral Kinkaid supplied a strong surface force, consisting of 5 battleships, 5 heavy and 4 light cruisers, 21 destroyers, and 1 escort carrier.

Unfortunately, this vessel, the Nassau, had little opportunity to show its capabilities. The weather was so bad that no launchings were possible between the time the vessel left the west coast and D-Day. Coupled with this was the fact that the main squadron aboard, a composite squadron of observation and fighter planes, had had only five days of training as a squadron before going into action.

After the initial bombardment by the surface forces, the Nassau operated to the northeast of Attu, usually at a distance of some forty miles, but sometimes as close to the island as ten miles. Weather seriously hampered flight operations every day.

Heavy fog and low overcast made air support difficult and at times impossible. The wind velocities were so slight that because of the relatively low speed of the vessel, it was necessary to catapult every plane that was launched. At no time were weather conditions such that it was felt advisable to launch an all-out attack.

In spite of the need for support for the ground troops, flights of four to eight planes were all that could be operated at any one time. On one occasion the elements were defied with disastrous results. In response to an urgent plea for assistance from the ground forces, two flights of four fighters (Grumman Wildcats, F4F's) were launched in weather so severe that four planes and four pilots were lost. In all, the squadron flew 179 sorties and lost, in the operation 8 planes and 5 pilots, one a Marine enlisted pilot. None of these losses was attributed to the human enemy, but to dangerous weather conditions.

The conquest of Attu was completed by the end of May. Work was rushed on the completion of air facilities, and, by 9 June, a fighter strip was operating. Attention could next be directed to the remaining Japanese stronghold, Kiska. For two and a half months, from 1 June to 15 August, Venturas made repeated daylight attacks on the island. Because army bombers also attacking the island were not equipped with radar, Venturas were used to guide the army attacks. During the same period, Catalinas made the usual patrols of the sea areas and conducted harassing night raids on Kiska.

After this extended period of "softening up" the enemy, the land assault was made on schedule, on 15 August — only to find that the enemy had fled. The only ships that had been sighted during the preliminary bombing phase had been two auxiliaries and two small freighters. Army B-25's had sunk two and damaged one of these. Two explanations of the method of Japanese withdrawal were possible. One was that the Japanese garrison had been taken from Kiska on submarines; the other was that the withdrawal had been accomplished during a few days of extremely bad weather that had grounded the search planes.

The capture of Kiska ended the Aleutian campaign. The Japanese had been driven out in slightly more than fourteen months from the time of their attack on Dutch Harbor. It had been necessary to build bases from nothing under the noses of the Japanese and then assemble the men, the supplies, and the ships to expel the enemy.

Although weather, a lack of detailed knowledge of the Aleutian chain, and total absence of facilities were often greater obstacles than the enemy, the presence of the Japanese presented an ever-constant threat that had to be guarded against in all operations. The Navy's pilots merit the highest praise for their courage and flying skill in the face of such obstacles.

The final phase of the North Pacific campaign was begun during the second year of war and continued throughout the duration of the conflict. This was an attack on the northern fringe of the Japanese empire, the Kuriles. The first attempted attack was a combined effort by army B-25's and navy Catalinas. In this and later raids made during the year, the Navy was handicapped by the fact that once again the Catalina was called upon to perform a task for which it had not been constructed. The best bombing altitude for protection from antiaircraft fire was 9,000 feet. With its load, the Catalina could barely reach this altitude.

Furthermore, the low speed of the Catalinas meant that an attack on the Kuriles necessitated flights lasting from twelve to fourteen hours. Flights of this duration placed a heavy strain on the officers and men participating in the raids. The inclement weather not only hampered flying conditions but affected bombing operations. Frequently, bombing attempts were unsuccessful because bombs could not be released as a result of the icing of the exterior bomb racks.

In spite of these handicaps, Catalinas continued to raid the Kuriles until February 1944, when they were replaced by faster and more heavily armed Venturas.

U-BOAT COMMANDER OTTO KRETSCHMER (LEFT) SANK THE MOST
ALLIED VESSELS IN WORLD WAR 2; A TOTAL OF 47 SHIPS.

20 – THE MARSHALLS CAMPAIGN

THE LONG PERIOD OF TRAINING, building, and testing our naval aviation forces began to bear full results in 1944. For the next year and a half, a series of staggering blows from our combined forces was to beat back the Japanese from one strategic stronghold to another until the home islands themselves were reeling under the attack, and with the fleet destroyed and merchant shipping decimated, the Japanese chose surrender to the alternative of complete destruction.

A foretaste of events to come had been handed the Japanese in the Gilberts in 1943. The series of strikes started in earnest with the invasion and conquest of the Marshalls early in 1944.

The efficiency with which the Marshall Islands were neutralized from the air, bombarded from the sea, and captured by our soldiers and Marines made it a model amphibious operation. The air, sea, and land conditions of the Marshall and Gilbert groups were so closely parallel that the experience gained in the seizure of Makin, Tarawa, and Apamama could be applied directly to the campaign against the strategic atolls to the north. None of our amphibious attempts was made at such small cost and with such glittering results. And in no campaign did out naval air arm play a more dominating role in respect to air opposition by the enemy.

The operation plan as prepared by the Commander, Fifth Fleet, Vice Admiral (later Admiral) R. A. Spruance, involved the occupation and defense of the Kwajalein Atoll, the largest of the Pacific atolls and a cornerstone in the Japanese outer defense system, of the lightly defended islands of Majuro Atoll, 250 miles to the southwest, the establishment of airstrips and naval bases on the conquered islands from which the remainder of the Marshall atolls could be dominated, and, finally, the seizure of Eniwetok, westernmost of the Marshall group and a steppingstone on the way to the Marianas.

D-Day for the landings on the Kwajalein and Majuro was 31 January 1944; in view of the initial success of the operation the date for the attempt on Eniwetok was moved up to 17 February.

Since the Japanese Fleet made no effort to interfere in the campaign, our air program was divided, in general, into three main phases: the neutralization of all enemy-held bases from which aircraft might operate against our forces; the direct support of the amphibious assaults; and the first carrier raids against Truk and the Marianas.

During the period required to execute these tasks, carrier planes flew 6,407 combat sorties, dropped 1,854 tons of bombs, and destroyed approximately 244 enemy planes in the air and on the ground. The scope of actions of our fast carrier task force exceeded those of any comparable period in the past.

The plan of neutralization began as soon as land-based naval aircraft and the heavy bombers of the Seventh Army Air Force, operating in Task Force 57 under Rear Admiral (later Admiral) J. H. Hoover, could take off from the newly won airfields on Tarawa, Makin, and Apamama.

During the weeks preceding the invasion, these planes flew reconnaissance missions on radial lines extending 850 miles from their Gilberts bases, and attacked, with ever-increasing intensity, the Japanese installations at Mille, Jaluit, Wotje, Maloelap, and Kwajalein.

These raids destroyed an estimated 50 enemy aircraft in the air and an additional 24 on the ground, but they did not succeed in completely neutralizing any of the Japanese airfields with the exception of Mille.

Their most successful functions were to keep the enemy off balance, provide photographic information, and protect, with offensive action, the expanding stock piles in the Gilberts. The complete neutralization of Japanese air power in the Marshall Islands awaited the arrival of the carrier planes of Task Force 58.

As in the Gilbert Islands campaign, the fast carrier task force, now under the command of Rear Admiral (later Admiral) M. A. Mitscher, operated, tactically, in four separate groups. And as before it was a formidable armada: 6 large carriers (Yorktown, Enterprise, Intrepid, Essex, Bunker Hill, and Saratoga), 6 light carriers (Belleau Wood, Cabot, Monterey, Cowpens, Princeton, and Langley), 8 fast battleships, 6 cruisers, and 36 destroyers.

Three groups assembled in Pearl Harbor and the fourth group (Task Group 58.3) formed at Funafuti in the Ellice Islands. Shortly before D-Day the force stood poised for its overwhelming assault on the enemy airfields.

With approximately seven hundred planes at his disposal, it became Admiral Mitscher's strategy to deal the Japanese air strength in the area an initial blow from which it could not recover. All four groups began launching aircraft at 0500 on 29 January. Task Group 58.1 (Rear Admiral J. W. Reeves) struck Maloelap; Task Groups 58.2 (Rear Admiral A. E. Montgomery) and 58.3 (Rear Admiral F. C. Sherman) attacked the Kwajalein fields and installations; and Task Group 58.4 (Rear Admiral S. P. Ginder) hit Wotje. Surprise was complete — in every instance enemy planes were caught on the ground. So successful were these tactics that by the end of the day not a single Japanese plane remained operational east of Eniwetok.

The first phase of the air campaign — the neutralization of bases from which the enemy might send planes against our amphibious forces — was completed during the following three days when Rear Admiral Sherman led his task group against Eniwetok. At 0450 on 30 January the Bunker Hill, Monterey, and Cowpens launched a bombing and strafing attack which burned fifteen Betty's and four float planes and removed any immediate threat from that isolated atoll.

Thereafter, the task group concentrated on buildings, gun positions, and small shipping; when it was relieved on 3 February by Task Group 58.4 (the Saratoga, Princeton, and Langley) it left the principal islands smoking and helpless to resist aerial bombardment. Rear Admiral Ginder's carriers continued the program of destruction.

In the meantime, at Kwajalein, the air campaign had entered its second phase — the support of landing troops. On the morning of D-Day, while Task Groups 58.3 and 58.4 continued the neutralization of Eniwetok, Wotje, and Maloelap, Task Groups 58.1 and 58.2 operated in direct support of the ground assaults on Roi and Kwajalein, the principal islands of the Kwajalein Atoll.

In addition, the CVE's of Task Group 52.9 (Coral Sea, Corregidor, and Manila Bay under Rear Admiral R. E. Davison) and of Task Group 53.6 (Sangamon, Suwannee, and Chenango under Rear Admiral V. H. Ragsdale) were stationed off the northern and southern ends of the atoll, their planes under the tactical command of coordinators who were present at the scene of action.

The experience gained in the Gilberts was applied with the most satisfying results; from an airman's standpoint the support phase, in which air attacks were "carefully and successfully timed to accord with ships' gunfire and the actual assault on the beaches," was ideally conducted. On 4 February, having been unmercifully beaten by an overwhelming combination of air, sea, and land power, the enemy ceased all organized resistance.

With the fall of Kwajalein there followed a lull in air operations. Task Group 58.4, however, continued the interdiction program at Eniwetok, and a neutralization group of four cruisers, two escort carriers (Nassau and Natoma Bay), and screening destroyers, under the command of Rear Admiral Small, was assigned the task of denying to the enemy his airfields at Wotje and Maloelap.

The rest of Task Force 58 and various of the amphibious support units entered Majuro lagoon, used by the Germans as a fleet anchorage during their period of control (1899-1919), and captured by our forces on D-Day without opposition. There followed a week of replenishment and planning. The very encouraging success of the operation to this point permitted D-Day for Eniwetok to be moved up from 10 May to 17 February; and as a powerful corollary to this attack, the fast carrier task force planned a great raid on the enemy stronghold of Truk.

Truk had always been a symbol of Japanese strength, and it was not without anxiety that Vice Admiral Mitscher led his three groups in the sortie from Majuro on 12 February. Intelligence information concerning the target was limited, a fact which heightened its reputation as a formidable bastion, difficult and dangerous to attack. For those who participated in the raid, and for those who study it in relation to the development of our carrier tactics, it is a climax in the central Pacific campaign.

At 0650 on 16 February, having completed an uneventful approach, the task force stood ninety miles off Truk, its presence apparently unknown to the enemy. A fighter sweep of 70 planes took off at this time and completed a very successful action in which, for a loss of 4 planes, our pilots shot 56 Japanese planes out of the air and destroyed an estimated 72 on the ground. This upset the balance of air power in our favor, and thereafter the enemy antiaircraft batteries provided the only serious opposition.

After the initial sweep, launches were staggered between carriers to provide a continuous flow into the target area. With the neutralization of enemy air power, shipping became the primary objective. Contrary to expectation, few Japanese men-of-war were found, but the anchorage was particularly productive of auxiliaries.

The Commander, Fifth Fleet's final report listed 41 of 55 enemy ships sunk or damaged — an impressive total and one which fulfilled the mission of the striking force. Our losses were 25 aircraft and the carrier Intrepid damaged by aerial torpedo during a brief enemy action on the night of 16-17 February.

The first carrier attack on Truk, when viewed in retrospect, was a psychological as well as a tactical victory. The myth of Truk's impregnability exploded with the four hundred tons of bombs which were dropped over the target; the Japanese battle force which had been spotted in the Truk lagoon some days before the attack retreated in the face of a threat; and our naval airmen gained new confidence in their weapons and in their ability to wield them.

In its effect on overall strategy, the raid was equally important. It proved that Truk could be eliminated as a menace to our advance by air neutralization alone, a fact which saved lives, time, and trouble. And, when considered in conjunction with the first raid on the Marianas a few days before, it helped secure the new bases at Kwajalein and Majuro from the possibility of an enemy counterthrust.

The immediate purpose of the Truk attack, of course, was to cover our landings on the main islands of Eniwetok Atoll. From 30 January to 6 February these islands had been subjected to daily poundings by Task Groups 58.3 and 58.4 in an effort to prevent their being used as forward staging bases against our operations to the east. Task Group 58.4 struck them again on 10, 11, 12, and 16 February, concentrating on gun positions, supply areas, and bivouacs. The enemy was never able to offer air opposition, his planes having been destroyed and the airfield cratered in the strikes of 30 January. When the landings began on 17 February the Japanese could resist only with fanaticism and the careless sacrifice of their lives.

Air support of the ground fighting followed previously conceived tactics. The Saratoga, Princeton, and Langley (Task Group 58.4) operated to the north of the atoll until 28 February, launching air strikes, searches, and patrols. Task Group 51.18 (the Sangamon, Suwannee, and Chenango) was stationed to the east and conducted call strikes and most of the anti-submarine patrols until relieved on 25 February by the Manila Bay. This air support work was as tedious as it was unspectacular; but because it was vital to the success of the landings, it deserves the commendation of naval historians.

The final aerial blow delivered in support of the Marshalls operation was aimed at the Marianas on 21 and 22 February 1944. The torpedoing of the Intrepid at Truk had caused a regrouping of the fast carrier task force and only two groups — 58.2 (Essex, Yorktown, and Belleau Wood) and 58.3 (Bunker Hill, Monterey, and Cowpens) — participated in the raid. The purpose of the attack was twofold: to cover the campaign in progress on Eniwetok by destroying enemy shipping, air facilities, and installations at Saipan and Tinian; and to make photographic reconnaissance of those islands, and of Rota and Guam, for use in the contemplated seizure of the Marianas.

These purposes were not accomplished without opposition. The Japanese responded with the first large-scale and determined air assault on our carriers since their dusk attacks off Makin and Tarawa.

During its refueling and approach periods the task force was spotted on two occasions by enemy search planes — incidents which removed the advantage of surprise at the same time that they added the factor of counterattack.

The situation was not promising, especially when the unknown character of the target and its apparent security within an inner circle of Japanese-controlled sea were considered. Admiral Mitscher determined not to be diverted, however, and sent word to his group commanders, Rear Admirals Montgomery and F. C. Sherman, to "prepare for a fight" during the run-in.

The Japanese assault was spectacular both in its execution and in its failure. Beginning at 2100 on 21 February, "bogies" were present near the disposition for thirteen hours; and during that period three distinct attacks were pressed to within gunfire range. Of a total of approximately forty raiders, fourteen to seventeen were knocked down and the rest driven off by the intense volume of our antiaircraft barrage.

Most of the Japanese bombers and fighters were surprised on the ground, where Hellcats, in repeated strafing runs, burned or damaged almost a hundred of them. The alarm, however, had caused a hasty dispersion of shipping units, and only a few freighters and patrol craft were damaged. The scarcity of this type of target, the early destruction of enemy aircraft, and the excellent photographic coverage already procured, influenced Admiral Mitscher's decision to retire after a third strike.

The force therefore set an eastward course after sending 281 sorties over the islands. Our loss of six aircraft is an accurate gauge of the air superiority obtained.

The strikes at Saipan, Tinian, and Guam marked the end of the operation to occupy and defend the Marshall Islands. The conquest of the key atolls in that group was a victory of air, sea, and land co-ordination which extended our control of the sea to the Japanese secondary line of defense. The campaign proved to the enemy that our fast carrier task force could be delayed only by the hazard of a fleet engagement, and, conversely, it revealed that Task Force 58 could roam at will in the central Pacific, subject only to logistic limitations.

During almost a month of supporting operations, carrier aircraft destroyed the remnants of Japanese air power east of the Marianas, discovered the vulnerability of those islands, and reduced Truk to the status of a minor outpost. The air weapon which had been forged in the Gilberts was tempered in the Marshalls; it was ready for greater tests to the westward.

21 – CONQUEST OF THE MARIANAS CAMPAIGN

FOUR MONTHS ELAPSED BEFORE THE NEXT large-scale amphibious operation took place — the campaign to wrest the Marianas from the Japanese. During that period naval aircraft, flying from newly commissioned fields on Kwajalein and Eniwetok, and from bases in the Gilberts, went about the business of reducing Truk, Ponape, Wake, and the enemy-held atolls in the Marshalls, to impotence.

One of the principal features of this land-based air campaign was its correlation with attacks by aircraft under the control of the South and southwest Pacific commands. The northwestward movement of these commands and the westward progress of the Central Pacific Force meant that Truk, for example, was now threatened from two sides. The air noose was tightening around thousands of miles of enemy sea, and the spheres of Japanese influence in the central Pacific were being reduced to the shattered remains of their once-strategic bases.

In the meantime, the fast carriers were not idle. On 18 March the Lexington, having repaired its battle damage, returned to conduct an "exercise" strike on Mille Atoll. By 22 March the bulk of the carrier force was assembled in the Majuro lagoon where it had replenished and rearmed for operations in support of the Hollandia invasion by General MacArthur's troops.

On 23 March this force sortied in two groups; on 26 and 27 March it rendezvoused with a third group which had previously been assigned support tasks in the South Pacific area; and on 29 March the combined force of 5 large carriers (Enterprise, Bunker Hill, Hornet, Yorktown, and Lexington), 6 light carriers (Belleau Wood, Cowpens, Monterey, Cabot, Princeton, and Langley), and screening battleships, cruisers, and destroyers, began a high-speed run toward the Palau Islands for a preliminary raid to aid the projected Hollandia assault.

The hope that heavy units of the enemy fleet might be found at Palau disappeared when search planes discovered and reported the force on 25 and 28 March, thus eliminating the element of surprise and alerting shipping in the area. As at Truk, however, the dawn fighter sweeps of 30 March displayed a potency which destroyed virtually all enemy airborne aircraft and created havoc among planes caught on the ground.

Unlike Truk, enemy aerial opposition did not end after the first day's fighter sweep. During the night of 30 March, the Japanese flew in reinforcements which engaged succeeding strikes in heavy combat. Our Hellcats won control of the air, however, and thereby exposed the islands to the bombs of the Helldivers and Avengers. Japanese plane losses in the approximate ratio of 19 to 1 proved once again the superiority of our equipment and pilot training.

The 115 sorties flown against ground installations accomplished results in conformity with previous experience; if they did not leave the bases of the Palau group utterly neutralized, they at least revealed the precariousness of their situation in respect to the advancing battle front.

The attacks on shipping targets produced an interesting conclusion: a preference for masthead rather than glide tactics resulted in increased bombing efficiency and was largely responsible for the sinking of 28 vessels and the damaging of 18 others. Masthead or "skip" bombing had been developed by the Army Air Force and applied with devastating results in the Battle of the Bismarck Sea, 6-9 February 1943. The Navy's use of the technique confirmed its superiority to high level or glide methods.

Finally, the raid introduced mine laying into the repertoire of the fast carrier task force. The mining operations carried out at Palau constituted the largest tactical use of mines made up to that time by United States air forces, and the first in history from carriers. A total of seventy-eight mines were dropped — an important contribution to the immobilization of enemy shipping in that area.

The operation plan did not limit the raid to the Palau Islands and their adjacent waters. On 31 March, Task Group 58.1 sent 142 sorties against Ulithi Atoll and Yap Island, and on 1 April the combined force struck Woleai with 150 tons of bombs. The cost of the four-day operation was forty-three aircraft from all causes, an impressive total at first glance but one which was not out of proportion to the destruction obtained.

Despite this deepest penetration to date of the Japanese defense system, air attacks on the task force were not as severe as those experienced off the Marianas in February. Excellent fighter direction, efficient combat air patrol, and, one suspects, a Japanese hesitancy born of recent experience, combined to protect our carriers and screening ships from damage. The withdrawal to Majuro was made without incident.

On 13 April 1944 Task Force 58 departed the central Pacific in three groups for operations in support of army landings in the Hollandia-Aitape area. Upon completion of that campaign the force retired by way of the Carolines and on 29-30 April conducted the second carrier raid on Truk. By this time a carrier attempt on that once-formidable fortresses had become anticlimactic.

But the Japanese were still strong enough to offer major resistance, and approximately 120 planes had to be destroyed before our pilots won control of the air. Thereafter, pre-established patterns of attack were employed, with staggered strikes pounding buildings, airfields, and defense guns, and ferreting out the few ships which necessity had forced the enemy to retain in the anchorage.

If the second Truk raid was an anticlimactic blow in the central Pacific offensive, it was nevertheless distinguished by individual heroism and the precision of fighter tactics. A study of the initial fighter sweep, for example, provides material for stories of high adventure. Two divisions from the Langley, while orbiting above the heavy cloud layer over Truk, spotted an estimated thirty Japanese fighters approaching from slightly below. The eight pilots attacked at once and in a few minutes of tense and expert fighting destroyed 21 planes without loss to themselves.

Some minutes later, a Lexington group of 11 Hellcats encountered 15 to 18 Japanese planes, of which they shot down 9 for the loss of one. Such examples of our air superiority produced that confidence which is the key to victory, and chastened the enemy with losses that he could not afford.

Although this was the last carrier raid in force before the invasion of the Marianas, the central Pacific air war continued with increasing tempo, waged by army, navy, and marine land-based aircraft. Routine tasks — long-range searches, anti-submarine patrols, transport flights — occupied the time and energies of naval pilots who were denied the thrill of carrier warfare. Others maintained the neutralization of by-passed islands with strikes that lost their adventure through repetition.

The luckier ones flew their Liberator on attack missions to Truk and Ponape, where excitement could still be had, or on reconnaissance flights to the Marianas. In the meantime, under cover of this activity, air, sea, and land forces were being assembled at Eniwetok for the assault on Saipan.

Late in May, a final carrier raid was made to indoctrinate new air groups and to strike a blow at airfields on Marcus and Wake islands, from which the Japanese might threaten the northern flank of our supply line to the Marianas. Three carriers participated — Essex, Wasp, and San Jacinto — under the tactical command of Rear Admiral A. E. Montgomery, Commander, Task Group 58.6. The group sortied from 3 Majuro on 15 May, opened the attack with a night fighter sweep on Marcus in the pre-dawn darkness of 19 May, and, in the face of intense antiaircraft fire, strafed and bombed that hapless island for two consecutive days. On 23 May the raid moved eastward where it struck Wake Island with 354 combat sorties.

With the return of the task group to Majuro on 25 May, Vice Admiral Mitscher's Task Force 58 began active preparations for its next great role in the central Pacific offensive. During the occupation and defense of the Marianas the naval air arm, operating at a new peak of efficiency, offered a challenge to the enemy which could be ignored no longer; and the Japanese, for the first time since the desperate days of the Solomon Islands campaign, accepted the challenge with their own fast carrier fleet.

As our amphibious ambitions brought us closer to Japan the power of the enemy to resist in the air increased accordingly. The campaign for the Marianas put the air war on a more equal basis than it had ever been in the central Pacific, and although American control of the air was won in a series of shattering defeats for the enemy, the example of quick and easy victory which had been established in the Gilberts and Marshalls could not be applied to the conquest of Saipan, Tinian, and Guam.

The reasons for this are obvious both to the military strategist and the geographer: loss of the Marianas portended consequences that demanded the commitment of Japanese air strength; and the comparative ease with which the islands could be reinforced from the empire by way of the Bonin-Volcano group supported that demand. And so, the Japanese Air Force expended its aircraft with lavish hand. Such strategy impeded our advance only momentarily and presented us with a solid triumph over enemy air power.

The Marianas campaign imposed tasks on the fast carrier force which increased its responsibility, its operating difficulties, and the importance of its role in amphibious warfare. The four groups under Vice Admiral Mitscher contained 8 large carriers (Hornet, Yorktown, Franklin, Bunker Hill, Wasp, Enterprise, Lexington and Essex), 8 light carriers (Bataan, Cabot, Belleau Wood, Monterey, Princeton, San Jacinto, Cowpens, and Langley), 7 new battleships, 13 cruisers, and 58 destroyers.

These ships could provide an air fleet of 900 planes to oppose local Japanese air power, support ground troops, and meet any threat from the enemy fleet. The full significance of this ability is only understood when it is realized that Eniwetok, the closest base for logistic support, was a thousand miles from the combat zone.

Task Force 58 sortied from Majuro on 6 June 1944 after the concentration of ships in that anchorage had been spotted by a high-altitude Japanese search plane. The approach to the Marianas, however, was uneventful and on 11 June at 1300, while two hundred miles east of the islands, a fighter sweep was launched which began the campaign with the advantage of surprise and destroyed 150 planes in the air and on the ground. This crippling blow prevented the enemy from reacting in strength, and that first and most anxious night in the vicinity was passed without attack on our carriers.

On 12 June the bombing began, and the islands of Guam, Rota, Saipan, Tinian, and Pagan felt the weight of high explosives on airfields and installations. With most of their airplanes burned on the ground or missing in action, the Japanese responded only with sporadic dusk and night attacks during this preliminary phase of the operation. These were unsuccessful and there was reason for optimism.

The inability of the enemy to foresee the danger to the Marianas left him with two convoys in the area which were found on the twelfth by carrier search planes of Task Groups 58.1 and 58.4. Rear Admiral W. K. Harrill's group (58.4) struck a formation of twenty ships which was fleeing on a northerly course, 125 miles west of Pagan Island. In an afternoon of merciless destruction, the enemy convoy was reduced to wreckage and a few survivors. One destroyer was sunk by strafing alone, a rare example, but one which suggests the character of the attack.

Search planes on the morning of the thirteenth found only a single ship, a few abandoned hulks, and the evidence of catastrophe. It is doubtful if more than one or two ships escaped. To the south, 135 miles west of Guam, another convoy of six ships became the target for a special attack mission sent by Rear Admiral J. J. Clark (CTG 58.1). This strike failed to locate the convoy on 12 June, however, and it was not until the next day that it was caught and strafed by twenty fighter-bombers. All of the ships were damaged, and two destroyers left trailing oil and a cargo vessel on fire.

Air strikes on island targets continued on 13 June in accordance with the program of neutralization. On 14 June Task Groups 58.1 and 58.4 rendezvoused and commenced an approach to the Bonin Islands, which they were to attack on 15 and 16 June in support of the D-Day landings on Saipan. This was the deepest penetration of empire waters ever made by a carrier striking force up to this time.

In the teeth of a growing gale, fighter sweeps and bombing missions were launched against Iwo Jima, Chichi Jima, and Haha Jima; and despite airborne opposition, heavy antiaircraft defense, and the most unfavorable flying conditions, planes from the Essex, Hornet, and Yorktown reduced the ability of the Japanese to resist and left the area in poor condition to stage aircraft into the Marianas. High seas, wet and pitching decks, and formidable cloud fronts made carrier landings so hazardous that the loss of only one plane operationally is a tribute to the skill of pilots and flight deck crews.

On the afternoon of 16 June, there being indications of threatening movements by the Japanese Fleet, the two task groups retired southward toward a rendezvous with the rest of the force.

In the meantime, in the early morning of 15 June, marine and army ground forces began their bitter month-long struggle to conquer Saipan. Throughout that grim ordeal, naval aircraft controlled the air over the troops and aided them in the bloody business of extermination. In addition to the fast carrier task force — which supported ground operations when it was not occupied with the sinking of ships, the neutralization of adjacent islands, and its duel with the Japanese Fleet — an impressive fleet of escort carriers hovered about the islands, performing the routine and necessary tasks of service and support.

The roster of these ships reveals by its length that the experimental use of escort carriers in the Gilbert Islands campaign had proved the value of the type in amphibious operations. Twelve CVE's supported the seizure of Saipan, Tinian, and Guam: Suwannee, Sangamon, Chenango, Kitkun Bay, Gambier Bay, Coral Sea, Corregidor, Fanshaw Bay, White Plains, Kalinin Bay, Nehenta Bay, and Midway.

And for the first time in the central Pacific, a division of escort carriers (Copahee, Breton, Manila Bay, and Natoma Bay) was used, not only for replacement plane service, but for the protection of oiler groups. The word "invaluable" has perhaps become hackneyed in its constant association with our escort carriers, but they deserved the adjective in the Marianas, where their absence could only have prolonged the war.

The story of the air war in the Marianas during this first phase cannot be restricted to a recital of the exploits of carrier aircraft. Facilities for operating planes from Saipan developed rapidly as our troops gained control of the southern part of the island.

The primitive airstrip at Charan-Kanoa was secured on 16 June, and the following afternoon a TBF crash-landed there. By the twentieth, this strip was operational for emergency landings and artillery observation planes. On 18 June, Aslito airfield was captured and a few days later army Thunderbolts moved in and assumed responsibility for local combat air patrol. The threat of the Japanese Fleet made long-range searches to the westward necessary, and five Mariners arrived at Saipan on 17 June.

On the sixteenth the tender Ballard had begun laying buoys for these planes in the open sea about two and a half miles from the reef off Tanapag Harbor. Long swells made operations here difficult, but planes and tenders were fairly free from harassment by enemy guns ashore or by enemy aircraft.

Before Saipan had been captured, seven aircraft tenders (Pocomoke, Chandeleur, Ballard, Casco, Mackinac, Yakutat, and Onslow) had been formed into a search, reconnaissance, and photographic command which became a principal source of naval intelligence. Nor should the cruiser and battleship observation planes be forgotten in any analysis of the part played by naval aircraft in the Marianas.

Unwieldy, unprotected, and unsung, these planes spotted for ship to shore bombardments, rescued downed personnel, and, in one amazing instance, shot down a Japanese fighter! If the pilots and crewmen of these spotting, search, rescue, and patrol planes did not receive glorification in newsprint, they earned the respect of those who profited by their labors.

It must not be supposed, however, that our overwhelming air strength accomplished, as it had in the Gilberts and Marshalls, the emasculation of enemy air power. Although airfields throughout the Marianas had been subjected to heavy attack beginning on 12 June, they were by no means completely neutralized by the fifteenth.

Tinian air facilities were covered by surface ships and combat air patrols from the fifteenth on, and there were heavy and repeated strikes on the fields of Guam, Rota, and Pagan throughout the remainder of June and into July; but these attacks were not sufficient to prevent the fields from being used as staging bases for Japanese aircraft. It is therefore not surprising that enemy planes carried out a series of raids on our ships in the vicinity of Saipan and on the forces ashore during this first phase of the campaign.

None of the raids was of serious proportions, and they were more of a nuisance than a threat to the success of the Marianas operation. Damage to the Fanshaw Bay, the battleship Maryland, and several auxiliaries, put these ships out of action but did not alter the fact that, for most practical purposes, our control of the air was complete.

This inability of the Japanese to turn the tide of our advance with land-based aircraft led them into a decision to bring their carriers into the action. This could be, and was, no surprise to our strategists. No matter with what bravery or fanaticism the enemy troops fought across the green fields and stony hills of Saipan, their defeat was but a matter of time if our domination of the skies was not challenged and overthrown.

And since American possession of the Marianas foreshadowed consequences of extreme danger to the homeland, it became evident to the Japanese that the gamble of another carrier battle, so long delayed, could no longer be avoided. The alternative of certain defeat under existing conditions made the possibility of victory through desperate action more attractive than it might otherwise have been.

The enemy's choice, therefore, was simple and logical, both to himself and to the American admirals. Only the outcome could be in doubt.

During the first phase of the Marianas campaign, the Japanese battle fleet was attacked by our submarines from its anchor at Tawi, in the southern Philippines, northward into the Sulu Sea, through the Straits of San Bernardino and into the Philippine Sea. Air searches launched from southwest Pacific and central Pacific bases groped for this force as it maneuvered in preparation for its attack, but were unable to locate it until 19 June. Our submarine pickets, however, had more success.

On the seventeenth, the Cavalla sighted an oiler group at 13°29' North, 130°45' East, and about seventeen hours later made contact with a force of fifteen or more ships in position 12°23' North, 132°20' East.

On the eighteenth, several reports and a number of sightings of small single-engined planes gave indication of the approach of the enemy fleet, but no actual contact was made. The situation on the morning of the nineteenth, then, was one of suspense and anxiety. The enemy battle force was known to be heading eastward and it was known to be of formidable proportions: the Japanese Fleet striking force which had been present in the Philippine area in early June was estimated to contain 9 carriers, 5 battleships, 16 cruisers, and 32 destroyers. Until it had been definitely located and brought within range of our aircraft, this fleet possessed strategical possibilities which increased its formidableness and our apprehension.

In the meantime, Admiral Spruance (Commander, Fifth Fleet) regrouped his forces for the coming battle. Task Force 58 was reinforced with units from Task Force 51 (Joint Expeditionary Force) and on 19 June it contained 7 large carriers (Hornet, Yorktown, Bunker

Hill, Wasp, Enterprise, Lexington, and Essex), 8 light carriers (Bataan, Belleau Wood, Monterey, Cabot, San Jacinto, Princeton, Cowpens, and Langley), 7 fast battleships, 8 heavy cruisers, 13 light cruisers, and 67 destroyers. Aircraft carried 450 fighters, 250 bombs, 200 torpedo planes, and approximately 80 scout-observation aircraft. In point of air and surface striking power, this was the greatest armada that had ever been assembled.

Commencing on 17 June — all groups of the fleet having rendezvoused — the general plan was to operate in an area west of the southern Marianas, advancing westward during the day, retiring eastward during the night and remaining continually within supporting distance of the amphibious forces at Saipan.

On 18 June, a formation was assumed which placed three carrier groups on a north-south line, 12 to 15 miles apart, a battle-line group of fast battleships 12 to 15 miles westward of the center carrier group, and a fourth carrier group (designated the battle line carrier group) stationed to the north of the battle line. Although Admiral Spruance was present and issued general policies, plans, and orders, Vice Admirals Mitscher and W. A. Lee, Jr. kept tactical command of the carrier and battle-line groups. On the morning of the nineteenth the entire fleet was fueled, armed, and ready for action.

At 0115 on 19 June a Mariner search plane out of Saipan made contact with a large enemy force of about 40 ships in a position 470 miles west of Guam. Because of transmission failure, however, this important information was not received by our waiting forces until seven hours later, a circumstance which not only improved the tactical position of the Japanese but deprived us of an opportunity which might have altered the course of the Battle of the Philippine Sea and resulted in much greater destruction of enemy forces.

Searches flown from our carriers, meanwhile, had negative results. The great air battle began, then, with the Japanese launching their attack on Task Force 58 in the early morning daylight of the nineteenth, without our knowing the position of the enemy carrier striking force.

The enemy attack which continued throughout the nineteenth involved the largest number of planes that had ever been sent against an opposing fleet. Fortunately, the attack was poorly coordinated and instead of being concentrated it was extended over a period of many hours.

This circumstance, combined with the excellence of our radar technique, aircraft equipment, and pilot performance overcame the difficulty of position and presented us with the greatest triumph in air to air combat achieved in the Pacific war.

No one can doubt that the Japanese strategy was cunning and that it contained the elements of victory. With our fleet between his carriers and the operational airfields on Guam and Rota, the enemy could launch strikes from a position outside the range of our planes and, after attacking, land on his island bases; then, to complete this squeeze-play strategy, these planes could be shuttled back to the carriers, attacking again en route.

It was a plan which forced us into a defensive role and therefore offered a measure of protection to the Japanese Fleet. It failed for reasons already suggested and because the enemy did not readjust his strategy to meet new problems and circumstances.

The Japanese attack developed in four stages and covered the entire daylight period of 19 June. The first evidence of air attack appeared at 0530 when several "bogies" were picked up by radar in the general direction of Guam, and, soon after, the evidence became conclusive when Belleau Wood fighters, in a request for help, reported many planes taking off from Agana Field. Until 1000 the air over that island was filled with the clash of opposing aircraft.

Approximately thirty-three of our Hellcats made contact with a slightly larger number of enemy fighters and established the pattern for the day by shooting down thirty-five planes for the loss of one. The destruction of these planes upset the enemy's plan to catch our carriers in an aerial vise, and warned us of events to come.

During this first stage of the action, Task Force 58 was eighty miles northwest of Guam. At 1000 the Alabama reported the first large "bogey" of the day bearing 265° true, distant 125 miles, at 24,000 feet or above, and closing.

Admiral Mitscher immediately ordered all planes over Guam to return, and directed his group commanders to launch additional fighters. In a matter of minutes all flight decks had been cleared and interception was being made at a distance of sixty miles from the fleet. The fifty to seventy Japanese attackers were cut to pieces in a running fight that brought the remnants of the group into antiaircraft range.

These desperate planes, singly and in small groups, pushed through an attack which hit the South Dakota with a small bomb, sent one plane crash-diving against the Indiana at the water line, and scored a near miss on the Minneapolis. Ships' gunners knocked down nine planes, which completed the action with the virtual annihilation of the enemy raiding force.

Shortly after 1100, as the few survivors of the first raid disappeared from radar screens, a new large "bogey" was detected. Another interception at sixty miles discovered a second raid of sixty to seventy fighters, dive bombers, and torpedo planes which was broken up and, to a large extent, destroyed in the same manner as the first. About four dive bombers broke through to score near misses on the Bunker Hill and Wasp but three of these were shot down by antiaircraft fire.

From 1130 to 1430 several raids by isolated planes or small attack groups were intercepted at distances from the force of two to sixty miles. These last hopeless attempts to reverse the fortunes of war failed through lack of co-ordination and strength, and our fighters splashed the Japanese planes with a determination increased by the day's success.

By 1400 the attacks on Task Force 58 had tapered off, and in the intervening lull our fighters paused to refuel and rearm. During the spectacular action over the fleet the situation at Guam had not remained static. With the alarm of the first raid on the fleet, all bombers and torpedo planes had been ordered into the air to eliminate the hazard of fire and explosion in the event of an enemy success against our carriers.

These planes, which had been armed and made ready for an attack on the Japanese Fleet, were diverted to Guam and Rota in an attempt to neutralize the airfields on those islands. This maneuver, while it contributed to the discomfort of the enemy, did not succeed in rendering the fields inoperational, and by late afternoon there were numerous reports of enemy aircraft in the Guam area.

As a result, fighter sweeps were vectored there which engaged the enemy on a large scale, and, by shooting down about seventy-five planes, completed the last stage of the air battle. Following the destruction of these planes at Guam, a few snoopers remained in the area of the task force until 2200, but the Japanese offensive was spent, and no further attacks were made.

The "Marianas Turkey Shoot" will be remembered as a climax in the history of naval aviation. No aerial defeat could have been more crushing for the enemy or more decisive in its relationship to our amphibious efforts. The 385 Japanese planes which were shot into the sea and the 17 others destroyed on the ground might well have turned the tide of battle on Saipan had our pilots been less skillful, our equipment less efficient, and our fighting spirit less determined.

Final statistics revealed that losses in air combat favored our Hellcats in a ratio of 13 to 1 when compared to enemy fighter plane losses, and 21 to 1 if other types are included.

If this demonstrated superiority in the air did not reveal to the Japanese that the war was lost, it proved to them that excellent strategy could not overcome the defects of equipment and pilot training. The only possible action left them now was retreat; and in the fact of their retreat lay the second challenge to the fast carrier task force.

Although the enemy inflicted only minor casualties on our force during the first day of the Battle of the Philippine Sea, our continued loss of position seriously handicapped any attempt at counterattack. In order to keep a constant fighter patrol in the air it was necessary to head the task force into an easterly wind, which consequently took our ships away from the Japanese carriers. By nightfall, Task Force 58 was only forty miles west of Rota — farther east than it had been in the morning and impotent to strike back.

With the conclusion of the day's action, therefore, the force immediately took a westerly course at twenty-three knots in an effort to close the enemy. For the purpose of covering the Marianas area, Task Group 58.4 (Essex, Cowpens, and Langley) was detached, which reduced the carrier strength of the main body to six CV's and six CVL's.

Search missions on the morning of 20 June were negative, and it was not until 1518 that contact was made with the retiring enemy fleet. By this time Task Force 58 had reached a position 370 miles west of Rota and was over 275 miles from the closest enemy group. It thus became evident that if the enemy was to be hit it would be necessary to launch a strike at extreme range and recover the planes in darkness.

At 1553 Admiral Mitscher advised Admiral Spruance that he intended to launch a deckload of 85 Hellcats, 77 Helldivers, and 54 Avengers. This flight was off at 1630 on a mission made hazardous not only by the defensive capabilities of the Japanese, but also by the lateness of the hour and the distance involved. All air groups made contact with the enemy fleet slightly before sunset after flights of from 300 to 330 miles.

The Japanese had disposed their fleet in three groups, a small group to the northeast, the main body in the center, and an oiler group to the southeast. It was an impressive force; returning airmen estimated that 6 carriers were present, screened by 4 battleships, 11 cruisers, and 22 destroyers. Despite the difficulties of coordinating an attack at that time and at that distance, the distribution of aircraft between the enemy carriers was well handled by the officers in tactical command of the various units. The action was tense and dramatic.

Through the varicolored bursts of the Japanese flak our planes hit the carrier targets hard, and although skillful and aggressive enemy fighters were in the air, they could not prevent the attack from being pressed home with vigor and determination. Two carriers were sunk and the rest damaged. A relatively small proportion of the total of bombs and torpedoes was expended on other types, but at least two destroyers and one oiler were sunk among the several hit. If the Japanese Fleet was not broken, it was removed as a threat to the Marianas operation.

This gallant strike cost us eighteen planes to antiaircraft fire and enemy interceptors but the battle against distance and the gathering darkness produced many more casualties. With the darkness came difficulties in navigation which forced our carriers to reveal their position with searchlights, star shells, and white truck lights in an effort to home the returning planes.

The scene was a nightmare of desperate efforts. Flight-deck crews worked feverishly to recover planes which were coming in on their last gallon of gas. Confusion caused by the darkness, and an eagerness to land caused some pilots to make approaches on cruisers and destroyers.

Others fouled the carrier decks with barrier crashes. And during the evening many more, lost or simply out of fuel, made water landings, raising the total of operational losses from all causes to 80. The subsequent rescue of 77 percent of the downed personnel did much to relieve the desperation of that night in the minds of those who were participants, but the event will be remembered by many as the most spectacular phase of the Battle of the Philippine Sea.

While this dramatic recovery of planes was taking place, the wounded Japanese Fleet was limping off toward the safer waters of the empire. A PBM located it the next morning, but by that time it was 350 miles from our force. Nevertheless, a decision to launch a strike was made and at 0550 on 21 June our planes took off in the hope of dealing the enemy a last, shattering blow. The mission, however, was unsuccessful; the flight proceeded to the limit of its combat range, and, having made no contact, jettisoned bombs and returned to base.

Searches during the remainder of the day were negative and at 1920 Admiral Spruance directed the retirement of Task Force 58 toward Saipan. At that point our carriers had reached a position only 545 miles from the Philippine island of Samar.

The Battle of the Philippine Sea was won, but it was a victory which brought criticism, as well as congratulations. The enemy fleet, for the first time since the naval decisions in the Solomon Islands, had come out to do battle and it had escaped destruction. Since the formation of the fast carrier task force, a fleet engagement had been one of the objects of our strategy and our hope. When the chance came it found our carriers committed to a defensive role by a proper hesitation to leave the amphibious operations at Saipan uncovered.

When the opportunity to strike back was grasped, the factors of time and distance prevented a decisive blow. The campaign in the Marianas was saved, but the Japanese Fleet escaped with its major strength, to fight again.

WITH THE REPULSE OF THE JAPANESE FLEET our air effort in the Marianas returned to the less spectacular tasks of patrols, searches, and field support missions. The only air threat that the enemy could muster came from his operational but battered airfields in the Caroline, Bonin, and Volcano Islands. Such a threat did not demand the attention of all the fast carrier groups, and a system of rotation consequently enabled our carriers to return to Eniwetok for rest and replenishment.

The first of the post-battle raids to involve more than routine operations was an attempted strike against the Volcano-Bonin group which the Japanese turned into an air battle of impressive proportions. At 0600 on 24 June, Task Group 58.1 (Hornet, Yorktown, and Bataan) launched a long-range fighter sweep against Iwo Jima which was intercepted by a large number of enemy fighters. In the resulting action our 48 Hellcats lost 4 of their number while destroying an estimated 68 Japanese fighters and bombers.

This defeat, however, did not discourage the enemy from attempting an attack against the task group with the remnants of his local air strength. This attempt proved equally disastrous: our interceptors shot down 46 more aircraft, to raise the day's total to 114. Its mission completed without bombing the airfield, Task Group 58.1 retired to Eniwetok without incident.

After rearming, Task Groups 58.1 and 58.2 rendezvoused on 1 July and set course for Iwo Jima on another raid to neutralize that flanking base. On 3 July a snooper reported the approaching carriers and forced the decision to launch a long-range sweep in an effort to obtain the advantage of some surprise. This stratagem was successful and the sweep of 63 fighters shot down 50 Zekes [Mitsubishi A6M; also called Zeroes] and destroyed an undetermined number on the ground. On 4 July a heavy flight schedule was carried out against Iwo, Chichi, and Haha islands which removed their immediate threat to the progress of operations in the Marianas.

These operations, after the defeat of the Japanese Fleet had permitted our aircraft to concentrate on supporting the troops, proceeded efficiently. On 9 July all organized resistance on Saipan ceased and the invasion of Guam, long postponed by the fanatic resistance on the former island, became the new object of our plans. Since the Japanese offered no organized air opposition, little could be said about the role of naval aircraft at Guam which distinguishes it from the similar role at Kwajalein and Eniwetok.

Air operations were devoted almost exclusively to an initial softening up of the landing beaches, followed by direct support of the combat teams. Fast carriers and escort carriers combined to punish the Japanese positions and exposed personnel with such severity that the landing on 21 July had all the characteristics of a well-rehearsed play.

Thereafter, Task Group 58.4 (Essex, Langley, and Princeton) and the Carrier Support Group (Sangamon, Suwannee, Chenango, Corregidor, and Kalinin Bay) remained to cover and assist the advance of our soldiers and Marines until the capture of the island on 10 August.

In a concurrent operation, landings were made on the island of Tinian to complete the amphibious phase of the campaign. Air support was furnished by Task Group 58.4, five escort carriers of Task Group 52.14, and land-based planes operating from the newly won fields on Saipan. Nine days after the landings on 24 July the island was declared secure. Naval aircraft, flying unopposed except for ground fire, supported the invasion troops with 1,775 sorties and 457 tons of bombs. The co-ordination of air, sea, and land forces, now made efficient by much practice, contributed to the success with which Tinian was neutralized, invaded, and captured.

During the assaults on Guam and Tinian the fast carrier task force roamed north and south, gathering photographic information and neutralizing those Japanese bases which might still send long-range bombers against our footholds in the Marianas. With a plan for the seizure of Yap, Palau, and Ulithi under consideration by the high command, it became necessary to reconnoiter the western Carolines again as well as to strike facilities which remained a threat in the central Pacific.

Vice Admiral Mitscher led three groups to this area and, beginning on 25 July, raided the islands for three days. This done, two groups of three carriers each steamed north for the third strike of the campaign against the Volcano-Bonin islands. Enemy air strength at Iwo, Chichi, and Haha had by this time come under the neutralizing power of land planes based on Saipan, and as a result of negligible air opposition, attacks were made on shipping targets with gratifying results. Our loss of 7 aircraft was greatly overbalanced by a Japanese loss of 32 ships sunk, others damaged, and 10 planes destroyed.

These raids guaranteed the security of Saipan, Tinian, and Guam. By the latter part of August, the defense of the islands could be turned over to aircraft operating from the newly commissioned fields, and the fast carriers and escort carriers were relieved of support duties in order to participate in the next westward movement.

This movement followed quickly and became a springboard into the Philippines. The success of the Palau-Ulithi operation can be attributed, in large measure, to the experience gained and the lessons learned in the air war for the Marianas.

22 – Operations in the Western Carolines

THE OPERATIONS AGAINST THE WESTERN CAROLINES — Peleliu, Anguar, Ngesebus [or Ngedbus], and Ulithi — covered the period of three months from August through October 1944. Every major command in the Pacific area was involved, and operations extended over the entire central and western Pacific. Nearly 800 vessels, 1,600 aircraft, and an estimated 250,000 navy, marine, and army personnel, exclusive of garrison forces, participated. Of the personnel, 202,000 were Navy, 28,400 were Marines and 19,000 were Army.

During the period from August 1942, to the completion of the western Carolines campaign, Allied forces in the Pacific Ocean had been operating from two directions at once. One of these campaigns, the South Pacific-Southwest Pacific, had moved steadily westward and northwestward. The other, the central Pacific campaign, had moved westward through the Gilberts and Marshalls.

In order to make an unbroken front between our two forces it was necessary to establish the Marianas-Palau line. Southwest Pacific forces concurrently were to establish a continuation of that line by the capture of Morotai. With this line firmly in our hands, further advances to the westward could be supported.

As early as May 1944, Admiral Nimitz directed Admiral Halsey to initiate planning for the seizure of Palau, and on the fifteenth of June, Halsey's headquarters at Noumea were closed, to be reopened at Pearl Harbor. In late July, ships which had participated in the Marianas campaign and new ships began to pass to the control of Admiral Halsey as Commander, Western Pacific Task Forces. In the meantime, the planning for the Marianas campaign was carefully scrutinized for any lessons of value which might be of use in Palau, and observers were sent to the operation itself. How many lessons had been learned is clearly evidenced when the Palau and Gilbert amphibious attacks are compared.

Although the two landings took place only ten months apart, the contrast between them is striking. In spite of thorough planning and the assembly of a powerful force, the attackers at Tarawa ran into considerable trouble. Conditions at Peleliu were, if anything, more difficult than at Tarawa. Despite this, however, fewer troops were killed at Peleliu than at Tarawa, and the issue was never in doubt for a moment.

The bombardment, minesweeping, beach preparation, air support, ship-to-shore movement, progress ashore, and supply of troops went almost entirely according to plan. Insofar as wise planning and painstaking preparation could make for success, the operation at Peleliu and Anguar showed gratifying improvement over the operations which preceded them.

In order to facilitate the completion of his mission, Admiral Halsey divided his forces into three principal parts: the first, the joint expeditionary force, under Vice Admiral T. S. Wilkinson, had as its main task the capture of the objective islands; the second, the Forward Area Force under Vice Admiral (later Admiral) J. H. Hoover had as its principal task support of Admiral Wilkinson's force and defense and development of captured islands; and third, the Covering Force, directly under Admiral Halsey, had tasks assigned which were bold in concept and require some elaboration.

The last named force consisted of all available fast battleships and carriers, plus escorts. Its job was "to utilize every opportunity which may be presented or created to destroy major portions of the enemy fleet." To create such an opportunity, the plan for use of the fast carrier force undertook a type of operation which had not been risked in the past, namely, a comparatively sustained attack on a substantial enemy deployed in depth over a system of many coordinated airfields.

Never had our force penetrated close to a large enemy land mass having up to 63 operational airfields dispersed over a wide area, upon which were based an estimated 650 planes which could be readily reinforced. Such an operation was deliberately undertaken in order to create an opportunity to engage the enemy fleet, and in order to gain and maintain control of the eastern sea approaches to the Philippine-Formosa-China coast areas.

Halsey's Covering Force, known as Task Force 30, contained various units of which the major was Vice Admiral Mitscher's Task Force 38. This was broken up into the usual four groups. Group 38.1, commanded by Vice Admiral J. S. McCain, contained the Wasp, Hornet, Cowpens, and Monterey. Group 38.2 under Rear Admiral G. F. Bogan, included the Bunker Hill, Intrepid, Cabot, and Independence. Group 38.3, headed by Rear Admiral (later Vice Admiral) F. C. Sherman, contained the Essex, Enterprise, Langley, and Princeton.

Group 38.4, commanded by Rear Admiral R. E. Davison, was constituted around the Lexington, Franklin, San Jacinto, and later the Belleau Wood. In addition to these large and medium carriers, eight CVE's, Barnes, Nassau, Nehenta Bay, Sargent Bay, Steamer

Bay, Sitkoh Bay, Rudyerd Bay, and Hoggatt Bay were in Halsey's force, the first seven being attached to Group 30.8, the Oiler and Transport Carrier Group, and the last to Group 30.7.

In the Joint Expeditionary Force, known as Task Force 31, were eleven CVE's, attached to Rear Admiral R. A. Ofstie's Task Group 31.2. These were the Marcus Island, Kadashan Bay, Savo Island, Ommaney Bay, Sargent Bay, Petrof Bay, Kalinin Bay, Saginaw Bay, Gambier Bay, Kitkun Bay, and White Plains.

The movement from the embarkation area in the Solomons to the objective was made in echelons and was uneventful. As far as is known, all task forces and task groups arrived undetected by the Japs, and, it is believed, gained complete strategical surprise. Task Groups 38.1, 38.2, and 38.3 sortied from Eniwetok on 29 August, and by 6 September were off Palau ready to attack aircraft installations and defenses. Meanwhile, in order to deceive, divert, and destroy the Japs, Group 38.4 struck Chichi Jima and Iwo Jima, 31 August to 2 September.

On the afternoon of the sixth the three task groups launched a preliminary fighter sweep against Palau. No airborne opposition was encountered, and many of the targets were found already badly damaged by the attacks of southwest Pacific B-24's which had been pounding the area since June. On the seventh and eighth, other air attacks were carried out. All told, during the three days 1,470 sorties were flown over Palau. Our losses were six planes to antiaircraft fire.

On 10 September, Group 38.4, which had struck Yap and Ulithi after its diversionary attacks on the Bonins and Volcano islands, took over the neutralization of Palau, and the other three groups of Task Force 38 withdrew to undertake operations against the Philippines.

Prior to the landings on Peleliu and Anguar the islands were given the usual air and surface bombardment. In addition to the carriers of Group 32.4 (Lexington, Franklin, Belleau Wood, and San Jacinto), four CVE's aided in this work. In these operations, and those which followed, Napalm fire bombs were used for the first time in quantity in carrier operations. For the support of the actual Peleliu landings, 15 September, there were available four CV's and four CVL's of Groups 38.2 and 38.4 and ten CVE's of Group 31.2. Landings went according to plan, and on D plus 1 Day the Peleliu airfield, one of our main objectives, was captured.

The fighting ashore, however, was extremely bitter, and it was not until the end of November that all Peleliu was in our hands. Air support after the landings was provided by the CVE's and Group

38.4 until the eighteenth of September, and thereafter the CVE's carried on alone until the end of the month. One of the notable features of the air support was the fact that this was the first occasion in which CVE's were employed throughout as a group and in such strength (ten carriers).

As a result of these operations, it was evident that escort carriers could completely handle the full support of a major operation once the beachhead was established. From 6 September to 1 October, 6,021 sorties were flown and only eighteen aircraft (0.3 percent) were lost to antiaircraft fire. There was almost complete lack of enemy air opposition.

The next move was the occupation of other islands in the group. Two days after the initial landings on Peleliu came the assault on Anguar, southernmost of the Palau group. No large concentration of the enemy was reported, and on the twentieth it was announced that island was secured. With the occupation of the northern portions of Peleliu completed by the twenty-seventh, it was decided to move on to the occupation of Ngesebus.

On the twenty-eighth this was accomplished by a shore-to-shore operation from Peleliu. At the same time, two other small islands were taken. Our casualties in the southern Palaus as of 14 October were 1,069 killed as compared with 11,586 of the enemy.

According to our original plans in the western Carolines, Ulithi Atoll (midway between Palau and Guam) and Yap were to be occupied in early October. As our campaign progressed, however, it was decided to abandon the attack on Yap and to undertake the Ulithi operation sooner than planned. Ulithi could be of real value to us since it possessed one of the best anchorages for large ships in the central western Pacific, a lagoon 19 miles long and 5 to 10 miles wide.

The forces involved in the occupation of Ulithi were under the direct command of Rear Admiral W. H. P. Blandy. Included in his task organization was an escort carrier unit, Task Unit 33.12.2, consisting of the Kitkun Bay, Gambier Bay, and White Plains and four destroyers. Preliminary minesweeping, reconnaissance, and fire support commenced on the twenty-first of September and continued for two days.

On the twenty-third, landings were made successively on Sorlen, Falalop, Osor, Mogmog, and Potengeras. No Japs were found, and the natives were friendly. The carrier force, during its support of the operation, maintained CAP and anti-submarine patrols. Since opposition was lacking, aircraft were for the most part in a stand-by status.

While the occupation of southern Palau and Ulithi was being undertaken, carrier strikes were directed at the Philippines in order to neutralize the many air bases there capable of hitting our landings in the western Carolines, as well as to obtain photographic coverage of enemy installations and of possible landing areas. Mindanao, roughly as large as the British Isles, was selected as the first target.

The basic plan was similar to prior operations. To effect surprise, the approach was to be made at night; to establish air superiority, the attack was to open with a dawn fighter sweep, and thereafter repeated attacks were to be launched at successive intervals while the task groups remained in the area. Group 38.3 (Essex, Enterprise, Langley, and Princeton) was assigned the central and northern airfields, while Davao and the southern fields, where the greatest opposition was expected, were assigned to Groups 38.1 (Wasp, Hornet, Cowpens, Monterey) and 38.2 (Bunker Hill, Intrepid, Cabot, Independence).

In accordance with the plan, a fighter sweep was sent out on the morning of the ninth of September. Everywhere opposition was negligible, and it became apparent that Mindanao was a sparse target for three carrier air groups. Hence, after sweeps on the tenth, the groups withdrew. Total Jap plane losses during the two days were 44 on the ground and 14 in the air. In addition, the enemy lost 83 ships destroyed, 80 others probably destroyed, and 34 damaged by our air and surface attacks. Our combat losses were 6 aircraft and 4 crews.

The most important feature of the carrier air attacks on Mindanao was the lack of opposition. The failure of the enemy to attack our carriers in force, the lack of air opposition at the targets, and the limited nature of the facilities at airfields indicated Jap weakness. The threat to our amphibious operations at Palau and Morotai from aircraft based at Mindanao, Palmas, or Talaud islands ceased to be important.

Upon retirement of the three task groups, Admiral Mitscher announced his decision to attack Leyte, Samar, Cebu, Negros, Panay, and Bohol, as well as airfields at Bulan on the southern tip of Luzon. Some two hundred Jap aircraft and substantial shipping were believed to be in this area.

After refueling and receiving replacement aircraft the groups set course for the initial launching point about forty to sixty miles east of the southern tip of Samar. No enemy attacks developed during the approach. At 0800 on the twelfth the fighter sweep took off. Over Leyte, Samar, and Bulan no airborne opposition was encountered, while on the islands fewer airstrips were found than had been anticipated.

The sweep over Cebu met with more success; about 38 Jap planes were destroyed in the air and 40 on the ground, at a cost of 2 planes to us. Subsequent attacks were made by our forces on the thirteenth and fourteenth. During all these attacks, 12 Sept. to 14 Sept., over 2,000 combat sorties were flown. We lost 22 planes, 12 pilots, and 14 crewmen, but the Japs lost 174 airborne planes and at least 200 on the ground, besides 71 vessels of various types sunk and 37 more probably sunk or damaged.

One result of these attacks, discovery of the Jap weakness, caused the abandonment of the plan to attack Yap and the recommendation by Admiral Halsey that the contemplated landings on Leyte and Samar be staged at an early date.

Following the second day's attack on the Visayans (central Philippines) Group 38.1 (Wasp, Hornet, Cowpens, Belleau Wood) was temporarily detached from Task Force 38 to furnish air support for the Morotai landings, on 15 September. While en route to Morotai, strikes were carried out against Mindanao.

One of these, against Zamboanga, meant a flight of three hundred miles to the target, yet the thirty-two F6F's making the strike completed the mission successfully. Very little enemy activity was discovered on any of these sweeps. Early on the fifteenth the group was about fifty miles south of Morotai. In accordance with plans, a fighter sweep was launched which destroyed twenty-eight grounded aircraft at Langoan without any loss to us. The landings at Morotai were unopposed, so 38.1 was relieved on the sixteenth. The next day Monterey reported to the group, relieving Belleau Wood, which reported to 38.4.

With the favorable termination of the Mindanao strikes, 9-10 September, preliminary plans were made to conduct a carrier air attacks on Luzon, the first one in the war. Luzon was a vital base and shipping center for the Japs, and hence formidable opposition was expected.

Following the operations in support of the landings at Morotai and Peleliu, and in preparation for the attack on Luzon, Task Groups 38.1, 38.2, and 38.3, on 19-20 September, effected a rendezvous and completed refueling and replenishment. By midnight of the twentieth, the three groups had reached a position somewhat north of the latitude of Manila, and three hundred miles east of it.

From that point, a high-speed run was made to the launching position, some seventy miles east of central Luzon. The approach was made in absolute radio silence under cover of a severe equatorial front. This front, although it screened our approach, held up the launching of planes by two hours and hindered their rendezvous when they were in the air.

At 0800 on the twenty-first, a fighter sweep of ninety-six F6F's was launched against Nichols and Clark fields. When the planes arrived over the fields, as many as fifty Jap planes were in the air. Once again, the score of planes destroyed was lopsided, with the Japs losing 66 in the air and others on the ground, while we lost only 6.

Throughout the day other strikes were launched at frequent intervals. The first strike by Group 38.1 arrived over Manila Bay at 0930 and found at least fifty worthwhile ship targets, but the Jap warships were not present. One notable feature of this strike was the effectiveness of our torpedo attacks, seven out of eight torpedoes dropped running hot, straight, and normal. By the end of the day the group had sunk 16 freighters, 1 large oiler, and 2 destroyers.

Group 38.2, during the same period, sank 1 medium freighter ninety miles north of Manila, 2 medium freighters and a large transport off the Luzon west coast, and a medium freighter in Subic Bay. Group 38.3 sank 5 merchant ships and a destroyer. Numerous other craft were damaged. The toll of Jap vessels alone reached proportions which well justified the risk taken in approaching the strong enemy base.

On the twenty-second, a fighter sweep and two strikes were all that were launched as there was indication of the approach of a typhoon from the north. After withdrawal, fueling was completed on the twenty-third and the three groups proceeded to their next assignment.

Information had been received by Vice Admiral Mitscher indicating that some of the Jap shipping which had been withdrawn from the Luzon area had proceeded south to Coron Bay in the Calamian group, north of Palawan. In view of the targets presented, Admiral Mitscher directed that strikes be made on shipping in Coron Bay on the twenty-fourth, and that three additional strikes be made on other Visayans targets. The strike against Coron Bay was launched at 0600 on the twenty-fourth from a point fifty miles northwest of Samar Island.

From the launching point to Coron Bay was a distance of nearly 350 miles, yet no one was lost operationally during the flight. Following the Coron Bay strike, others were launched against suspected troop concentrations on Leyte, against aircraft and ground installation on Cebu, Negros, Masbate, and Panay, and against numerous

Jap vessels. There was almost complete lack of enemy airborne op-position, clear evidence of the effectiveness of prior strikes. The most important result of the strike was the sinking of 16 large ships and 10 smaller craft and the damaging of numerous others. Following the second Visayans attack, the three groups retired, 38.2 to Saipan, 38.3 to Palau, and 38.1 to Manus.

Damage inflicted on the Japs by the groups (9 September to 24 September) was estimated at 880 planes destroyed and 241 vessels of various types sunk. We lost 54 aircraft in combat and 59 pilots and air crewmen. In addition to the extensive damage inflicted on the enemy, an interesting feature of the operation was the use, for the first time with the fast carrier task force, of a fully equipped night carrier, the Independence. Its complement of planes was 16 night fighters and 9 night torpedo planes.

23 – FROM NEW BRITAIN TO MOROTAI

THE STARTLING DEVELOPMENTS ACROSS THE central Pacific have tended to overshadow a parallel movement across the southwest Pacific. While the Marshalls were being taken, the Marianas conquered, and the Carolines won, the armed forces of the Allies were beating their way from the Solomons up through New Guinea until, by the time the moment was ripe for the retaking of the Philippines, they had advanced to control Morotai, standing between the Philippines and the Netherlands East Indies.

In this movement naval aviation played an important assisting role to the advancing army air and ground forces, which carried the brunt of the burden. In general, this assistance followed three lines of activity. In the first place, as in the early stages of the war, there were raids of varying intensity upon the Japanese in the region.

By 15 February 1943, the southern Solomons campaign had been brought to a virtual close by our landings on Green Island. From this time on, Rabaul-bound convoys were the main targets of army bombs and Fairwing 17's "Black Cats." The latter, during 1943 and early 1944, greatly improved their technique of attack, and as a result came to be called upon more and more for special missions against Japanese convoys, rather than for routine nightly marauding patrols.

One such special mission took place on the night of 15 January 1944, when six PBY's were sent out to intercept a convoy en route from Truk to Rabaul before it could reach the safety of a weather front off New Ireland. A special communications plan was arranged, and contact was made about 0130, but it took some time to get all the planes together as the work had to be done largely by correlating radar "blips" with an occasional sight contact.

The enemy was traveling in two separate convoys about eight miles apart, but during the attack became so confused that all ships were milling about together, and the antiaircraft fire set up, although heavy, was so dispersed that it was ineffectual.

On the first run-in, one of our planes scored two hits on what turned out to be a 10,000-ton oiler that flared like a torch. The light from this doomed craft made it easier to make attacks on the rest of the convoy, and two merchantmen were hit and left dead in the water.

Not only was Japanese shipping struck, but enemy strongholds were also the subject of attack. Principal targets were Rabaul and Kavieng. Big navy PB4Y's (Liberators) of VB-106 patrolled the area, and in February 1944, in addition to their regular duties, were especially effective as "spotters" for destroyer attacks. Carrier groups also entered the picture attacking both Kavieng and Rabaul as well as raiding shipping bound for these ports.

From early in 1944 to the surrender of the Japanese in 1945, land-based navy and marine TBF's and SBD's, accompanied by F4U's and F6F's, gave Rabaul and Kavieng a regular going over, from raids two or three times a week to daily attacks.

These operations contributed heavily to keeping down any Japanese attempts to make effective use of these bases as our forces swept along the New Guinea coast and into the Admiralties.

An important aspect of naval air activity in the southwest Pacific was the extension of bases to keep pace with the advance of Allied ground forces. During the latter part of 1943 and the early months of 1944, the outstanding New Guinea seaplane base was Samarai, on Milne Bay. The tide of conquest, however, opened new bases along the northern New Guinea coast and on near-by islands.

The most important development of base facilities in the southwest Pacific began in March 1944. This was the construction of a major naval repair and supply base on Manus, in the Admiralty Islands. This island has one of the finest harbors of the entire area and commands the Bismarck Sea and the northeast coast of New Guinea. It was found to be lightly defended and was quickly occupied. PBY's of VP-34 assisted in the first landings by slipping in a reconnaissance party and picking it up on the following morning from a rubber boat. On 25 March, only ten days after the invasion of Manus, ComAir Seventh Fleet transferred his headquarters and Fairwing 17's operations into the forward area, and operated under the Thirteenth Air Force.

Along the northern New Guinea coast, Japanese shipping could not make the short runs from island to island at night which had been possible earlier. The ever-increasing Army Air Force daytime sweeps were thus reducing the "Black Cats" to searches for slim pickings. Furthermore, by 12 April, long-range PB4Y Liberators of VB-106 were in the area. In their long flights to spy out enemy activity, they even reached the Japanese-dominated Philippines. The fact that the "Black Cats" were being curtailed in their activities to special missions, did not mean that the Catalina was no longer useful. As night operations declined, air-sea rescue became more important.

One remarkable rescue was made at Kavieng on 15 February. The pilot of a PBY received a call that a B-25 was down in the water. Securing fighter cover, the pilot went to the area and found not one, but several planes downed in the harbor. Undeterred by the heavy swell, the navy pilot made four separate landings and effected the rescue of the crews of five B-25's. The last take-off was with twenty-four persons aboard.

In addition to attacking enemy shipping and bases in the southwest Pacific, naval aviation contributed to the effort by carrying out supporting operations in the Netherlands East Indies. During 1943 and well into 1944, PBY's of PatWing 10, as well as other planes of the Allies made raids from western Australia, doing considerable damage and keeping the Japanese invaders on the defensive.

Prior to 1944 there had been little in the way of offensive operations by the Allies in the Indian Ocean area, and, indeed, there had been a good deal of criticism of the lack of such operations by the British sea forces available in the Indian Ocean. It will be pointed out, however, that the British at this time had only one carrier, the Illustrious, with a complement of forty planes, available for fleet operations. The other carriers in the region, including CVE's of the U.S. Navy at various times, were primarily assigned to the protection of convoys carrying land-based army planes and supplies to the India-Burma theater.

In March 1944, however, there was a new development. The carrier Saratoga, with two destroyers in attendance, left our naval base at Majuro in the Gilberts on the fourth, with orders to join the British Eastern Fleet. Stopping first at ports in Australia, the Saratoga continued on its way and made rendezvous with the British on 27 March about a thousand miles south of Ceylon, and participated in fleet exercises for almost a month preparatory to making diversionary strikes at the Netherlands East Indies at the same time that the Hollandia landings were to be made.

The first of the strikes was made on 19 April against Sabang, a Japanese base at the northern tip of Sumatra. A feint, in the form of a preliminary raid on the Andaman Islands had thrown the Japanese off guard, and the main raid was a marked success. Saratoga fliers sank a transport and left another smoking, in addition to strafing and causing considerable damage to other shipping and shore installations, including near-by airfields and parked aircraft. The lone American pilot shot down was picked up by a British submarine which maneuvered under the point-blank range of a coastal gun, while twelve fighter pilots ruthlessly strafed the decks of a Japanese destroyer that attempted to interfere in the rescue operations.

The second strike was made at Soerabaja, in Java (which had become the most important Japanese base in the Netherlands East Indies) and did considerable damage. This strike was made to co-ordinate with a southwest Pacific command advance in New Guinea, where landings were being made on Wadke. The Allies were fighting a coordinated war; the left hand knew what the right hand was doing.

In addition to harassing raids on enemy bases and shipping in the southwest Pacific and attacks in the Netherlands East Indies, naval aviation in April 1944, returned to its familiar task of giving air support to amphibious operations. This time the target was the north coast of New Guinea, and landings were to be made at both Hollandia and Aitape. The amphibious force assembled for this operation was the largest group of the sort yet brought together in this area. The landing force was supported by strong elements of the United States and Australian fleets, including five task groups of aircraft carriers, with a total of nineteen flattops.

The principal prizes in the Hollandia area were the network of three airfields at the base of a rugged chain of mountains, the former Dutch seaplane base on Lake Sentani, and the best anchorage in that part of New Guinea. Our forces, feinting at Palau before turning abruptly south toward Hollandia at night, met hardly any opposition either on land or in the air. A single enemy plane attacked one of the Humboldt Bay beachheads on the night of 23 April, and on the following night one of our destroyers was attacked by twelve torpedo planes without damage to the vessel.

In addition to the work done by the Fifth Air Force in the pre-invasion softening of the area, it was estimated that navy carrier planes of Task Force 58 destroyed 132 aircraft — 16 as they attempted to take off in the target area, 13 downed by combat air patrols over a ten-day period, and the remainder destroyed on the ground. From 21 to 24 April, the task force planes flew over 3,000 sorties in support of the Hollandia landings. Nearly 750 tons of bombs were dropped, mainly in advance of troop landings, and rockets were used by TBF planes from the Hornet and the Bunker Hill.

In addition to demonstrating the potent effect of carrier air support of amphibious operations, in this case far beyond what turned out to be actually needed, the Hollandia-Aitape campaign gave an excellent dress rehearsal for the composite operations of many ships and air components all working together to make one large and intricately coordinated attack. Such tremendous undertakings were not long in coming in another area of the Pacific war; within six weeks we were to invade the Marianas.

In comparison with the Marianas campaign and other operations in the central Pacific, naval aviation's support of the continuing amphibious operations in the southwest Pacific was relatively minor. An exception was the conquest of Morotai, the last steppingstone to the Philippines. Strategically located, Morotai in our hands would place the Philippines, Borneo, and the Netherlands East Indies within range of our growing air power. The island, furthermore, straddled the Japanese supply lanes, and in our hands could be used to cut off vast territories under enemy control, not the least among which were the oil fields of Balikpapan and Tarakan, on Borneo.

The conquest of Morotai was, as it turned out, merely a part of a larger campaign aimed at placing our forces in a position to advance on the Philippines. Task Force 38 was assigned the role of giving air coverage for the landings on Morotai, which were affected in September 1944. As in the case of the Hollandia invasion, the air support was far more than was needed. The landings on Morotai were virtually unopposed, and as a result our full carrier air strength was never employed.

Nevertheless, the mere presence of such a potent force affected the Japanese reaction in the area, impelling them to withdraw their forces, limit their supplies, hide their shipping, and finally give up a strategic position without even fighting for it.

24 – WHIRLWIND IN THE PHILIPPINES

IN DECEMBER 1941, the Japanese sowed the wind in the Philippines. Before it went the stubbornly resisting but far outmatched forces of the Americans and their allies. Nearly three years later, the Japanese forces in this area reaped the whirlwind. And what a whirlwind it was!

The years between the attack on the seaplane tender William B. Preston on 8 December 1941 and the opening raids of our carrier task force on 10 October 1944 had seen United States naval aviation develop into the greatest naval air striking force in existence. Our materials of war had been wrought on an ever-increasing scale by the willing hands of a free people. Our pilots and other aviation personnel had been trained under improved techniques on sprawling fields throughout the United States and had been tested on the field of battle.

Our carrier task forces had been thoroughly tried out under a variety of combat conditions and were well prepared to meet the novel and difficult situations that were to confront them in the Philippines. Our maintenance and supply units had developed a performance that was commensurate with that of their combat brethren.

The successful occupation of Peleliu in the Palau Islands and Morotai in the Halmaheras opened the way to a deeper thrust across enemy lines of communication — the reoccupation of the Philippines. Here was an opportunity not only to recapture territory formerly in our possession, but to gain strategic control of Japanese supply lines in the South China Sea, and secure harbors and depots needed for the concentration of men and supplies for operations against the home islands of Japan.

Support of the Philippines operation required that the personnel of Task Force 38 extend themselves to the limit. No other period of the war in the Pacific up to this time had included as much intensive air action. No previous operation involved quite the same problem. Past operations had been against relatively small islands which were easily isolated.

Once the fast carrier task force had gained control of the air over the island by the simple process of destroying its defensive air power, the force needed only to intercept reinforcements which could come in either under their own power, which was seldom possible because of the distances involved, or by a carrier force which the enemy seldom dared to risk.

In the Philippines, however, this situation was changed. The land mass, with its widely dispersed airfields and great size, presented a difficult problem for the carrier task forces. To maintain control over even a part of the area, they would have had to remain relatively fixed and to pick off reinforcements as they came in.

In view of the long campaign that loomed ahead, it appeared necessary to slow down the rate at which the Japanese Air Force could bring up its reserves by crippling the staging bases in the Ryukyus and Formosa. That control of the air ultimately gained was due to the distances at which the carrier force could strike against reinforcing and staging bases without leaving the landing operation unprotected.

During this period, the task force roamed almost at will, successfully overcoming all opposition by the Japanese Air Force and making considerable progress toward its ultimate destruction. Attacks were made on every operational airfield within range of its planes from Amami O Shima in the Nansei Shoto to Leyte Gulf in the Philippines and through the South China Sea to Saigon.

The campaign for the recapture of the Philippines can be divided into the following phases:

1. Preparatory phase — raids on Nansei Shoto and Formosa 10-15 October.
2. Support of Leyte Landings 18-23 October.
3. Battle for Leyte Gulf 24-26 October.
4. Support of occupation of Leyte 29 October-20 December.
5. Support of Mindoro landings 13-21 December.
6. Support of occupation of central Luzon 30 December 1944-19 January 1945.
7. Completion of occupation of Philippines 17 January-15 August.

Strikes on the central Philippines, carried out in support of the occupation of Peleliu and Morotai islands, revealed the weakness of the Japanese defenses in that area. The rescue of an American pilot who had been with friendly guerrillas on Leyte, revealed the small number of Japanese troops on the island. This information, forwarded to higher echelons of command by Commander, Third Fleet, with a recommendation for the cancellation of the planned operations in the western Carolines in favor of the immediate seizure of Leyte, resulted in a change of plans by the Joint Chiefs of Staff.

With the date for the landing on Leyte set for 20 October, and with the nearest Allied air base nearly five hundred miles distant, the task of establishing initial control of the air over Leyte fell to the fast carrier task force. The establishment of this control required the destruction of large numbers of enemy aircraft in the Philippines and the neutralization of surrounding bases through which reinforcing aircraft from the empire would pass. It was, therefore, necessary that the task force penetrate, for the first time, the enemy's inner defense ring.

The opening raids of the first phase of the Philippines operations were made by Task Force 38 on 10 October 1944. The fast carrier task force, operating together as a unit for the first time since the Marianas campaign in June, was divided into four task groups and comprised a total strength of 9 carriers, 8 light carriers, 6 battleships, 4 heavy and 7 light cruisers, 3 antiaircraft cruisers, and 57 destroyers. It was commanded, during October, by Vice Admiral M. A. Mitscher, Commander, First Carrier Task Force, Pacific, and operated with the Third Fleet under the command of Admiral W. F. Halsey.

The first objective was the Nansei Shoto — a six hundred-mile chain of islands which extends from the southern tip of Japan to Formosa and forms the inner defense ring composed of strongly protected naval bases and numerous air bases through which reinforcements for the Philippines could be staged. The main effort of the force on the first day was directed against Okinawa, but small attacks were also made against Amami Shima, Daito Jima, Kume Shima, Kerama Retto and Miyako Jima.

The approach of the task force was undetected due to the use made of covering weather conditions and the destruction of picket boats and snoopers by supporting land-based planes from Saipan. The first attack caught the Japanese planes on the ground. Aircraft, airfields, shipping, and ground installations were covered thoroughly. The scale of the attack was one of the heaviest ever delivered by the fast carrier task force in a single day. Strike sorties totaled 1,356, tons of bombs dropped amounted to 541, and 652 rockets and 21 torpedoes were expended. The effectiveness of Okinawa as a staging base was interrupted, at least temporarily.

On 10 October 1944, while some Third Fleet carrier planes attacked airfields over the 600-mile long Ryukyus, others destroyed harbor installations and burned the heart out of this provincial capital.

While the task force fueled next day, a fighter sweep was sent against Aparri on northern Luzon, and on the night of 11 October, under cover of darkness, the entire force began a high-speed approach on Formosa, stronghold of Japanese air power.

In spite of the fact that the enemy was by now aware of the presence of the task force in the area, there were no attacks until after each of the task groups had launched dawn fighter sweeps. These sweeps met strong opposition over the target. The opposition continued to be strong on the strikes that followed the sweep, but fell off considerably during the day.

At dusk and after dark, sporadic air attacks were launched against the task force, but night fighters and antiaircraft fire were adequate protection, eleven of the attacking planes being shot down. No damage was inflicted on the ships of the force.

The neutralization of air bases was carried out on the twelfth and thirteenth by bombing attacks on hangars, fuel dumps, shops, and other major servicing facilities. Over half the 772 tons of bombs dropped during the period were expended on these targets alone. Weather conditions hampered attacks on shipping, but several profitable targets were found. A total of 26 ships, 8 of which were large or medium transports, were sunk, 14 probably sunk, and 41 damaged. No important naval units were found.

Enemy air attacks during the period 12-16 October, while the task force was operating off Formosa, were the strongest encountered since the Marianas. During this period nearly one thousand airborne enemy planes were engaged, 43 percent of which were encountered en route to, or near, the task force. More enemy planes actually reached our surface forces than ever before, and enemy raiding tactics were good.

At dusk on the thirteenth, the cruiser Canberra was torpedoed, on the afternoon of the fourteenth the carrier Hancock was slightly damaged by a near miss, two hours later the cruiser Reno was hit by a damaged plane, and at dusk on the same day the cruiser Houston was torpedoed. Both of the successful torpedo attacks were made by the fast new Japanese torpedo bomber "Frances."

Salvage Group 30.3 was formed on the fourteenth to escort the damaged Canberra to safety. The group, made up of ships from Task Groups 38.1 and 38.3, included a towing unit composed of 2 heavy and 3 light cruisers, 8 destroyers, and a covering unit composed of 2 light carriers, 1 heavy and 1 light cruiser, and 5 destroyers.

On the fourteenth, in the last action against Formosa, a dawn fighter sweep was made against the airfields, after which the force retired toward the Philippines. Task Groups 38.2 and 38.3 fueled, while 38.4 launched strikes to neutralize the airfields on northern Luzon. Task Group 38.1 protected the cripples. In spite of the severe pounding which the task force had given Formosa on the first three days, the scale of enemy effort on the fifteenth was greater than ever, suggesting that reinforcements had been flown in from surrounding bases. On this day the Hornet was damaged.

The action off Formosa was the cause for much celebration by the Japanese. The Japanese radio broadcast the news that the greater part of our aircraft carriers had been sunk, and that a large number of escorting ships had either both sunk or damaged. It was also reported that orders had been issued to the Japanese Fleet to attack and annihilate the remnants of our force.

Contrary to these jubilant Jap announcements, the facts were that, except for the two cruisers which had been torpedoed, there was little damage to the ships of our force, and the score of the air battle was 658 enemy planes destroyed against our combat loss of 63. In addition, considerable damage had been inflicted on airfields and ground installations, and large numbers of transports and small coastal vessels had been sunk and damaged.

Strikes against Formosa were strategically important in that they proved that a fast carrier force could approach the strongest enemy air base outside Japan itself at a time when the enemy was aware of the approach and could then not only protect itself against the greatest aerial opposition which the enemy could mount, but could deliver damaging attacks against shipping and ground installations.

In this respect, Formosa was a stunning defeat for the Japanese and indicated that the enemy's strategic plan for the control of areas south of Formosa was in jeopardy.

The next three days were spent in retiring from the Formosa area and in escorting the damaged cruisers toward Ulithi. In view of the Japanese claims of damage to the task force, Groups 38.2 and 38.3 operated within range but off to the east of the crippled division (Task Group 30.3), in an attempt to lure the Japanese into an attack on the "remnants of the force." In effect, the crippled ships and their escorts were offered as "bait" to the Japanese Fleet. A heavy attack by land-based planes was made on this group on the sixteenth at about 1400 in the area northeast of Luzon. Fighters from the Cabot and Cowpens, the two light carriers operating with the crippled cruisers, destroyed approximately fifty of the sixty attacking planes. The attackers were successful, however, in putting another torpedo in the already crippled Houston. This second hit caused her to list badly, but despite some difficulty in towing, she continued to make progress toward Ulithi.

On the sixteenth, a search plane made contact with an enemy surface force of cruisers and destroyers but due to communication difficulties and the coming of darkness, the enemy escaped. On the eighteenth, attacks were launched on central and southern Luzon by Task Groups 38.1 and 38.4, and on Aparri and Laog in northern Luzon by Task Group 38.2. Task Group 38.3 fueled within supporting distance of 30.3. On the nineteenth, additional strikes were launched on the airfields of Luzon and 30.3 passed beyond the range of enemy land-based aircraft and proceeded to Ulithi.

The Third and Seventh amphibious forces, covered by the surface forces and carrier aircraft of the Seventh Fleet, began initial landing operations on 17 October with the occupation of the islands controlling the eastern entrance to Leyte Gulf. Air support for the Seventh Fleet was provided by the escort carriers under the command of Rear Admiral T. L. Sprague. These, organized in three units under Task Group 77.4 were made up of a total of 16 escort carriers, screened by 9 destroyers and 11 destroyer escorts.

On the morning of the twentieth, landings were made on Leyte under cover of a heavy fire from the surface ships and air bombardment of the CVE's and Task Groups 38.1 and 38.4 of the fast carrier task force. Air opposition was negligible. The preliminary strikes by Task Force 38 against airfields on Formosa and the central Philippines, aided by the destruction of some seventy-six enemy aircraft by CVE planes, was effective in reducing enemy air opposition so that we enjoyed complete but temporary air supremacy.

Task groups of Task Force 38 alternated in supporting ground operations on Leyte and in searching for enemy fleet units. On the twenty-first, Task Groups 38.2 and 38.3 launched strikes against southern Luzon and the Visayans. Strikes on shipping, particularly at Manila, produced good results. Twenty-two ships were sunk, including 2 large and 5 medium transports, and 2 oilers. An additional 51 ships were reported as probably sunk.

On 23 October, Task Group 38.1 and the carrier Hancock from Task Group 38.2 retired to Ulithi, and the Bunker Hill from Task Group 38.2 departed for Manus to pick up a replacement air group.

All indications pointed to the fact that the Japanese Navy would take vigorous action in opposition to the serious strategic threat to its position brought about by our landings at Leyte. In addition, the Japanese in their broadcasts had minimized their defeats by boasting of the complete destruction awaiting our forces when they had been lured to the westward, and their Navy could not afford to wait. There was also some evidence that they were influenced by their own propaganda and believed that their air attacks at Formosa had been successful in destroying a large part of our Third Fleet. Our position at Leyte favored the Japanese strategically. Our lines of communication were stretched a tremendous distance while theirs had been appreciably shortened.

The reaction of the Japanese Fleet resulted in a series of air and surface actions for the control of Leyte Gulf which developed in the following separate stages: (1) location of enemy fleet units, (2) air attacks on enemy forces in the Sibuyan Sea, (3) the Battle of Surigao Strait, (4) the battle off Samar, and (5) the battle off Cape Engano.

On the basis of submarine reports, Commander, Third Fleet, disposed his forces so that information regarding future movements of the enemy fleet would be learned at the earliest possible moment. Task Group 38.3, with CTF 38, proceeded to the vicinity of Polillo Island off the southern end of Luzon, Task Group 38.2, with ComThirdFleet, operated off Surigao Strait, and Task Group 38.4 was situated about sixty miles off the southern coast of Samar. Each group launched reinforced search teams at dawn which fanned out to the westward and covered the entire westerly side of the Philippines from northern Luzon to southern Mindanao and its western sea approaches.

At about 0822, CTF 38 received an emergency contact report from a Task Group 38.2 search plane, that an enemy force of 4 battleships, 8 cruisers and 13 destroyers, without transports or carriers, had been sighted in the Sibuyan Sea off the southern tip of Mindoro, on course 050, speed 15 knots. An amplifying report received approximately one hour later, stated that the enemy force was in two groups, the first consisting of 2 battleships, 4 heavy cruisers, and 7 destroyers, and the second, five miles astern, of 2 battleships, 4 heavy and 1 light cruiser and 6 destroyers. Both groups of this force, designated as the Central Force, were reported to be steaming up the east coast of Mindoro Island.

This represented a major portion of the Japanese Fleet strength and constituted a serious threat to the success of our landings on Leyte, which it could approach through any one of the numerous inter-island channels in the central Philippines. Commander, Third Fleet, therefore, directed the immediate concentration of all task groups around Task Group 38.2 off Surigao Strait and recalled Task Group 38.1, then on its way to Ulithi. All task groups were directed to launch strikes on the Central Force as soon as possible.

At about 0905, a search-strike group from Task Group 38.4, the most southerly of the groups, sighted an enemy force of 2 battleships, believed to be the Fuso and the Yamashiro, 1 heavy cruiser and 4 destroyers on a northwesterly course to the southwest of Negros Island. This force, designated in action reports as the Southern Force, was about 215 miles west of Surigao Strait and was also in a position to reach Leyte Gulf during the night of 24-25 October. The search-strike group attacked immediately and reported at least three 500-pound bomb, plus several rocket, hits on each of the battleships and rocket hits on the cruiser and destroyers.

All ships were heavily strafed. Before additional strikes could be launched against this force, Task Group 38.4, under orders from ComThirdFleet, moved north and passed out of range.

Task Group 38.3, operating to the north, made immediate preparations to launch a strike against the Central Force but before it could do so three large raids of forty to sixty enemy aircraft each were reported approaching the group. Additional fighter planes were launched, and the task group maneuvered to keep under cover of rain squalls as much as possible.

The largest raid was the third — a group of fifty or sixty planes about equally divided between dive bombers, torpedo planes and fighters. First intercepting fighters to reach this group were led by Commander David McCampbell, Essex air group Commander, who accounted for nine planes on his own, while his wing man got six. Approximately 150 enemy planes were shot down during the attacks.

As a result of the efficient performance of the fighters and the use made of the cover of rain squalls, no organized formation of enemy planes was successful in reaching the ships. A few single planes did break through, however, and one of these was successful in placing a bomb in the Princeton. The bomb started large fires and heavy explosions on the hangar deck caused the ship to drop astern.

Later, a tremendous explosion, undoubtedly in the after bomb and warhead magazine, blew the stern off the ship. The light cruiser Birmingham, alongside at the time, suffered considerable damage and very heavy personnel casualties. Damage to the Princeton was such that it was necessary to destroy her.

The large number of carrier-type planes present in the raids on Task Group 38.3 led to the conclusion that a carrier force was in the vicinity, and a search was launched at 1230 to the north and northeast, the direction in which it was supposed the carrier force might be. At 1540, a search plane reported contact with an enemy force of

3 battleships, 4 to 6 heavy cruisers, and 6 destroyers in position 18-10N, 123-30E, on course 210 at speed 15. A second contact reported one hour later, identified the force as containing 2 carriers, 1 light carrier, 3 light cruisers, and 3 destroyers on course 270, speed 15, in position 18-25N, 125-28E.

Other reports established the total of 17 ships in the enemy force, consisting of 2 battleships of the Ise Class (with flight decks), 1 Zuikaku carrier class career, 3 light carriers, 5 light cruisers, and 6 destroyers. Distance to the enemy and the fact that a strike group launched on the Central Force had not yet returned, made it impossible to strike this carrier force before dark. The major burden for the attack of the Central Force fell on Task Group 38.2 which was not only nearest to the enemy but was unhampered by other attack missions or by major air attack. Three strikes were launched by this group, totaling 146 planes which delivered attacks at 1026, 1245, and 1550. One attack was carried out by planes of Task Group 38.4 and two were made by Task Group 38.3.

Returning pilots reported 2 light cruisers, 1 seaplane carrier and 1 destroyer sunk, 1 heavy cruiser and 1 destroyer probably sunk, and 5 battleships, 5 heavy and light cruisers, and 8 to 12 destroyers damaged. Of the larger ships, one was reported as being afire and down at the bow, and another appeared badly damaged. (It was later established that the heavy cruiser Chokai was sunk in this action, and the battles Musashi damaged so severely that she sank later in the day off the southern tip of Mindoro Island.) At the end of the attack the enemy force was milling around aimlessly and when last seen had reversed its direction and was on a westerly course.

The general situation was now as follows:

Three enemy forces had been discovered moving in the direction of Leyte Gulf at a speed which might result in a coordinated dawn attack on Leyte on the twenty-fifth. The combined strength of the three forces was almost equal to the assumed total strength of the Japanese Fleet and constituted a serious threat to our operations at Leyte. Surface forces were apparently moving up from the south and west through the Philippines, while a carrier task force was moving from the north on the eastern side of the Philippines.

Both surface forces had been attacked, the Central Force much more heavily than the Southern, but the extent of actual damage and its effect on further movements of the forces was not actually known. Opposed to the enemy and in a position between the approaching enemy forces, we had the Seventh Fleet with its old battleships and escort carriers, and the Third Fleet composed of the fast battleships and carriers of Task Force 38. Except in heavy cruisers, we had an overwhelming superiority over the Japanese.

It should be noted that there was no single unit of command over all the forces for the operation in the Philippines short of the Joint Chiefs of Staff in Washington. The Seventh Fleet, commanded by Vice Admiral Kinkaid, operated under General MacArthur, Supreme Allied Commander, Southwest Pacific area, with the following specific responsibilities assigned in the Operation Plan:

This force will, by a ship to shore amphibious operation, transport, protect, land, and support elements of the Sixth Army in order to assist in the seizure, occupation, and development of the Leyte Area. The Third Fleet under Admiral Halsey was operating under Admiral Nimitz, Commander in Chief, U. S. Pacific Fleet and Pacific Ocean areas. Its responsibilities, as shown in the Operation Plan were:

. . . Furnish necessary Fleet support to operations by forces of the Southwest Pacific. . . . Forces of Pacific Ocean Areas will cover and support forces of Southwest Pacific. . . . Western Pacific Task Forces (Third Fleet) will destroy enemy naval and air forces in or threatening the Philippines Area, and protect the air and sea communications along the Central Pacific Axis. In case opportunity for destruction of major portions of the enemy fleet offers or can be created, such destruction becomes the primary task. . . . Necessary measures for detailed coordination of operations between the Western Pacific Task Forces and forces of the Southwest Pacific will be arranged by their respective Commanders.

The Third Fleet was therefore operating in the Philippines area for the purpose of destroying enemy air and naval forces, and was available to the Supreme Allied Commander, Southwest Pacific, for direct support of the occupation upon request. While so operating it was directly responsible to, and under orders of, Commander in Chief, Pacific Fleet. This lack of a single unit of command may or may not have been responsible for further developments resulting in the initial success of the Japanese forces in transiting San Bernardino Strait. It is the first operation in the Pacific war to be carried out without a single unit of command, and in that respect is worthy of note.

Commander, Seventh Fleet, Vice Admiral T. C. Kinkaid, assumed that it was his responsibility to protect the forces at Leyte against the Southern Force approaching through Surigao Strait and that the Third Fleet was providing similar service at San Bernardino Strait. The surface forces under his command, 6 old battleships, 4 heavy and 4 light cruisers, 25 destroyers, and the PT boats of Task Group 70.1, were more than adequate to deal with the Southern Force. These ships took position in the lower Leyte Gulf across Surigao Strait to await the approach of the enemy. Task Group 77.4, the escort carriers, remained under way in the open sea to the eastward of Leyte Gulf.

Our battle forces, under the command of Rear Admiral (later Vice Admiral) J. B. Oldendorf, Commander, Bombardment and Support Groups, were deployed so that the battleships and cruisers were in line across Surigao Strait, with destroyers and PT boats stationed on patrol in the Sunrise. The Southern Force, approaching through the strait, was divided into two groups. The landing, or attack group was composed of the battleships Fuso and Yamashiro and the heavy cruiser Mogami in column, screened by four destroyers. The transport group, following approximately four miles behind, was composed of four or five cruisers and six destroyers.

Our forces awaited the approach of the enemy and when the range closed, cut them to ribbons. The first attack was made at 0259 by the PT boats and destroyers in the Strait, and the battleships and cruisers opened fire at 0350. When the cease fire order was given at 0410, 2 battleships and 3 destroyers had been sunk and the cruiser and 1 destroyer, though damaged, had escaped. The trailing group turned away from the trap into which they were moving and evidently escaped undamaged. Thus, in a very brief but intensive action, our battleships, which had been "sunk" at Pearl Harbor, removed the threat of the Southern Force without damage to themselves.

At 2025 on the night of the twenty-fourth, Admiral Halsey directed the concentration of the fast carriers in preparation for an attack on the Japanese force reported to be approaching from the north. Search planes, which were watching the Central Force in the Sibuyan Sea, were called in and the concentration of Task Groups 38.2, 38.3 and 38.4 (Task Group 38.1 was still hurrying back from the direction of Ulithi) was affected about 150 miles northeast of San Bernardino Strait. The carrier task force, under Vice Admiral Mitscher, then proceeded north at 25 knots to attack the enemy at dawn.

First contact with the enemy was made at 0205 by a night radar search plane, and was reported as three large and three small ships on course 110, speed 15. Another group of six ships was reported by the same plane about forty miles from the first. The nearest of these groups was only eighty miles from Task Force 38.

Because the proximity of the two forces and the fact that they were on approaching courses presented a possibility of surface action, Task Force 34, composed of battleships, cruisers and destroyers from the task groups, was formed and assigned a position in advance and to the right of the carrier groups. Engine trouble forced the tracking plane to return to the force and a replacement was not able to regain contact with the enemy force.

Preparations for attack were made prior to dawn by all task groups, and at 0555 a search was launched followed immediately by the attack groups from each of the carriers. The attack air groups orbited fifty miles to the north of the task force to await contact by the search planes.

Contact was regained at 0735 with an enemy force bearing 015 true and 140 miles distant from Task Force 38. Obviously, the enemy had changed course and run at high speed after the search plane had lost contact during the night. The weather at the target was perfect. At 0840, the air groups, which had been orbiting, struck.

The enemy was taken by surprise. The fifteen or twenty planes he had in the air were quickly shot down or driven off, and because the planes which had attacked Task Group 38.3 the day before had evidently not returned from Luzon, our planes attacked without air interference during the remainder of the day. Carriers were the primary targets for the first strike.

The light carrier which had launched planes received several direct hits with 1,000-pound bombs, followed by three torpedo hits, whereupon she exploded violently and sank. Several hits left another light carrier dead in the water. Damage was also reported on the large carrier, one battleship, one cruiser and one destroyer.

While this strike was in progress, Commander, Third Fleet, received a series of urgent messages from Commander, Seventh Fleet, regarding the plight of the escort carriers off Samar. The first of these, received at 0822, reported enemy battleships and cruisers firing on the escort carriers from a position fifteen miles astern.

At 0848, Commander, Third Fleet, ordered Task Group 38.1, then fueling in a position nearly within striking range of the enemy, to proceed at the best possible speed and to launch strikes on the enemy fleet. At 1115, Task Force 34, with its fast battleships and Carrier Task Group 38.2, were detached from the fast carrier task force to go to the aid of the forces in the Leyte area.

Before the planes of the first strike against the Northern Force returned, another was launched. This, although consisting of fewer planes than the first, was successful in making several additional torpedo hits on the light carrier. The enemy formation had now broken up into two groups, one standing to the north making good speed away from the area, and the other circling around the crippled ships.

At 1300, a third strike, which included 150 of the planes that had participated in the first strike, concentrated on the undamaged ships in the group attempting a getaway, in an effort to create more cripples which could be disposed of later. The large carrier was hit hard with many bombs and was observed to sink at 1430.

The light carrier in this group succeeded in evading serious damage by maneuvering radically at high speed. The fourth and fifth strikes which attacked between 1500 and 1710 finished off another light carrier and seriously damaged a battleship with a torpedo and several bomb hits which slowed her temporarily. A light cruiser was stopped dead in the water by several bomb hits, and a destroyer was reported sunk by strafing.

As the attack developed, it became apparent that there would be many crippled ships which could be overthrown by surface forces and sunk by gunfire. A special group was formed for this purpose with the four heavy cruisers of Cruiser Division 13, and twelve destroyers. In the only instance of the Pacific war in which our surface forces participated in the destruction of a Japanese aircraft carrier, this group had a crippled light carrier, the Zuiho, under fire at 1630, and before the destroyers could close to deliver a torpedo attack, she rolled over and sank.

Later that night, at 2059, a large destroyer of the Terutsuki class was also sunk. Submarines which had been disposed in the probable path of retirement, accounted for one cruiser sunk and one probably sunk during the night.

No damage was sustained by any of the ships of the task force during this engagement. Although heavy antiaircraft fire was encountered from all of the enemy ships during the entire engagement, only ten of the attacking planes were lost by gunfire during the day.

This engagement was the ultimate in the development of carrier warfare which had been foreshadowed in the Coral Sea, at Midway, and with less success, at the Marianas. With the opposing fleets entirely out of sight and range of each other, one was able to wreak devastating destruction on the other by air power alone. The situation was entirely against the Japanese.

Without the protection of their planes, which had not returned from the airfields on Luzon where they had landed after their unsuccessful attack on Task Group 38.3 the day before, they were helpless against the crushing blows delivered by our air groups. Because of this condition, they were also unable to launch counterattacks against our carriers. This situation made it possible for us to release most of our protective forces for the more lucrative business of attack.

By no means the least important factor in the destruction of the Japanese force was the organization of our attack forces. An air coordinator, whose duty involved the assessment of the damage being inflicted and the assignment of attacks on specific targets to each of the striking groups as they came in, was kept over the enemy during the entire attack. Three air group commanders carried out the co-ordination assignment at different periods of the attack, together completing the best organized aerial attack ever delivered by our carrier planes against a concentrated enemy naval force.

The following summary indicated the devastating damage inflicted on the Japanese Fleet in this engagement:

SUNK
1 large carrier (Zuikaku)
2 light carriers (Chitose, Chiyoda)
1 destroyer

POSSIBLY SUNK
1 light cruiser (Tama) Damaged
1 light carrier (Zuiho; later sunk by cruisers)
2 battleships (Ise, Hyuga)
1 heavy cruiser (Ashigara)
2 light cruisers (Natori, Oyodo; Natori later sunk by submarines)
4 destroyers (1 later sunk by cruisers)

Only two ships of the enemy force, a cruiser and a destroyer, escaped without damage.

At 2200, Task Groups 38.3 and 38.4 reversed course and departed for a fueling rendezvous.

Early on the morning of 25 October, the escort carriers and screen of Task Group 77.4 were moving forward from their night positions to their stations nearer Leyte Gulf, where routine patrols and strikes in support of groups of troops and operations in Surigao Strait were to be flown. The three task units were disposed from a point ninety miles southeast, to a point sixty miles northeast of Suluan Island, off Leyte Gulf. By 0530 the Leyte CAP (combat air patrol) and the local CAP and ASP (anti-submarine patrol) had been launched. At 0658 the CVE's to the south launched searches for enemy surface ships.

Before this search had departed, however, the carriers in Task Unit 77.4.3, the northernmost group, made an alarming discovery — the stacks and pagoda masts of Japanese battleships were sighted coming up over the horizon. Simultaneously with this discovery, an ASP plane reported the presence of the force. There had been no advance radar or visual contact with this force.

Task Unit 77.4.3 immediately changed course to due east, which was close enough to the wind to permit launching without closing the enemy too rapidly, and all available planes were launched. Flank speed was ordered, and all ships began making smoke. An urgent contact report was broadcast in plain language giving the distance and bearing of the enemy and requesting immediate assistance.

The Japanese force, which had been attacked in Sibuyan Sea the afternoon before, had reversed its course under attack. Apparently, after Task Force 38 planes had withdrawn, it had again changed course and resumed its progress toward San Bernardino Strait. This fact had been reported at 2115 by carrier search planes but no further sightings were made until it appeared over the horizon off Samar. It approached in three columns.

In the center column were 4 battleships, to the port were 2 heavy cruisers and 2 destroyers, while the starboard column contained 4 heavy and 1 light cruiser and 1 destroyer. The enemy opened fire at 0659, and his heavy gun salvos began falling near the CVE's. The enemy closed rapidly, and the volume and intensity of his fire increased. Shells were falling all around the helpless CVE's, every ship being straddled repeatedly.

The situation was extremely critical. Task Unit 77.4.3 composed of 6 of the relatively slow escort or "baby" carriers screened by 3 destroyers and 4 destroyer escorts, was exposed to an enemy force of fast battleships and cruisers which had approached within range of its big guns before being discovered. Surface forces of the Seventh Fleet were deep in southern Surigao Strait, too far away to be of help, and, due to five days of bombardment plus their brief but destructive engagement with the enemy Southern Force, extremely low on ammunition.

The fast battleships of the Third Fleet were operating with Task Force 38, then engaged with the enemy carrier force to the north. It was clear that the CVE's were "on their own" and must protect themselves with their own weapons — one 5-inch gun each for which only antiaircraft ammunition was available, their screening ships, destroyers and destroyer escorts, and their aircraft.

At 0706 the enemy was closing the range with disconcerting rapidity, and it did not appear that any of the ships could survive another five minutes of the volume of heavy caliber fire that was being put out by the Japanese force.

Planes of Task Unit 77.4.3 had been launched as soon as the attack began. All available planes were scrambled under conditions which had never been met before. Because the wind was from the direction of the Japanese force, planes were launched partly crosswind while the carriers were under heavy fire from the enemy.

The air groups let go with everything they had. Fighters strafed and bombed heavy cruisers and battleships, and torpedo planes attacked with bombs and torpedoes.

When the ammunition had been expended, the pilots made dummy runs in the face of heavy antiaircraft fire. Any tactic which might draw the attention of the enemy away from the "jeep" carriers was used.

As the planes ran out of fuel, and it became impossible to land on the carriers, the army airstrip at Tacloban on Leyte, which was not yet operational, was used. Some ammunition was available there, as well as a supply of aviation gas.

Immediate retaliation against the enemy force was also carried out by the destroyers and little destroyer escorts in a series of torpedo attacks.

In one of the most gallant and heroic acts of the war, these little ships advanced against the heavy units of the Japanese Fleet, entirely unprotected by covering fire except that from their own 5-inch guns, and launched their torpedoes at ranges varying from 9,000 to 4,000 yards. The Johnston and the Samuel B. Roberts were sunk by a concentration of enemy gunfire, and the Hoel went down an hour later after having received about forty hits by 5-, 8-, and 14-inch projectiles.

These determined attacks by the destroyers and destroyer escorts were successful in creating a temporary diversion from the

CVE's, and their gunfire inflicted some damage to the superstructure of the Japanese ships. In addition, one torpedo hit was observed on a Kongo-class battleship, and two cruisers may have received fatal torpedo damage.

At about the same time that the torpedo attack was launched, all available aircraft from other units were ordered to strike the Japanese force. Planes from the CVE's to the south were recalled from support strikes over Leyte to a rendezvous point over Suluan Island. Reinforced by additional planes from the CVE's, they joined in the attack on the enemy at 0830.

As the engagement developed, the enemy deployed his force so that the cruisers and destroyers flanked the CVE's, and the battleships followed directly astern. This placed the escort carriers in a position exposed to fire from three directions, that from the flanking cruisers proving particularly damaging. The Gambier Bay took her first hit at 0810 and ten minutes later a hit in the engine room reduced her speed so that she fell out of formation.

As he fell back, enemy fire concentrated on her, and in five minutes she was dead in the water. Two or three cruisers poured in their fire, making at least twenty-six hits in the next twenty minutes, sending her down at 0911. The Kalinin Bay received fifteen hits and the Fanshaw Bay took four. Men on the White Plains were wounded by flying shrapnel from shells exploding all around her.

The CVE's, which had begun their retirement on a due east course, had been forced by the oncoming enemy to swing first to the south and then around to the west until they were being driven directly into the coast of Samar.

The spirit of the personnel manning the CVE's faced with those overwhelming odds was characterized by a gunnery officer on one of them. During the worst of the engagement, when the cruisers were closing the range, he told his crew, "Just wait a little longer boys, we're sucking them in to 40-millimeter range."

At about 0930, just as the complete destruction of the CVE's seemed inevitable, the Japanese suddenly broke off action and appeared to retire from the area. However much of a surprise and relief this might have been to the men on the CVE's, the enemy's next move was even more inexplicable. Their force maneuvered aimlessly, remaining in the vicinity of the battle for the next three and a half hours, taking no further action against our forces.

During that time, it had swung from north to east, then around in a complete circle until it was heading north again. At 1310, it had reached a position only thirteen miles west of the point at which the battle had started. At that time, however, the Japanese seemed to acquire purpose and retired toward San Bernardino Strait.

With the material now available, only surmise can be made as to the reasoning of the Japanese commander in breaking off the action at a time when everything seemed to be going his way. He held a decided advantage in speed, gun power, and position which placed the entire force of CVE's at his mercy. His decision certainly saved the escort carrier force from destruction and can only be regarded as a lost opportunity for the Japanese.

It can be reasoned, however, that the factors which must have influenced the Japanese commander's decision made it less surprising than it first appeared. The Japanese plan of attack, while it violated a fundamental of naval strategy in that it divided the weaker force into three attacking groups, was cunningly conceived and based on a concentration of these groups at Leyte Gulf at dawn on the twenty-fifth. Air attacks on the Southern and Central forces on the twenty-fourth, although not stopping their approach, did succeed in slowing down the Central Force so that it was behind schedule in arriving at Leyte Gulf.

How much information the Japanese admiral may have had about the destruction of the other enemy forces or about the disposition of our units that were hastening to the aid of the CVE's is, of course, unknown; and it is safer simply to state that the enemy commander withdrew for reasons best known to himself, unless one chooses to accept the explanation put forth by the commanding officers of several of the CVE's in their official action reports that it was the intervention of divine providence.

During the entire period the Japanese had been under air attack by the planes of Task Group 77.4. In addition to the diversionary effect of these attacks, serious damage was inflicted on enemy ships. In 252 fighter- and 201 torpedo-plane sorties, 191 tons of bombs and 83 torpedoes had been expended on the enemy force. The confusion of the action and the periods of poor visibility made it difficult to assess the damage to the enemy or to determine the individual effort responsible for it. The results, however, were most gratifying.

The combined air and surface attacks sank 2 heavy cruisers (the Suzuya and Chikuma) and 1 destroyer, heavily damaged 1 battleship, 1 heavy cruiser and 1 destroyer, and caused some damage to 3 cruisers and 1 destroyer.

Our losses were the destroyers Hoel and Johnston, the destroyer escort Samuel B. Roberts, and the CVE Gambier Bay, all sunk by gunfire. In addition, 2 destroyers and 2 CVE's were damaged by hits. Planes lost during the engagement totaled 17 fighter and 18 torpedo planes.

Considering the superior speed and gun power of the enemy, it is surprising that no more damage was sustained. The enemy not only failed to close aggressively, but his rate of fire was slow. Although his salvo patterns were small and often unfortunately close, the number of hits in relation to the number fired was very low. The fact that the Japanese were using armor-piercing ammunition, which in some cases passed entirely through the unarmored merchant hulls of the escort carriers without exploding, undoubtedly saved the two which were hit but not sunk.

Shortly after Task Unit 77.4.3 was engaged by the enemy surface force, Task Unit 77.4.1 to the south was subjected to a determined aerial attack. The attack was not only unexpected, having come in without detection by radar, it was different from any that had previously been experienced.

The gravity of the military situation brought about by our landings at Leyte and the pressure of air strikes against islands near the homelands had driven the Japanese to desperation tactics. Some strange quirk of the Japanese mind had conceived the violent and flaming death of a plane crash on the deck of an enemy carrier as a glorious and desirable end for a true warrior of the Emperor. Significantly, this tactic was called the "Divine Wind" after the storm which had once succored the Japanese homeland from invasion by the Mongols.

In theory the exchange of a plane and pilot for the destruction of one of our ships was profitable for the Japanese. In practice, however, although the suicide attack succeeded in sinking and damaging many ships, it was not effective enough to stop our forces or even to relieve the pressure being brought to bear on the enemy. Prior to this engagement there had been instances of pilots crashing or attempting to crash on carriers and warships after they had been severely hit, but this is the first instance in which the attack seemed to be organized and planned as a suicide maneuver.

Although all methods were used by later pilots, the general pattern of these early attacks was the same. A low approach to escape detection by radar, a fast climb on arriving at the surface force, and a sudden dive into the ship releasing the bomb just before the impact formed the usual method employed. The bomb usually penetrated the ship, and the plane tore up the deck, both exploding and starting fires. The location of the hit and the condition of the planes on the flight and hangar decks generally determined the seriousness of the damage. If our planes were fueled and armed in readiness for a strike, a fire among them spread very fast, and the exploding bombs, torpedoes, ammunition, and fuel tanks resulted in a serious situation.

The first attack occurred at 0740 when the CVE's were landing planes returning from strikes. The antiaircraft batteries opened up on three enemy planes, identified as Zekes, coming in at 8,000 feet. One peeled off in a 20-degree dive, passed through a seemingly impenetrable hail of fire, and exploded violently on the forward flight deck of the Santee. Serious fires broke out on the flight and hangar decks. Accurate antiaircraft fire destroyed suicide planes just before they hit the Petrof Bay and Sangamon. Later, a suicide bomber hit the Suwannee causing a heavy explosion, fire, and many personnel casualties. None of the ships was seriously damaged and all were back in action before the end of the day.

At 1050 the already badly shaken CVE's of Task Unit 77.4.3 were hit by a series of suicide attacks just as they were landing their aircraft. During the attack, eight planes made suicide dives on the five remaining CVE's. The Kalinin Bay was hit, and the White Plains, Fanshaw Bay and Kitkun Bay received minor damage from near misses.

One plane penetrated the center of the flight deck of the St. Lô, and it sank without further enemy action. All escorts stood by, leaving the remaining CVE's without screen until late in the day when the task unit made rendezvous with Task Unit 77.4.1 to the south.

The work of an already exceedingly busy day was not yet finished. Remnants of the Japanese forces, fleeing in several directions, were not to escape without further damage. Planes of Task Group 38.1, which had been directed by ComThirdFleet to the aid of the CVE's, hit the retiring enemy forces at 1330. As the enemy ships retreated northward toward San Bernardino Straits, they were continually attacked and harassed by the planes of Task Group 38.1 and those from the CVE's. When the enemy force entered the Strait, several cripples trailed behind and were finished off by the planes of Task Force 38 the next day.

Except for the pursuit phase, this ended the Battle for Leyte Gulf, in which the Japanese Navy had been "beaten and routed and broken by the Third and Seventh Fleets." A total of 3 battleships, 1 large carrier, 3 light carriers, 1 heavy cruiser (XCVS), 3 heavy and 6 light cruisers, and 10 destroyers were reported sunk by the commanders of the Third and Seven fleets, as well as 21 additional combatant ships damaged.

Of these, 1 battleship, 1 carrier, 2 light carriers, 4 heavy cruisers, 1 light cruiser, and 3 destroyers were sunk by carrier airplanes alone. In addition, other units except those sunk at Surigao Strait by Seventh Fleet surface forces, were initial damaged and crippled by air attack to the extent that they could be overtaken by surface forces and sunk.

Our losses were 1 CVL (Princeton), 2 CVE's (Gambier Bay and St. Lô), 2 destroyers, 1 destroyer escort, and 1 PT sunk, and 1 light cruiser, 6 CVE's, 3 destroyers, and 1 destroyer escort damaged. The Japanese had paid a heavy price for their all-out attempt to stop our landing in the Philippines, and in addition had failed to accomplish their mission. The destruction and damage inflicted on the Japanese Fleet had reduced its potency as an offensive weapon.

By the evening of 26 October, Third and Seventh Fleet forces were operating in the vicinity of Leyte Gulf. Land operations were progressing so well that the strategic air support mission of the Leyte-Samar area by the Third Fleet appeared almost completed.

On 27 October, the 5th Army Air Force assumed responsibility for the support of land operations and Supreme Allied Commander, Southwest Pacific, directed that Task Force 38 refrain from striking land targets without specific request.

It developed, however, that due to the delays in airfield development, land-based aircraft could not perform their task without help from the carrier planes which therefore remained in direct support until 25 November. Task Group 38.1 had departed for Ulithi on the twenty-sixth, and Task Group 38.3, with CTF 38, departed on the twenty-eighth. The CVE's of Task Group 77.4 had withdrawn in order to rearm and replenish their aircraft. This left Task Groups 38.2 and 38.4 with the entire burden of protecting the ground and surface forces at Leyte. On 30 October, Vice Admiral J. S. McCain, Commander, Second Carrier Task Force, Pacific, relieved Vice Admiral M. A. Mitscher of command of Task Force 38.

Many enemy airfields in the Visayans, Luzon, and Mindanao were still operational and capable of launching air attacks on our ground positions and supply convoys. Enemy planes were also capable of providing cover for their reinforcements moving up through the inter-island channels to Leyte.

Attacks on the staging bases through which reinforcements were being flown seemed necessary and the planes of Task Group 38.2, although a much smaller force than needed for the job, struck hard on 28 October.

In spite of the destruction of enemy planes in the air and on the ground, air opposition could not be entirely eliminated because of the number of airfields in the area and because of their proximity to bases which linked them to the large concentrations of aircraft in and around the empire.

A series of suicide attacks on 29-30 October and 1 November was successful in damaging some of the ships of the force. On the twenty-ninth the Intrepid was hit in a 20mm gun gallery causing a fire and some personnel casualties but not affecting her battle efficiency. On the thirtieth, a suicide plane tore a 40-foot hole in the flight deck of the Franklin causing considerable damage to the hangar deck and destroying many planes. The Belleau Wood was hit in the same attack by a crash on the after end of the flight deck.

Both these ships were forced to retire to Ulithi and later to a continental navy yard for extensive repairs. On 1 November, one destroyer was sunk and three damaged in Leyte Gulf as a result of suicide attacks.

It soon became evident that only attacks on these staging bases could reduce the flow of Japanese air power toward Luzon, and that Task Group 38.2 could not of itself muster sufficient force for the job. A long-planned strike by the fast carriers against the Japanese homeland was therefore regretfully postponed and Task Group 38.2 retired to Ulithi to join the remainder of Task Force 38 and prepare for raids on the staging bases.

On 3 November, the newly replenished groups, 38.1, 38.2, and 38.3, rendezvoused at a point about midway between Ulithi and Luzon and began a high-speed run on Luzon. (Task Group 38.4 was now retiring to Ulithi for rearming.) While the task groups were assembling, the antiaircraft cruiser Reno was torpedoed by a submarine and so seriously damaged that she was forced to return to Ulithi under escort.

The three task groups launched destructive fighter sweeps and strikes against the airfields on Luzon and the Manila Bay areas for two full days on 5 and 6 November. Fourteen different fields were hit, the largest concentrations of enemy aircraft being found at Mabalacat, Clark, and Lipa fields. A total of 729 enemy aircraft were destroyed in the two-day attack.

Damage to shipping in Manila Harbor was also extensive. The heavy cruiser Nachi was sunk leaving the harbor, and a Natori-class cruiser, several large and medium cargo ships, and seven destroyers were damaged. As a result of these strikes there was an immediate improvement in the air situation over Leyte.

From 7 to 11 November the task force retired from the area to fuel and reorganize. Task Groups 38.1 and 38.3 fueled and 38.4

relieved 38.2 which retired to Ulithi. The Wasp with a two-destroyer escort proceeded to Guam to pick up a replacement air group.

Japanese efforts to reinforce their Leyte garrison continued in spite of the decreased effectiveness of their air power in the protection of convoys. On 11 November, Task Force 38, now operating under the tactical command of Rear Admiral (later Vice Admiral) F. C. Sherman, caught and virtually eliminated an enemy convoy of 3 large and 1 medium transport, 5 destroyers, and 1 destroyer escort. The attacking planes sank the entire convoy (with the exception of one destroyer) just as it entered Ormoc Bay.

Strikes on Luzon were made in the next two days. On 14 November, Vice Admiral McCain resumed command, and the task force, now consisting of two groups, renewed attacks on targets in the Manila Bay area. Much shipping destruction was accomplished, and the installations on Manila docks and at Cavite Navy Yard were given a good going over.

After fueling at sea and being joined by Task Group 38.2, our forces again attacked the Luzon-Mindoro area on 19 November but, except for the destruction of a hundred enemy planes, good targets were lacking, and further strikes did not appear advisable. The force withdrew from the area, Task Group 38.1 joining 38.4 in Ulithi. The two remaining groups remained at sea in order to continue support for the Leyte operation as the need arose.

On 25 November these two groups again launched strikes on the Luzon area seeking to destroy crippled enemy combatant ships and to damage further the reinforcement pipe line to Leyte. One heavy cruiser, the Kumano, was found and destroyed along with other important shipping, and many other components of the enemy's merchant and transport strength were damaged.

During this period of operations, the suicide tactics of the Japanese pilots were successful in damaging several ships. On 5 November, a suicide bomber crashed the signal bridge of the Lexington impairing her battle efficiency only slightly but causing relatively high personnel casualties. As a result, she was routed to Ulithi for repairs.

On 25 November, a well-executed and deceptive attack developed shortly after noon which resulted in heavy damage to ships of 38.2 and some damage to 38.3. The Hancock caught fire from the debris of a suicide plane which was exploded by antiaircraft fire just above her deck. The Intrepid's flight deck was pierced by a crashing plane which started a serious fire, and shortly afterward she was hit a second time. The Cabot also took a hit. A second plane crashed close aboard and the exploding bomb caused additional damage. The Essex received only superficial damage from a hit by a suicide diver.

During the period 3-27 November, which was the closing period of operations in direct support of Leyte-Samar operations, planes of Task Force 38 made 6,062 sorties of which 4,198 were strikes. During this period 54 enemy warships, totaling 140,600 tons, had been sunk, more than three times that tonnage damaged; and 768 enemy aircraft had been destroyed.

The fast carrier force had beaten down the Japanese effort to maintain air supremacy over the Philippines and had crushed enemy attempts to reinforce the troops opposing our occupation of Leyte.

By the end of November our troops controlled all of Leyte except for the area around Ormoc Bay, at which point the Japanese, aided by the reinforcements which they put ashore in spite of air attacks, were offering strong resistance. Land-based air, which at the beginning of the month had consisted of only two squadrons of day fighters and one squadron of night fighters, had been reinforced so that by the end of the month the Army had nearly three day-fighter groups, two night-fighter squadrons, a light bomber group, and a photographic unit operating from airfields on Leyte.

Navy units also operating in this area were three VPB (patrol bombing) squadrons, a Liberator (PB4Y) squadron and a Ventura (PV) squadron. Heavy rains and enemy air raids had handicapped the preparation of airstrips on Leyte, but in spite of these, airfields had been completed in the Tacloban, Burauen, and Dulag areas. Land-based air made 2,640 sorties against enemy targets during the month, most of which were directed against air strength and shipping in the Visayans area and the enemy reinforcement pipe line to Ormoc Bay.

Toward the latter part of the month, the CVE's of Carrier Division 29 (later relieved by Carrier Division 27) operated in the area a hundred miles east of Leyte Gulf providing combat air and anti-submarine patrols for the convoys proceeding in and out of Leyte.

Our land-based air offensive against Leyte targets during December was flown entirely by fighters and fighter-bombers. Marine Corsairs of Squadrons 211, 218, and 313 (Marine Air Group 12) were added early in the month to the army P-38's, P-40's and P-47's. Interdiction of the Japanese supply route to Ormoc resulted in the destruction of 2 destroyers, 4 destroyer escorts, 14 supply and personnel transports of various sizes, 3 oilers, and 14 coastal vessels and luggers.

The airfields in the Visayan area and on southern Luzon were pounded steadily by the army, navy, and marine fighters, and by bombers from Leyte, accounting for some 145 enemy planes destroyed on the ground and 150 in the air. The ground action for the occupation of Leyte was completed successfully on 20 December.

IN ORDER TO ESTABLISH AIR BASES within supporting distance of future operations on Luzon, General MacArthur had determined on the occupation of Mindoro early in December. The establishment of a beachhead on this island required that the landing force approach from Leyte through enemy controlled water of the Mindanao and Sulu seas, and near the islands still under Japanese control. Due to the delays in the development of air facilities on Leyte, this venture was delayed until 15 December.

Air support for the occupation was provided by Third Fleet forces with Task Force 38, which had the responsibility for the neutralization of the hundred airfields on Luzon, and the Seventh Fleet forces which provided cover for the transports and attack groups during passage through the Visayan area, as well as direct support of the landings. The Heavy Covering and Carrier Group (77.12), under command of Rear Admiral T. D. Ruddock, Jr., consisted of 3 old battleships, three light cruisers, 6 CVE's and 18 destroyers.

This combination set up here for the first time, paralleled the organization of the fast carrier task force and provided for the protection of the force with the gun power of the battleships and cruisers against surface raiding forces and heavy antiaircraft fire power and air cover against air attack. Further protection for the force was provided by land-based air, particularly at dusk each day.

Considerable enemy air opposition developed during the passage to the beaches. The light cruiser Nashville was hit by a suicide plane on the afternoon of 13 December while passing between Negros and Mindanao and forced to retire to Leyte Gulf. Later the same day, the destroyer Haraden was also hit. Eleven enemy aircraft were shot down during this attack by combat air patrols and antiaircraft fire.

Opposition on the beaches was negligible. Troops were landed on schedule with no casualties. Twenty-two enemy planes were destroyed during the unloading phase of the landings on 15 December which resulted in damage to two LST's, one destroyer, and slight damage from near misses to the CVE Marcus Island.

Immediate construction of airfields was undertaken, and by 22 December, San Jose Field was operational for army fighters and bombers.

By far the most amazing performance in connection with the landings on Mindoro was displayed by Task Force 38. Now reorganized and consolidated into three task groups for more adequate protection against suicide attacks, the task force sortied from Ulithi, and after training exercises at sea began a high-speed run on Luzon on 13 December.

Armed with a completely new set of tactics — designated by such colorful names as Tomcat, Zipper, Jack Patrol, Moosetrap, and others — designed to provide more adequate protection against suicide attacks, the task force proceeded to spread an air blanket over the airfields of Luzon. So effective was this blanket that not one navy plane was lost as a result of air combat with enemy aircraft during the three-day period it was spread.

The air blanket, one of the new tactics devised for the Philippines, was essentially a method of maintaining combat air patrols over enemy airfields in sufficient strength throughout the day and night to prevent the take-off or landing of any enemy aircraft. It began with a strong fighter sweep to clear the sky of enemy fighters.

Additional waves of fighters were then sent in to relieve those on station so that a continuous patrol was kept over the field. Once the enemy air power was beaten down only small patrols were needed to keep the fields immobilized, and all remaining aircraft were utilized in attacking ground and shipping targets.

The problem was not simple. There were from ninety to a hundred enemy airfields in the area assigned to the Third Fleet forces. The entire island of Luzon was allotted by areas to the three task groups and further divided within the task groups into assignments for individual carriers and air groups. This insured the complete coverage of all fields, which was essential to the success of the mission.

So successful was this air blanket that, with the exception of one flight of enemy aircraft which took off before the first strike on 14 December, no Luzon-based aircraft attacked the Mindoro-bound convoys, no enemy air attacks on Task Force 38 penetrated the defensive patrols, and none of the ships of Task Force 38 was damaged.

With enemy air power effectively pinned down, the remainder of the striking groups roamed up and down the coast in search of shipping and ground targets. During the three-day period, 62 enemy planes were destroyed in the air and 208 on the ground, 16 ships (exclusive of luggers [small boats] and barges) were sunk, and 37 were damaged. In addition, gas and fuel dumps, ammunition dumps, warehouses, hangars and buildings were destroyed, truck convoys and locomotives were strafed, and other serious damage inflicted on enemy ground installations.

At the conclusion of the strikes on 16 December, Task Force 38 began its withdrawal toward a fueling rendezvous to the eastward, intending to return to Luzon on the nineteenth for a continuation of the tactics just completed. Plans were nullified, however, by an unusually severe typhoon which formed near the force on 18 December.

As a result of the high-speed approach to the Luzon area and the continuous operations for the following three days, many destroyers were low on fuel, and it was imperative that fueling operations be carried out. Efforts were made on 17 December in spite of increasingly heavy weather, but without much success. A different rendezvous was designated for the following date at a point which was estimated to be south of the track of the typhoon, but a change in the course of the storm placed some of the units of the force directly in its path. The average strength of the wind ranged from 50 to 75 knots, with gusts as high as 120 knots.

Mountainous and confused seas built up. Some of the destroyers found themselves unable to change course by any combination of engines and rudder, and experienced rolls in excess of 70 degrees. Those with considerable free liquid surface in their tanks and bilges, as a result of being low on fuel, exceeded the stability range. As a result, the Hull, Spence, and Monaghan capsized, and at least two others, the Dewey and Aylwin, had narrow escapes from the same fate.

The lighter aircraft carriers also suffered considerably. A total of 146 airplanes were lost, including 8 blown overboard from the battleships and 11 from the cruisers. Fires broke out on the Monterey, Cowpens, and San Jacinto. These vessels also received considerable structural damage to their hangar decks from planes and other material adrift.

After the storm had subsided on 19 December, fueling was completed, and after spending the twentieth and twenty-first sweeping the area in search of survivors from the lost destroyers, the force retired to Ulithi for repairs and replenishment in preparation for the support of landings at Lingayen Gulf.

The operation for the invasion of Luzon was a natural sequence to the landings on Leyte and Mindoro. It was by far the largest operation carried out in the Pacific war to this date and involved co-operation between forces not under the same command to a greater degree than had been the case in most previous operations. The Seventh Fleet, which was attached to the Southwest Pacific command, was expected to provide air-surface cover and support and, along with Pacific Fleet submarines, submarine reconnaissance and interdiction of sea routes from Singapore and the empire.

The Fourteenth and Twentieth Army Air forces undertook to attack enemy bases in China, Formosa, and Japan, while navy search planes from the southwest and central Pacific and army planes from China scoured all possible approaches. So far as the Japanese installations in the Philippines were concerned, they were to be struck by planes of the Southwest Pacific Force, which included army, navy, and marine units, and by the aircraft of Task Force 38.

All these air operations were to be coordinated with amphibious landings in Lingayen Gulf on 9 January 1945 under the control of General MacArthur.

Task Force 38, which had been in almost continuous operation since the opening of the Philippines campaign on 10 October, had undergone minor repairs at Ulithi and had made preparations for another extensive period of operation.

The task force, still under the command of Vice Admiral J. S. McCain, operating as a part of Admiral Halsey's Third Fleet, was now organized into three carrier task groups, one night-carrier task group, and a service group. The 3 carrier task groups were made up of 7 large and 4 light carriers, 6 battleships, 7 heavy and 6 light cruisers and 48 destroyers; the night carrier group was composed of 1 large and 1 light carrier and 6 destroyers; and the service group of 7 CVE's, 9 destroyers, 18 destroyer escorts and 25 oilers. The fighter plane complement of the large carriers had been increased considerably, and for the first time since prewar training cruises, two marine fighter squadrons (VMF 124 and 213), equipped with F4U's (Corsairs), were operating from one of the carriers of the fast carrier task force.

After training exercises and fueling en route, the task force arrived in launching position on 3 January to begin the operation for the invasion of Luzon. Its mission was the destruction of enemy aircraft on Formosa. With the attacks staged on 3 and 4 January in this area, the task force experienced the first of an almost continuous spell of bad weather that was to hamper its operations during the entire month of January. A considerable number of pilots on the first strike could not make their way through the heavy front that existed between the task force and Formosa and were forced to return to the carriers.

During the two days, air opposition was light at the target and practically nonexistent over force. Efforts to "blanket" the target with continuous air patrols were unsuccessful and perhaps unnecessary because of the heavy weather.

Plane sorties totaled 1,595 during this attack, of which approximately half were defensive flights over the force. Thirty enemy planes were shot down, 81 destroyed on the ground, 36 ships of varying sizes were sunk, and 43 damaged. Adverse weather conditions all contributed to the comparatively high aircraft losses suffered in the attack, which were 18 lost in combat and 14 lost operationally.

After a fueling operation, the attention of the task force, in response to a request from ComSeventhFleet, was switched to the airfields of Luzon. Here, also, flying conditions were found to be "from undesirable to bad." Air opposition was light, only 14 enemy planes being destroyed in the air and 18 on the ground. Good targets were reported in the Clark Field area, and afternoon strikes were diverted to this region. Enemy aircraft were difficult to locate, being widely dispersed and thoroughly camouflaged; however, results were satisfactory, 75 being destroyed on the ground and 4 shot down in the air. Shipping attacks resulted in the sinking of 4 medium cargo ships, 2 medium oilers, and numerous smaller craft.

On 9 January, the day set for the landings at Lingayen Gulf, the task force returned to Formosa to support the landings by attacking the bases from which fields on Luzon would be reinforced. Heavy frontal weather continued to hamper flight, but the force was successful in reaching Formosa and the Pescadores Island targets with all but a few of the scheduled strikes. Considering these conditions, results of the attack were good. Nine ships were sunk, including 3 destroyer escorts and 2 large oilers, no enemy planes approached the task force, and no heavy air opposition developed to the landings at Lingayen Gulf.

While Task Force 38 met relatively light opposition, which was possibly due to adverse weather conditions, the Seventh Fleet forces approaching Luzon for the amphibious landings were not so fortunate.

The invasion armada proceeded to the objective in four separate groups, divided according to the speed of the ships and the scheduled time of arrival. All groups were escorted by CVE's of Task Group 77.4 under Rear Admiral C. T. Durgin. The track followed by each of the groups was the same. From Leyte Gulf the route passed through Surigao Strait and the Mindanao Sea, northward through the Sulu Sea, through Mindoro Strait into the South China Sea, thence northward off the west coast of Luzon to Lingayen Gulf.

The first group, consisting of the Minesweeper and Hydrographic Group plus oilers, ammunition ships, salvage tugs, and screen, left

Leyte on 2 January and completed its sortie successfully on 6 January. Minor damage to four ships of the group resulted from enemy suicide attacks which were experienced each day except 4 January.

The Bombardment and Fire Support Group and the Escort Carrier Group, (12 CVE's) comprising the second group, sortied Leyte Gulf on 3 January and met serious enemy air opposition. On 4 January, the Ommaney Bay (CVE), with its planes on deck fully gassed and armed, was crashed by an undetected suicide plane. Fires and explosions broke out which could not be controlled, and the ship had to be abandoned and sunk. The Lunga Point narrowly escaped a similar fate.

Shortly after leaving Mindoro Strait on the following day, this group was the object of three enemy raids. Combat air patrols shot down 18 enemy aircraft during the day and antiaircraft fire destroyed 5 suiciders, but 7 others succeeded in crashing ships. The heavy cruiser Louisville and a destroyer in the van group, and the Australian heavy cruiser Australia, 2 CVE's, Manila Bay and Savo Island, and a destroyer and destroyer escort in the rear group, were hit. All ships were able to proceed with their groups, arriving off Lingayen Gulf on 6 January without further incident.

The third and largest group, composed of the cruisers and destroyers of the Close Covering Group, the transports and landing craft of the San Fabian Attack Force and two CVE's of This Unit 77.4.3, left Leyte on the evening of 4 January. First contact with the enemy was on 5 January when a midget Jap sub fired two torpedoes at the light cruiser Boise. The submarine was damaged by a plane on anti-submarine patrol and then was rammed and probably sunk by a destroyer of the screen.

The only determined air attack on this group developed early in the morning of 8 January. Although combat air patrol planes shot down six and damaged several of the attacking planes, a suicide crash against the side of the Kadashan Bay (CVE) tore a 15-foot hole at the water line, and the transport Callaway was also hit. Both ships continued with the formation, but the Marcus Island (CVE) had to land and service Kadashan Bay planes until repairs could be affected.

The fourth and last group, containing the transports and landing craft of the Lingayen Attack Force and two CVE's left Leyte Gulf on the morning of 6 January. It was attacked by six enemy planes, one of which dived suddenly on the Kitkun Bay (CVE) striking a glancing blow on the port side. Resulting fires were brought under control, but serious underwater damage caused flooding in the entire and fire rooms. She was taken under tow but was later able to proceed under her own power at ten knots. Four of the attacking planes were shot down by combat air patrols.

During the approach of the third and fourth groups, the first two conducted preliminary operations in Lingayen Gulf. The Escort Carrier Group (77.4) operating northwest of Leyte Gulf provided combat air and anti-submarine patrols over the ships of the force and carried out attack, reconnaissance, and photographic missions on Lingayen land targets according to plan. In spite of the loss of the Ommaney Bay and damage to the Manila Bay, which rendered it incapable of full flight operations, CVE aircraft flew a total of 788 sorties during the three days immediately preceding S-day.

Serious enemy attacks against the ships in Lingayen Gulf developed on 6 January. By wide dispersal and good camouflage, the enemy had succeeded in avoiding the complete destruction of his aircraft, and by taking full advantage of the large land masses surrounding the Gulf other local conditions which made radar detection almost impossible, the Japanese succeeded in sinking and damaging many ships during this preliminary phase. Although the CVE air groups struggled valiantly to intercept the attacking aircraft and shot down large numbers of them, many reached their targets.

In three attacks on 6 January, 16 ships were victims of Japanese suicide tactics. Only one of these, a minesweeper, was sunk, but 2 battleships, 2 cruisers, 2 destroyers, 1 fast transport, and 1 minesweeper were seriously damaged, and 2 cruisers, 4 destroyers, 1 seaplane tender, and 1 minesweeper received minor damage. More than 50 enemy suicide planes attacked the ships in Lingayen Gulf on this day.

It was apparent from this sudden show of enemy air strength that defensive combat air patrols were not adequate to cope with the situation and that offensive measures were required. Land-based air, due to weather and other conditions, had been unable to provide this offensive as had been planned, and the brunt of the load was borne by the CVE's.

The escort carriers and their squadrons performed brilliantly and surpassed themselves in defense of our forces, but lacked the number of planes and type of equipment which the maintenance of both local defensive and distant offensive patrols required.

In response to a request for help from Commander, Seventh Fleet, Commander, Third Fleet, directed Task Force 38 to cancel scheduled strikes on Formosa to attack the airfields of Luzon, particularly those in the Lingayen area. As a result of these strikes (already related), the intensity of enemy air attacks diminished sharply on and after 7 January.

Following heavy bombardments by the Surface and Air Support groups on S-Day, as well as on the three preceding days, landings on the Lingayen and San Fabian beaches were made on schedule against negligible enemy opposition. Smoke screens over the ships in the transport areas and combat air patrols by planes from the CVE's prevented enemy air from interfering with the assault and unloading operations.

At dusk on 10 January the battleships and cruisers of the Bombardment and Support groups were combined with the escort carrier group to form the Lingayen Defense Force under the command of Vice Admiral Oldendorf, as a measure against surprise interference with our invasion by units of the Japanese Fleet. The majority of the escort carriers maintained a covering position to the northwest, while the remainder protected the convoy routes approaching Lingayen Gulf. Tender-based seaplanes conducted barrier patrols and night searches from Lingayen Gulf beginning on the night of S-Day.

Heavy, medium and fighter bombers of the Far Eastern Air Force concentrated attacks on enemy lines of communication and pounded enemy airfields in the central plains of Luzon. In support of the invasion, 2,238 tons of bombs were dropped on Luzon during the month of January, while islands in the central Philippines received another 600 tons.

During the first half of the month, land-based air destroyed or damaged 79 locomotives (50 percent of the prewar total), 456 railway cars (25 percent of the prewar total), 468 motorcars, 67 staff cars, 18 tanks, 5 armored cars, 10 fieldpieces, and 3 caissons, in addition to bombing shipping, bridges, and important highways.

The penetration of the South China Sea by our fast carrier task force was an event of outstanding significance. It was the first challenge to Japanese control of these waters since the early days of the war. In this action the task force roamed the sea, sweeping everything before it and leaving destruction in its wake.

The operation was not without its dangers. The South China Sea is a body of water approximately 600 miles by 1,200 miles long, restricted on all sides by large land masses which at that time were still controlled by the enemy. It was estimated that approximately one thousand planes were based in the area, most of which were Japanese army aircraft.

The large number of enemy air bases surrounding the sea would be brought within striking range of our carrier planes by this sortie, but by coming within range a position would be reached at which the enemy could also attack the force. In addition, it was anticipated that the weather would be poor for flight operations.

On the night of 9-10 January, after striking Formosa during the day, the Third Fleet forces made a high-speed passage through Luzon Strait, followed by the fast fleet oilers and CVE's which passed though Balintang Channel to the south.

From that night until the morning of the twelfth, the force proceeded on a southwesterly course, fueling on the way. Long-range searches were kept in the air for the protection of the force and the detection of enemy naval or merchant shipping at sea.

Task Group 38.2, which completed fueling before the other task groups, left the force on the eleventh and proceeded at high speed toward the launching point off the Indo-China coast, the other groups joining later.

Strikes, launched on 12 January over 420 miles of the Indo-China coast reaching as far as Saigon, produced devastating results in spite of adverse weather. Forty ships were sunk, including the dismantled French cruiser Lamotte-Piquet, the Japanese cruiser Kashii, 9 destroyer escorts, 4 patrol craft, 4 large oilers, and 3 large freighters. Convoys caught at sea were entirely wiped out and shipping found in harbors was bombed and strafed with excellent results. Total tonnage sunk on this day was 127,000, and an additional 22 ships, totaling 70,000 tons, were damaged.

An aerial torpedo demolished the long dock at Camranh Bay, oil tanks and planes at Saigon were seriously damaged, and buildings, warehouses, and hangars all along the coast were fired and damaged. Successful blanketing of airfields resulted in little enemy air opposition. Fourteen enemy planes were shot down and ninety-seven destroyed on the ground. The over-all results of the day's work were among the best of the war to that time.

On completion of the attacks, the force retired to the northeast at high speed to escape a typhoon to the south. Heavy seas made fueling difficult and two full days were required for its completion. On 15 January, from a position 250 miles from Hong Kong and 160 miles from Formosa, strikes were launched at these targets and at points along the near-by China coast. Unfavorable flying weather caused the discontinuance of strikes by noon, but a fighter blanket was maintained over the airfields of Formosa.

The attacks were resumed the next day under continuing bad weather. The main assault was launched against Hong Kong, an area particularly rich in shipping and harbor targets. The six strikes launched against this target met the most intense antiaircraft fire ever experienced in carrier operations, one report describing it as varying "from intense to unbelievable." Enemy fire control and accuracy were effective, particularly on the first strike in which one out of every eight attacking planes was lost.

In the two-day period, 30 planes were lost in combat and 31 in operations. Enemy aircraft destroyed amounted to 26 in the air and 21 on the ground, making this one of the few instances of the war in which our losses exceeded those of the enemy. However, damage to enemy installations had been heavy in spite of enemy opposition and unfavorable weather, 12 ships, exclusive of luggers and barges, being sunk, and 27 damaged.

Fueling was again delayed by inclement weather, and it was not until 19 January that the operation was completed. Task Force 38, which the Japanese broadcasts described as "bottled up in the South China Sea," passed through Balintang Channel north of Luzon and into the Pacific, without enemy opposition, on the night of 20 January. By midnight the force had completed passage and was proceeding to a position from which it would attack Formosa the next day.

Aided by the first good flying weather of the month, attacks on shipping were pressed home with vigor. The first strike was launched before sunrise, and others followed regularly throughout the day. More than a thousand sorties were flown. Enemy air opposition was light over Formosa but the day's total of 104 planes destroyed on the ground was the highest achieved by the task force during the month of January. Nine large and medium oilers and freighters, totaling 53,000 tons, were sunk and 11 other ships were damaged. About noon, Task Force 38 was under air attack for the first time since November. While the force was in a position approximately a hundred miles east of the southern coast of Formosa, a bogey contact was reported but not confirmed.

Fighters, sent out on an intercepting course, did not make contact. Many friendly planes were returning to the force at the time, and identification was difficult. Suddenly a single-engined plane glided out of the sun and dropped two small bombs, one of which hit the Langley (CVE) on the forward part of the flight deck. Within two minutes another enemy plane, also undetected until the last minute, dove out of the clouds and crashed through the Ticonderoga's (CV) flight deck starting large fires among the planes on the hangar deck below. Eight Cowpens fighters intercepted 18 enemy fighters approaching the force from northern Luzon and destroyed 14 of them, saving the force from another attack.

At the same time, however, planes approaching from the opposite direction were successful in evading destruction by intercepting fighters and without warning a second plane crashed the island structure of the Ticonderoga, adding to her damage. Later in the afternoon, the destroyer Maddox, on picket duty thirty-five miles toward Formosa, was damaged by a surprise attack.

Although this ended the success of the enemy for the day, it was not the end of damage to our ships. Engaged in recovering aircraft returning from strikes, the Hancock (CV) was seriously damaged by a 500-pound bomb which dropped and exploded on the flight deck from the bomb bay of a torpedo plane which had just completed a landing.

The three carriers were successful in controlling the fires and the Langley and Hancock were able to resume the recovery of aircraft in a few hours. The Ticonderoga and Maddox were unable to continue operations, however, and withdrew to Ulithi.

The force continued operating in the Formosa area, launching strikes on the islands in the Nansei Shotos on the following day. Although attacks were carried out against shipping and aircraft, resulting in the sinking of 4 medium and 25 small cargo vessels and the destruction of 28 planes on the ground, the primary objective was photographic coverage of the islands in anticipation of subsequent landings. Forty-seven photographic sorties succeeded in obtaining an 80 percent coverage of the priority areas. In one case, a photographic team of four planes flying abreast covered an airstrip in one pass, saving repeated individual runs over a heavily defended area.

After midnight, having completed almost a full month at sea largely in waters hitherto considered to be controlled by the Japanese, Task Force 38 retired to Ulithi to make preparations for the impending conquest of Iwo Jima.

The success of the Third Fleet forces in support of the invasion of the Philippines, climaxed by their sortie into the South China Sea, was acclaimed by Fleet Admiral C. W. Nimitz, Commander-in-Chief, U. S. Pacific Fleet, as one which "cannot be measured alone in the tangible damage inflicted on Japanese ships, planes and shore installations, substantial as that was.

Like the first attack on Truk in February 1943, and the first assault on large masses in the Philippines in September 1944, the penetration beyond the islands off the Asiatic mainland to strike at China coast targets had considerable strategic significance. From this time onward, no area outside of the immediate Japanese homeland and northern China was safe from assault by our carrier force, and even Japan itself was to feel the weight of carrier raids during the next month. Third Fleet forces had traversed 3,800 miles during the ten days in the China Sea without battle damage; and the weakness of enemy air reaction had shown the entire region to be wide open for future attack."

While the fast carrier task force was operating in the South China Sea, the invasion of Luzon proceeded satisfactorily. Only slight opposition was met in the advance through the central plains, but increasing resistance developed on the left flank in the direction of the approaches to the mountain passes to northern Luzon. Guerrilla forces conducted widespread demolition and sabotage, contributing materially throughout the campaign.

Land-based air moved into the Lingayen area on 12 January, operating from the newly acquired and reconditioned air base at Lingayen. These forces relieved the escort carriers of their responsibility for direct support missions on 17 January, the date on which the Lingayen Defense Force was dissolved. Eight CVE's, with screen, departed from the area to Ulithi to prepare for the invasion of Iwo Jima. The six CVE's remaining were re-formed under Rear Admiral F. B. Stump and departed for Mindoro, to support minor amphibious landings on 29 and 30 January in the Manila Bay area.

The withdrawal of the CVE's ended the support of the Philippine campaign by carrier-based aircraft. Control of the air had been wrested from the Japanese and the strength of land-based air was now such that carrier support was no longer needed.

On the date of the dissolution of the Lingayen Defense Force (17 January) responsibility for the direct support of ground operations and the protection of convoys passed to the Commander, Allied Air Forces. The air forces under his command included those of the army Far Eastern Air Force, the Marine Corps First Marine Air Wing, and the navy patrol squadrons of the 10th and 17th fleet air wings.

Close support missions, pre-invasion bombings, cover for army landings, fighter sweeps, combat air patrols, and reconnaissance missions were flown by these forces in support of the advancing ground troops. Interdiction of the enemy's rear carried out by land-based air included the destruction of ammunition and fuel dumps, storage and supply depots, buildings, warehouses, and barracks, besides attacks on important highways, bridges and enemy troop movements.

Units followed the progress of the ground occupation, shifting from Luzon, Leyte, and Mindoro, through the Visayans and into

Mindanao, eventually extending to include support of the invasion of the islands of the Sulu Archipelago.

The success to airmen of the First Marine Wing in supporting ground operations was expressed by Major General V. D. Mudge, commander of the 1st Cavalry, U. S. Army, as follows:

"The Marine dive bombers of the First Wing have kept the enemy on the run. They have kept him underground and have enabled our troops to move up with fewer casualties and with greater speed. I cannot say enough in praise of these men of the dive bombers, and I am commending them through proper channels for the job they have done in giving my men close ground support in this operation."

The same high opinion of the work of the Marines was held by the men of the 41st Infantry Division, who presented a plaque to the Marine Air Group commemorating the efficiency and effectiveness of the work achieved. The letter accompanying the plaque stated:

It is the desire of the Commanding General of the 41st Infantry Division to present this plaque to the officers and men of the Marine Air Group at Zamboanga in appreciation of their outstanding performances in support of the operations at Zamboanga, Mindanao.

The readiness of the Marine Air Group to engage in any mission required of them, their skill and courage as airmen and their splendid spirit of cooperation in aiding ground troops have given this division the most effective air support yet received in any of its operations. The effectiveness and accuracy of the support given by this Group proved a great factor in reducing casualties within the Division. The work and cooperation of this Group had given the officers and men of the 41st Infantry Division the highest regard and respect for their courage and ability.

United States troops entered Manila on 26 January, and its fall was formally announced by General Douglas MacArthur on 6 February 1945. It required, however, nearly a month of hard fighting to free the city of the Japanese; this was accomplished on 24 February.

The campaign progressed southward, and on 21 April General MacArthur announced that U. S. and Filipino troops were in control of all major and most of the smaller islands of the Visayans. Landings were made on Mindanao in the Davao area on 17 April, and on 4 May the city of Davao was captured. Although the recapture of the Philippines was assured, strong but futile resistance by the Japanese continued in isolated sections of the islands on Luzon and Mindanao until the Japanese surrender on 15 August.

25 – Iwo Jima

THE PATTERN OF AMPHIBIOUS OPERATION was familiar by the time
the American forces set out to capture Iwo Jima. It had been tried at
Tarawa, perfected in the Marshall invasions and had reached ma-
turity in the operations which brought about the fall of the Marianas
and the Philippines. The vast fleet of troop transports, supply ships,
escorts, land bombardment units and carrier task forces would sortie from their respective bases, set sail for their objective and move
through the now familiar sequence: air strikes and shelling of shore
installations, carrier plane attacks on subsidiary Japanese airfields
to obviate the possibilities of Jap counteraction by air, destruction
by surface ships and aircraft of any Jap fleets barring the way, assault of the beaches, and ultimate capture of the island by land
action.

Preliminary air bombardment of Iwo Jima was delivered by
heavy horizontal bombers operating from Marianas' bases. A long series of area bombing missions was flown during the months
preceding the landing with an eye toward the neutralization of airfields and installations, destruction of gun positions and fixed
defenses, and the unmasking of additional targets. Although the tonnage of bombs was large, apparently no permanent results were
obtained. The southern portion of the island is soft volcanic sand
which easily craters but is just as easily smoothed out again. On the
north Iwo Jima is rocky and has many steep ravines. The Japs were
well dug in and only those bombs which hit the few exposed targets
did any lasting damage. Furthermore, as far as can be determined,
the psychological gain from the prolonged bombardment could not
be measured in terms of reduced efficiency on the part of the Japs
on the day of the landing. The immediate air support of the Iwo landings was entrusted to two units: Vice Admiral M. A. Mitscher's by
now world-famous Task Force 58 and Task Group 52.2, commanded
by Rear Admiral C. T. Durgin. The two forces supplemented each
other. Task Force 58 was to steam northward and lie off the Japanese coast while its planes attacked the Tokyo area. During the
initial phases of the assault on Iwo it would return to give direct aerial support to the invasion before turning its attention once more to
Tokyo.

Finally, it would make photo reconnaissance runs over Okinawa, Kerama Retto, and Amami Gunto — a presage of things to

come. Task Group 52.2, on the other hand, was primarily concerned with direct support of the landings. En route to Iwo it would provide submarine and combat air patrols and once the objective was reached it would bomb positions and spot targets for fire support ships and land artillery.

Task Force 58 consisted of 11 fleet carriers, 5 light carriers, 8 battleships, 6 heavy cruisers, 11 light cruisers, 81 destroyers and some 1200 planes when it sortied from Ulithi on February 10, 1945. Assuming a generally northward course, it arrived off the coast of Japan in the Tokyo area early on February 16. The operations for the first day's strike, which occurred exactly one year after the first raid on Truk, involved continuous fighter sweeps against airfields in the assigned target areas, with the planes taking off at dawn on the sixteenth. About 1130 the torpedo and bomber units were launched for raids on the Nakajima Ota and Koizumi aircraft plants. The strikes were only partially successful because weather conditions were poor. Many planes could not get through to their assigned targets and had to be content with dropping their bombs on airfields and smaller industrial plants.

During the day, the fighter sweeps continued, but by sunset all planes, and night groups were taking off. Night operations were seriously limited by weather which forced the planes to restrict their activities merely to defensive searches where offensive operations had been originally planned. This bad weather was not without recompense, however, for the poor conditions kept Japanese planes from making attacks on the task force.

On the morning of 17 February, the fighters again took up the burden of neutralizing the Tokyo airfields while the bombers turned their attention to aircraft engine plants in the area. In addition, shipping in Tokyo Bay was bombed, and a light carrier and several smaller units were sunk. By noon, since the weather had become much worse the planes were called back, and the task force retired toward Iwo Jima.

Between 19 and 23 February, Task Force 58 was occupied with rendering air support to the Iwo landings. During that period, constant strikes were conducted on ground installations, while morning and afternoon raids hit Chichi Jima and Haha Jima in the Bonin group to the north for the purpose of neutralizing enemy air operations. When these tasks were successfully completed, the task force steamed northward to deal once more with its primary target.

The second Tokyo strike was made on 25 February and attained a fair measure of tactical surprise in that the task force was not discovered on the run-in until about 0230 that day. This gave the

Japanese about five hours to prepare for the first attack, and apparently this preparation consisted of flying as many planes as possible out of the Tokyo area. Fields which had based large forces of planes during the previous attacks were practically deserted. Some nonaggressive fighters were encountered, but all these avoided combat and no attacks were launched against the task force.

Weather conditions were extremely unsatisfactory, but the bomber strikes were able to break through and severely damage the Koizumi aircraft plant. At noon the weather forced the planes back to their carriers, and shortly thereafter retirement from the Tokyo area was affected. The next day an attack on Nagoya was planned, but since the planes could not be launched the force retired to the south.

On 1 March the squadrons were busy again, this time over Okinawa. Task Force 58 had cruised to Iwo after the second Tokyo strike and then had moved westward towards the Ryukyu chain. The approach to Okinawa attained complete tactical surprise, and the task force was not attacked during the operations, while fighter sweeps and photographic missions scoured the islands, gaining 80 percent coverage of the required area. Following the first day's operations, the action was broken off and the force turned toward Ulithi, where it anchored on 5 March.

Despite consistently poor weather, the operations of Task Force 58 were a complete success. Over 5,500 sorties were flown in dropping almost 1,200 tons of bombs and firing 10,000 rockets. Jap losses included 393 planes destroyed in the air and 266 on the ground. Another 141 were probably destroyed and 354 were damaged. Our losses were 143 planes and 95 pilots and air crewmen. Pickings were slim on the fast disappearing Jap Fleet and merchant marine. One carrier and about seventeen small combat ships were destroyed, while seventy-nine merchant ships and small craft went to the bottom. About 126 assorted ships, including two destroyers, were damaged.

Ground targets in the Tokyo area received a heavy mauling. Two aircraft plants were rendered at least temporarily inoperative and two plane engine plants were hard hit. Twenty-three airfields around Tokyo received attention which left them in various stages of ruin; runways were pitted, planes destroyed, hangars burning, shops demolished, barracks damaged, and fuel dumps fired. The net result, in addition to the actual damage, was that the Japanese were unable to funnel planes to the Iwo area where they might well have presented serious opposition to our landings. For that reason alone, if for no other, the activities of the task force may be considered highly successful. Meanwhile, Task Group 52.2 was carrying out its operations off Iwo Jima. This force, although dwarfed by the giant strength of Task Force 58, was a powerful striking unit in its own right. Its main power lay with the carriers Saratoga and Enterprise and 12 escort carriers with their force of over 400 fighters and bombers. These were protected by 2 heavy cruisers, 1 antiaircraft cruiser, 12 destroyers, and 14 destroyer escorts. The group left Ulithi on 10 February following Task Force 58 and preceding the bombardment and fire support ships.

By 12 February the ships had arrived in the Saipan-Tinian area for scheduled rehearsal and then they proceeded north to Iwo, which was reached on the fifteenth and sixteenth. Attacks were first launched 16 February on Iwo Jima installations and on the airstrips of Chichi Jima and Haha Jima and continued without abatement until 8 March when the situation on Iwo no longer required air support. By 11 March most of the units had returned to Ulithi and the rest were underway. The strikes at Iwo were, in the main, direct support to land operations. The three days preceding D-Day (19 February) were used to bomb out as many emplacements as possible: gun, mortar, antiaircraft sites, rocket launchers, pillboxes, supply dumps, troop concentrations and coastal defense guns.

After the landings, the support became extremely specific as planes were used to eliminate individual pillboxes or to strafe cave entrances. The planes were also called on to direct artillery fire, an extremely hazardous mission because of their vulnerability to antiaircraft fire. The blows at Chichi Jima and Haha Jima were aimed primarily at enemy air power, for the Japs made some attempts to gather squadrons at those points for attacks on our amphibious forces.

Such attacks were not without success. During twilight and dusk on 21 February the Japanese pressed home a determined torpedo and suicide attack on the task group's carrier units. It was made at a time when many planes were returning from operations over Iwo and with the resultant chattering of radar screens, tracking of the Jap planes was very difficult. The Saratoga was the first to be hit, a suicide plunging into her flight deck about 1700. This started fires and made the landing of planes difficult, but the situation was in no way critical. The attack continued, however, and three more Jap fighters succeeded in crashing the huge ship, starting fires on the forward part of her flight deck and making it impossible for her to land her planes. She was badly damaged, although by no means fatally, and was forced to leave the action at once.

Planes had been sent aloft, meanwhile, to intercept the Japanese, but darkness, weather, and the cluttered radar scopes hampered their efforts and the attackers continued to come in. About 1945 the Lunga Point was hit a glancing blow by a torpedo bomber. The plane exploded just before impact with the ship and then skidded across the flight deck and plunged into the sea on the opposite side. Fire broke out but was quickly extinguished, and operations were resumed on schedule the following morning.

Not five minutes later the Bismarck Sea took a suicide plane near the after elevator, and this was almost immediately followed by another plane which hit just forward of the elevator. Fires were started and spread forward rapidly. Many small explosions followed as if from her antiaircraft ammunition, and then two large explosions occurred, coming from deep inside the ship. It was soon evident that the vessel could not be saved, and she was ordered abandoned and sank two hours later.

Planes of Task Group 52.2 flew over 8,800 sorties from 16 February to 11 March. Most of these flights were against land installations on Iwo, and it has been difficult to appraise accurately the number of ground targets damaged or the extent of such damage. Known damage or destruction, however, included enemy coast defense, antiaircraft, machine-gun positions, mortars, rocket launchers, tanks, trucks, pillboxes, trenches, buildings, supplies and troop concentrations.

Very few Jap surface craft were encountered, as the Japanese Fleet did not choose to oppose our landings. Three merchantmen were sunk, however, and 7 more probably sunk while 35 small craft were damaged. Seventeen planes were destroyed in the air and on the ground, with 6 more probably destroyed and 15 others damaged. Hunter-killer attacks on enemy submarines achieved a score of one possible sinking and damage to three others.

In comparison with the record set up by Task Force 58, the task group's losses seem excessively high. Sixty planes were shot down, most of them victims of antiaircraft fire. Eight pilots and 12 air crewmen were reported dead or missing. The Lunga Point received minor damage, the Saratoga was badly mauled, and the Bismarck Sea was sunk with about 350 of her men and officers.

These losses, however, had to be accepted as did the heavy casualties suffered by the Marine Corps landing forces in the capture of this bloody but vital bastion, Iwo Jima.

26 – OKINAWA

IN EARLY MARCH OF 1945 Task Force 58 was in Ulithi atoll. At anchor in the blue waters calmed by the protecting ring of coral reef and green islands, the carriers were huge, silent, and baking hot under the tropical sun. That sun felt good to the officers and men of Task Force 58. It made them forget the chill driven into their bones by the raw, wet winds of a few weeks before off Tokyo and Iwo Jima. After noon chow flight decks, gun galleries, and all available topside spaces were filled with half-nude figures sopping up the warm, pleasantly enervating sunshine.

Sluggish landing craft filled with khaki-clad officers or ships' crews in dungarees churned toward the palm-fringed beaches and thatch-roofed bars on the island of Mogmog. Lighters, barges, and tankers plowed around the anchorage glutting combatant ships with fuel, ammunition, and provisions. It was a period of rest and replenishment.

As always in port, scuttlebutt on the coming operation was the chief topic of conversation among the crews and junior officers. With native shrewdness the mess cook in the galley, the "airdale" in the rear cockpit of a Helldiver, the fighter pilot, the junior officer commanding a five-inch gun mount all spoke of Japan's inevitable doom. How to do the job best and at least cost was the question.

That something big was in the wind was evident from the constant shuttling of admirals' barges and captains' gigs between the carriers and battleships. Secret conferences were going on; big plans were being laid. Many guessed the answer. "Okinawa," they said, "there's no other place to go." Some argued for Formosa, but not many.

Most of the men with Task Force 58 at this time had observed the haphazard, hesitating development of the Japanese suicide dive, the body crash of the Kamikaze pilot. They had seen it first in the Philippines, then off Formosa. No pattern or calculated plan seemed behind the attacks. They were ill-organized and with notable exceptions missed fire. Much concern was felt before the Tokyo strikes, but concern turned to confidence and almost contempt when Japanese planes during those strikes had fled to the west, refusing to meet the American challenge.

Thus, the average man in the fleet guessed the locale of the next operation, and he faced the task before him with renewed confidence. What he did not guess was the extent to which the enemy was prepared to employ pilots and planes of the Kamikaze Corps — the factor that was to distinguish the Okinawa campaign from all other campaigns in naval history, as the longest, toughest, most costly battle ever fought and won by the men and planes and ships of the United States Navy. He did not guess, but he might have been forewarned by an incident that happened a few days before the fleet sortied.

The last dim glow of a tropical sun was just fading from the horizon, and in the mood of relaxation, most ships were showing movies. The large carrier USS Randolph was typical. Down on the steaming hot hangar decks officers and crew were sweating through the second reel of a mystery thriller. A few wisps of breeze strayed in the openings on the sides of the hangar deck. Men out on the catwalks for the cool evening air watched the dark waters lap at the ship's anchor chain.

Except for the occasional bright patch of a movie screen, the ships were dimmed out, but all around the atoll, lights gleamed from the shore depots working overtime to load the big ships. It was almost like any harbor back home in peacetime.

Then, without warning, the Randolph shuddered; the concussion of a heavy explosion drowned the smooth screen voices. Smoke poured in the after end of the hangar deck. The dark waters blazed with the reflected light of a welling flight deck fire. Stunned crew members automatically rushed for their battle stations at the insistent clang of general quarters above the surprised, indignant exclamations and the successive, smaller explosions from the stern of the ship.

General quarters sounded on all the other ships in the harbor. Amazed officers and crew streamed topside to watch the leaping flames and orange core of the gasoline fire enveloping the stern of the Randolph. On the USS Yorktown, a Marine in a gun sponson exclaimed, "I tell you it was an F6F. Came in low over us and crashed right into her." Theories flew fast, but not until the smoking, twisted stern of the Randolph had cooled sufficiently for examination the following day, was it definitely determined that a Kamikaze pilot had made the 800-mile one-way trip from the island of Minami Daito to crash his single-engine seaplane into a carrier in Ulithi Atoll.

Vaguely, it seemed unfair. This wasn't supposed to happen in port; port was for recreation and rest. Nerves were on edge the next few days as flag officers hurriedly re-formed the task groups to exclude the Randolph on which repair crews were already working.

The stage was set for the Okinawa campaign.

BY MARCH OF 1945, the enemy must have known his number was up. On paper, the Japanese goal of "Hakko Ichiu, or the eight corners of the world under one roof" still looked impressive. The enemy held Borneo, the Dutch East Indies, Singapore, Malaya, Burma, great chunks of China. All his big armies were still intact. But his power existed only in the boastful, self-deluding words of the Domei News Agency which prepared propaganda for export.

Paradoxically, his stolen territories representing some of the richest lands in the world were practically worthless to the enemy. His vital supply line to the southwest had been cut to a painfully small trickle by navy carrier and search planes, by submarines, and by army bombers.

A few three-hundred-foot craft — called "sugar dogs" by the aviators — managed to slip up the China coast by night and thence to the Jap mainland. Another trickle slipped along the overland rail route through China — highly touted but impractical for the volume of supplies needed. The fleet of combatant ships that should have been protecting Japan's supply line had been mauled in the Battle for Leyte Gulf, and what was left remained bottled up in the Inland Sea.

Among many strategic materials, Japan's most pressing need was for high octane fuel — aviation gas for fighters to meet the crushing power of the B-29's and the terrifying new actuality of carrier raids on the mainland. Japanese scientists in desperation tried with some small success to "crack" oil from pine-tree roots for the vital fuel. Yet these and other efforts were like trying to stop a herd of stampeding elephants with a peashooter.

Japan was throttled, gasping. Despite her fatal weakness, Japan was not yet beaten to the point of surrender. To deliver the final blows, we needed one more land base, a final steppingstone for which there were three basic requirements: first, location within easy fighter plane range of the mainland; second, enough space for the huge airfields and dispersal and storage areas to support B-29 operations; third, harbor facilities ample and diversified enough to be developed as an advance staging base for the invasion of the main Japanese islands.

Within the tight ring closing on Japan, only one island filled the bill. Stringing southwest of Kyushu parallel to the China coast and extending to Formosa to form the eastern boundary of the East China Sea is the Ryukyu, or Nansei Shoto chain of islands. Keystone and largest island of this chain is Okinawa. On aviators' charts, the air-distance circles spreading out from Okinawa brought Kyushu within 350 miles, Tokyo within 800. Okinawa already had five air-fields built by the Japanese.

There were enough additional level spots on Okinawa's rugged, mountainous terrain to provide the B-29 fields and dispersal area that were needed. The irregular coast line of the narrow, 60-mile-long island offered good anchorages. Nakagusuku Wan on the east coast was the harbor to which crippled enemy units had re-tired after the Battle of the Philippine Sea. Naha Harbor on the east coast was adequate for smaller vessels.

Okinawa Gunto — "a prefecture of Japan . . . a densely popu-lated chain of rugged islands" — had all the requirements. And the decision to wrest this island from the Japanese as a final base from which to overwhelm the enemy precipitated the greatest test in his-tory for the United States Navy. In this test the naval air force was to play a major role — a role which marked the culmination and fullest realization of the battle techniques developed during the relentless march across the Pacific.

H-Hour was 0700 on the fourteenth of March. Task Group 58.4 was elected to lead the units of Task Force 58 for the sortie marking the opening of the Okinawa campaign. The first ships to move in the morning light were the small, quick destroyers. They poked their bows out Ulithi Atoll's entrance channel to guard against the possi-bility of a Jap submarine lurking outside the coral reef waiting for such a moment. Then the heavy units began to slide through the waters which were unctuously smooth at this breezeless hour of the morning. As the big carriers gathered speed and formed in Indian file to pass through the narrow confines of the channel, they moved with a flowing ease that belied their size.

Task Force 58 was in full strength that morning with four regu-lar task groups and a night fighter task group. Despite the temporary loss of the Randolph, the carrier core of the force included 10 CV's and 6 CVL's. This was buttressed with 8 battleships, 2 new battle cruisers, and 14 heavy and light cruisers.

Admiral Raymond A. Spruance, as Commander, Fifth Fleet, was in charge of the entire Okinawa campaign until relieved by Admiral (later Fleet Admiral) W. F. Halsey late in May. His flagship, the cruiser Indianapolis, operated during the initial phase as part of Task Group 58.3, as did also the flagship of Vice Admiral Marc A. Mitscher, commander of Task Force 58.

Task Group 58.1, which included the CV's Hornet, Bennington, and Wasp, and the CVL Belleau Wood, was commanded by Rear Admiral J. J. Clark. Task Group 58.2 with the CVs' Franklin and Hancock and CVL's San Jacinto and Bataan was under the command of Rear Admiral R. E. Davison. With Rear Admiral F. C. Sherman as task group commander in the CV Essex, Task Group 58.3 also included the Bunker Hill and the CVL Cabot. Task Group 58.4, under the command of Rear Admiral A. W. Radford in the CV Yorktown, included the CV Intrepid and the CVL's Independence and Langley.

The widely known "Big E," the famed CV Enterprise, now loaded exclusively with night fighters and night torpedo planes and functioning as a night carrier, was to operate with Task Group 58.4 during the daytime and at night, with necessary screening units, was to operate independently as Task Group 58.5 under the command of Rear Admiral M. B. Gardner.

The aviation phase of the strategy for the Okinawa campaign worked out by this team of flag officers was infinitely complex in detail but direct and simple in its basic concepts. While the landing forces were mounted at Leyte, Guadalcanal and Saipan, Task Force 58 would strike a heavy neutralizing blow at airfields on Kyushu, Shikoku, and western Honshu.

Then as the troop ships and train of supply vessels converged on the objective, Task Force 58 would operate to the northeast of Okinawa providing an aerial barrier between the invasion forces and the major source of the enemy's air power on the mainland. While the fighters were holding off enemy planes, the bombers of the task force would knock out Okinawa's defenses and then together with planes from the escort carriers of Task Group 52.1 would provide air support for the troops until sufficient land-based air forces could be established to take care of the job.

Routine training exercises filled the first two days on the long trip from Ulithi to Kyushu. This training was undertaken seriously, almost doggedly, for as soon as the ships left port, the word of the coming operation had been passed on.

Now the die was cast. As an aviation radioman put it, "It's one thing to chop your gums in port about what's going to happen next, but, brother, when the captain comes on the bullhorn and says, 'Target, Kyushu,' you stop kidding and start thinking — serious thinking."

Pilots, who knew that once again they would be flying over the Jap mainland, dived with unusual ferocity on the target sleds towed behind each carrier. And antiaircraft gunners, who realized their ships would be ready targets within easy range of enemy planes swarming from a hundred fields, shot in tight-lipped earnestness at the red sleeves trailing far behind the tow-planes.

On 16 March, Task Force 58 rendezvoused with squat tankers and escort carriers loaded with replacement planes for a final "topping off" eight hundred miles to the southeast of Kyushu. Replenishment finished, the task force set course for Kyushu and began the high-speed run-in.

As the throb of the giant turbines stepped up, one thought was with practically every man in the fleet. "Would the force get in undetected? Could it slip under the Jap searches and past the picket boats?"

Surprise always has been a major factor in carrier operations. Achieve surprise and you catch the enemy planes before they can take to the air. You pin the enemy down before he can hit back. The carriers had achieved surprise on their first Tokyo strikes. And the enemy planes when they finally did get off the ground simply fled west.

Throughout the anxious day of the seventeenth of March when Task Force 58 was churning toward Kyushu at high speed, there were several reported bogies. They all turned out to be friendly planes or did not come close enough to spot the force. Shadows lengthened and light faded from the overcast sky. Top-ranking officers relaxed a bit; the worst day was over, and the force had not been spotted.

But their ease of mind was short-lived. Snoopers were detected shortly before ten o'clock — 2141 precisely. "Bogey bearing 142 degrees, distance 72 miles," rasped the voice over the loudspeaker from Combat Information Center where fighter director officers were tensed before the spoke of blue light spinning round the dial on their radars. Within a few minutes, seven more "raids," or groups of enemy planes, were detected. This time it was certain the force had been spotted.

The snoopers hung on, shadowed the force throughout the night. Two were shot down by Enterprise night fighters, but others took up the hunt. The enemy was fully alerted some eight hours in advance of our first scheduled strike! This probably meant trouble.

On the morning of the eighteenth, skies were overcast, and the pre-dawn launches of aircraft had to be held up for a few moments. But at 0545, the first fighter sweeps, and combat air patrol fighters roared off the decks. Task Force 58 was a hundred miles to the east of the south tip of Kyushu.

The invisible beam of the radar combed the skies, anxiously seconded by the eyes of ship lookouts. A light, low overcast made the lookout's job doubly hard. The enemy snoopers of the previous night had been reinforced.

Many bogies closed Task Group 58.3 at 0500, dropping clusters of flares. One strayed within range of the task group's antiaircraft guns and was shot down.

This was only the beginning. Although never in great force, enemy planes pecked away at our ships throughout the day. Our gunners were hot, and all the task groups splashed enemy planes.

Task Group 58.4 was attacked most persistently and was the only group to suffer damage. At 0737 a single-engined plane popped out of the overcast and dived at the Enterprise, releasing a 600-pound bomb that hit the deck at the forward elevator but failed to detonate. Damage was minor. Half an hour later, a twin-engined Betty made a low gliding run on the group and was exploded a few yards short of the Intrepid. Flaming debris from the plane smashed into the Intrepid's side, killing one man and doing minor damage.

Between 1300 and 1500, the Yorktown was subjected to three attacks by dive bombers screaming out of the overcast. The first two missed with their bombs and the third plane's 600-pound bomb struck the ship a glancing blow. The bomb hit in the island structure and grazed down the side, exploding at hangar deck level. Damage, however, did not hamper the operational capacity of the ship.

Meantime, planes at the target on this first day of strikes piled up a comfortable score. In the air our fighters shot down 102 enemy planes, and on the ground our strafing, rocketing, and bombing attacks destroyed or damaged some 275 aircraft. Heavy damage was inflicted on airfield installations.

During the morning of that first day, photo reconnaissance revealed enemy combatant shipping concentrations at Kure and Kobe, enemy naval bases in the Inland Sea. Vice Admiral Mitscher decided to send his strikes on the following day against these ships with the naval base installations at Kure and Kobe as alternate targets.

The night of 18-19 March passed quietly. Planes thundered from the decks at 0525 to begin the second day of sweeps and strikes. The planes found their targets and damaged in varying degrees an enemy battleship, a hybrid battleship-aircraft carrier, three CV's, a CVL, a CVE, a heavy cruiser, and a light cruiser.

But our planes paid heavily to the intense antiaircraft fire encountered. One task group alone lost thirteen aircraft to the murderous cone of flak thrown up over the Kure Harbor.

While task force planes were dodging flak at the target, the fleet was attacked by enemy aircraft. Shortly after 0700, a single-engined Jap plane poked out of the clouds over Task Group 58.2 and placed two bombs squarely on the deck of the Franklin to start a devastating series of explosions and fires. A few minutes later, the Wasp in Task Group 58.1, took a single bomb hit amidships on the flight deck. The bomb pierced flight deck, hangar deck, and another deck below before exploding. Despite considerable damage, flight operations on the Wasp were resumed within an hour.

With the morning strike planes back aboard, the task force retired in the early afternoon of the nineteenth. Air cover was provided for the stricken Franklin and fighter sweeps were launched against Kyushu airfields to smother any remaining enemy planes.

Because of the Jap propensity to concentrate on crippled ships, heavy air attacks were expected. Only one developed on the nineteenth when eight enemy planes were intercepted early in the evening by combat air patrol eighty miles from the force. Five were shot down and three escaped to their bases.

At the end of the second day, in addition to the ships and shipping installations damaged at Kure and Kobe, intelligence officers counted 97 planes destroyed in the air and 225 additional aircraft destroyed or damaged on the ground to bring the two-day total to 199 aircraft shot down and 500 destroyed or damaged on the ground.

No Japs appeared that night, and by morning of the third day, the twentieth, the Franklin was 160 miles from the nearest enemy base but proceeding at only five knots under tow of the cruiser Pittsburgh. Her rudder, jammed hard over, was causing trouble.

Complete absence of enemy attacks throughout that morning gave the surviving members of the Franklin's heroic crew an opportunity to get the ship moving under her own power at a speed of 15 knots.

Finally, at 1630 on the twentieth, the enemy tried again. Fifteen "bandits," approaching low and fast, split up for individual attacks on Task Group 58.2. The Franklin was unscathed as seven planes were shot down by fighters and seven by ships' gunners, but Enterprise was hit by our own gunfire and put out of action for flight operations, and the destroyer Halsey Powell was severely damaged by a plane shot down by the Hancock's guns.

Shadowers appeared again at dusk and persisted throughout the night of 20-21 March. Task Groups 58.1 and 58.3 smashed a torpedo plane attack shortly before midnight, and three additional snoopers were splashed as tracer shells from antiaircraft guns periodically lit up the night sky.

At 1400 the next day, 21 March, a large group of bogies was detected by radar a hundred miles to the northwest. Extra fighters were scrambled and within a few minutes 150 Hellcats were airborne. Twenty-four fighters from Task Group 58.1 were the first on the scene to intercept thirty-two twin-engined Bettys and sixteen single-engined fighters. In a short, wild melee our fighters shot down all the Bettys and all their fighter escort for a loss of two of our own Hellcats.

There was a significance to this attack that was not realized at the time. The Bettys shot down were unusually slow, and pilots reported seeing what seemed to be a small, strange, extra pair of wings extending below the fuselage and parallel to the wing on the enemy bombers. Gun camera films, when developed, confirmed the pilots' reports. What these weird instruments were did not dawn upon our forces until a few days later.

The decisive obliteration of this attack group ended Jap efforts against the Franklin and, in fact, terminated the first phase of the Okinawa campaign. We had taken some damage, but we had dealt heavy blows.

DESPITE THE RIGOROUS PRELIMINARIES AT KYUSHU, in one sense the Okinawa campaign did not really begin until the sixth of April.

Indeed, much had happened since the damaged Franklin, Wasp, and Enterprise had been sent safely on their way to Ulithi. On the twenty-third of March, the navy air force moved in on Okinawa in a familiar pattern of activity. Bombing attacks were carried out on all known defense installations as heavily loaded Avengers and Helldivers streamed from carrier decks with clock-like regularity.

Airfields along the Ryukyu chain to the northeast and southwest of Okinawa were strafed and bombed to deny them to the enemy. Sturdy Hellcats buzzed leisurely over the hundreds of ships en route to Okinawa — first the minesweepers, then the forces that made the initial landings in the Kerama Retto to the west of Okinawa, then the cruisers and battleships of the naval bombardment groups, and finally the transports loaded with troops for the frontal assault on Okinawa itself.

Enemy air opposition was sporadic; the strikes on Kyushu apparently had served the intended purpose. On 1 April, Easter Sunday morning, our landing forces waded into the beaches just off Yontan airfield after a scathing series of "pre-how-hour" aerial

attacks and naval bombardment. The bitter enemy opposition anticipated was not met. The troops walked across the island within three
days. And still no heavy air attacks. The operation proceeded ahead
of schedule; it was easy, almost ominously easy.

But on the sixth of April, there was an abrupt change; the enemy, at long last, hit back. On that day the Okinawa operation
became the toe to toe aerial slugging match that is burned into the
memory of all the men who fought in the campaign.

"Cloudy becoming overcast at dawn. Visibility 8-10 miles. Northerly winds 10-15 knots. Slight sea and low swell from NW." In this
matter-of-fact language the aerologist's report described 6 April — a
perfect day for the Kamikaze!

By the time the shrouded sun was full over the horizon, it was
evident this was not to be another quiet day. Reports of large groups
of bogies crackled over ship radios. Loaded torpedo planes and dive
bombers were struck below, de-gassed and de-armed. Fighters were
serviced and warmed up as Task Force 58 prepared to meet the
threat.

The attacks reached fullest intensity in the early afternoon. The
enemy came and kept coming in attack units ranging from single
planes to groups of 30 or more. All told, more than 400 enemy
planes — all suiciders — surged down the Ryukyu chain that day.

Our naval aviators on combat air patrol met the thrust as far
north as Amami O Shima, midway between Kyushu and Okinawa;
they met it at the protecting ring of radar picket destroyers thrown
out around the major concentration of ships at the Okinawa anchorage; and they met it disposed in an arc to the north of Task Force
58.

Many young navy pilots became aces on that day. Breathlessly
they told of the "sitting ducks" they had exploded out of the sky. As
their reports were pieced together, it was apparent that many of the
Kamikaze planes were the oldest of flyable models — obsolescent
dive bombers of 1937 vintage, ancient torpedo planes, outmoded
seaplanes. Mixed in the conglomeration of enemy types were twin-
engined bombers, sleek, new, high-speed interceptors of the latest
enemy fighter model, and many Zekes, the enemy's familiar work-
horse fighter.

At Okinawa and over the task force that day, carrier-based fighters shot down 236 enemy planes. In achieving this score, our losses
totaled two planes. Task Force 58 ships splashed an additional 13
planes. Escort carrier fighters added 55 enemy planes to the total

and antiaircraft fire from ships at Okinawa accounted for 35 more. The enemy's Sunday punch was effectively blocked.

This first large-scale suicide attack set the pattern for all the others that followed. With minor exceptions, total enemy planes participating in each of the attacks ranged from 100 to 300. The attacks were spaced generally from seven to ten days apart, apparently the minimum time in which the enemy could muster his forces for an all-out blow. In the interval between major smashes, smaller-scale attacks were launched, but these were scattered and haphazard.

For suicide purposes, the enemy scraped the bottom of his aviation barrel. The oldest models that could still take to the air, including many trainers, were dispatched on the missions with frightened, ill-trained pilots at the controls. For most of the Kamikaze pilots it probably was their first combat mission — and their last!

Our fighter pilots constantly reported in amazement the ease with which they had destroyed the enemy. A division of four fighter pilots on the Yorktown amassed a total of fifty enemy planes between themselves without so much as receiving a single enemy bullet in any of their planes. "No evasive action," became a stock phrase in pilot reports describing the reaction of enemy planes which were destroyed.

Along with the suicide planes, the enemy usually sent newer fighter models, apparently as a covering force to engage our combat air patrol while the Kamikaze planes slipped in to smash our ships. Better planes flown by better pilots still were no adequate match for our planes, pilots, and teamwork tactics.

As the ultimate in suicide attack, the Japanese introduced a new weapon during the Okinawa campaign. Our first hint of the weapon had come on 21 March with the destruction of the thirty-two Bettys, on each of which had been observed a mysterious extra set of wings below the fuselage. Ground forces storming ashore at Okinawa discovered the first positive evidence of the "piloted bomb." They found several of the weapons assembled.

The flying bomb had stubby wings 16 feet across, a long-pointed nose which contained a 2,000-pound warhead, a cockpit with elementary controls, a twin-rudder tail assembly, and four tubes behind for rocket propulsion. The gadget was fitted to be carried half inside the bomb bay of twin-engined bombers of the Betty or Peggy type.

Engineers estimated that a speed of more than five hundred miles an hour could be attained by the bomb, but because of the small control surfaces they felt it would not be very maneuverable. It was obvious the mother plane would have to have the target in sight before the piloted bomb could be released. Typically skeptical, our forces dubbed the new weapon "Baka" — Japanese for "fool."

After the discovery of the bomb ashore, reports of sightings were made from various ships under air attacks, but it was not until the twelfth of April that the enemy scored with the Baka bomb. The destroyer M. L. Abele, already damaged by a suicide plane, took a hit amidships from a Baka. The ship broke in two and sank. In the three months at Okinawa, only three other hits by Bakas were recorded, and these hits were considerably less destructive than the first.

By their own admission in propaganda broadcasts to our forces, the Japanese expended more than three hundred Baka bombs — and, of course, a similar number of pilots — to obtain these four hits. Perhaps the best tip-off on how the average Jap soldier felt about this weapon was provided by a scared young Japanese, a crew member of an enemy bomber shot down by Task Group 58.4 fighters and later fished out of the water by a destroyer of that group. Under interrogation on the admiral's flagship, the nineteen-year-old petty officer said that volunteers to fly the bomb were notably lacking. Orders were necessary to convince pilots they should die for the Emperor in this fashion. Baka was a most appropriate name for the weapon.

The enemy's attacks which were mounted from Kyushu — principally from Kanoya air base — usually split two ways. The major arm reached down the Ryukyu chain to the shipping concentrated at Okinawa. This thrust usually was stopped at the radar picket destroyer line by our aggressive combat air patrol. The minor arm of the attack reached out for the fast carriers operating to the northeast of Okinawa. The attacks on Task Force 58 invariably were preceded by search planes sent out to find the exact position of the carriers to be radioed back to suiciders standing by at Kanoya air base awaiting the word.

Supplementing the Kyushu-based elements of the attack were smaller scale fleets of suicide planes stemming from Formosa. These planes either made the trip directly from northern Formosa air bases or attempted to stage through the airfields on Ishigaki and Miyako Jima, at the southern anchor of the Ryukyu chain. The neutralization of these airfields during the first two months of the operation was assigned to the units of Task Force 57, the first British carrier force to operate in the Pacific. In addition to denying the enemy his

Sakishima Gunto airfields, the British absorbed many suicide attacks that otherwise would have descended upon Okinawa.

At first the apparent goal of the suiciders attacking our shipping at Okinawa was the inner anchorage — the heavy fire support ships and the transports loaded with troops and supplies. But our combat air patrol effectively smashed these attacks at the outer ring of radar picket destroyers. Only scattered, lone planes slipped through and scored an occasional hit. As a result, the harassed suiciders began to dive on the first ships they sighted. And the first ships they sighted were the radar picket destroyers and their cluster of supporting ships.

Thus, the radar picket destroyers served two heroic purposes. Primarily they were the first line of defense, and their tools in this function were two — radar, and a team of from four to sixteen fighter planes assigned to each destroyer. The fighter director officer working in the destroyer's Combat Information Center picked up the approaching enemy raids frequently more than a hundred miles from Okinawa. As he tracked the raid in on his radar screen, the fighter director vectored his team of fighters to meet the attack, which sometimes was obliterated completely and almost always scattered.

Secondarily, the radar picket destroyers assumed the grim function of absorbing the attacks of suiciders that slipped through the combat air patrol network. Their guns took over where the fighters on patrol left off. They became the shock absorbers and finally the prime target of the Kamikaze attacks.

Our forces used everything in the book to break the back of the suicide offensive. In addition to the combat air patrol planes controlled by the radar picket destroyers, a "barrier patrol" of fighters was thrown across the path of enemy suicide attacks two hundred miles north of Okinawa. As a collateral mission, these planes carried out daily bombing attacks to deny to the enemy fields at Tokuno and Kikai Jima which might be used as staging bases for planes coming down from Kyushu. Night fighters and night bombers extended these attacks round the clock.

The necessity for providing daily support bombing missions for the troops battling against an enemy dug into caves and every conceivable defensive position, held the fast carriers to a restricted operating area and prevented them from sending regular strikes against Kyushu to overwhelm the suiciders before they could start for Okinawa.

To fill this gap, the B-29's, although not ideally equipped to bomb airfields, were called in to help during the campaign. Between

16 April and 10 May, B-29's based in the Marianas flew 1,391 sorties over Kyushu airfields and further disrupted the enemy's attacks.

Altogether, seven mass suicide attacks were staged between 6 April and 28 May, the critical period of the Okinawa campaign. On 6 April, a total of 339 enemy planes were destroyed; the next attack, on 12 April, cost the Japanese 208 planes; on 16 April, 270 planes were smashed; on 28 April, 115 planes were destroyed; on 4 May, 137 planes were destroyed; on 11 May, 128 planes were destroyed; and from 24 to 28 May, in a persistent series of attacks, 316 planes were downed.

The grand total of enemy planes destroyed during this period was 3,594. Task Force 58's share of this total was 2,259, of which 1,640 were exploded out of the air.

Our total combat and operational plane losses for the period came to 880. This figure includes 266 planes destroyed on damaged carriers. Task Force 58 lost 251 planes in combat, bringing its personal score against the enemy to a 9 to 1 ratio.

At Okinawa the United States Navy suffered the heaviest punishment in its history. Out of approximately 1,500 ships engaged in the campaign, 30 combatant ships were sunk and 223 were damaged during the three-month struggle with the Japanese Kamikaze force. One out of every seven naval casualties suffered throughout the entire war resulted from the Okinawa fighting.

The great preponderance of damage and destruction was wrought by suicide planes. Of the 253 vessels sunk or damaged from all causes, suiciders were responsible in 189 of the cases.

The ships that held the radar picket line to the north and west of Okinawa suffered severely. Destroyers, as a type, were the most frequently hit. Twelve destroyers were sunk, and sixty-seven were damaged.

Although Task Force 58 did not have a single vessel sunk throughout the campaign, the fast carriers took more damage than during any previous operation. Of the ten CV's that sortied to begin the Okinawa operation, eight ultimately were hit, suffering varying degrees of damage. The Bennington and Hornet were the only two of the original ten to escape damage from enemy action. Damage to carriers and supporting ships kept the force down to three task groups instead of the original four that started the campaign.

The escort carrier forces which had taken the brunt of suicide attacks in the Philippines fared much better at Okinawa. Three CVE's were hit of which only one, the Sangamon, was seriously damaged.

The toll exacted by the suiciders cannot be measured alone in terms of casualties to men and ships. An utterly exhausting pace was demanded of ships' personnel in combat for days on end.

Ships on radar picket station were rotated in three-day periods. During their stay on station, men slept, when they slept at all, flung out on steel decks at battle stations. They ate K-rations when there was time between attacks. They fought off unending air attacks on a scale never before experienced. As many as fifty planes attacked a single ship; lone destroyers downed more than twenty planes in a single day with antiaircraft fire alone.

The fast carrier forces broke records for staying at sea und combat conditions. Without respite, Task Group 58.1 operated for 47 days, 58.4 for 62 days, and 58.3 for 77 days. Fighter pilots piled up unprecedented numbers of combat flying hours. Plane maintenance crews, alert all day at battle stations, worked half the night at servicing and repairing planes.

Men of the Navy met and exceeded the demands on their courage and durability. Implacably they fought the enemy but even more desperately they fought to save their ships when the Kamikazes scored hits. It was commonplace for destroyers hit by two or three suiciders to make it back to port. The USS Laffey, smashed by six suiciders and two direct bomb hits, steamed back to a repair base under her own power. Among the carriers, the Franklin and Bunker Hill were monumental examples of the ability of our crews to save badly wounded ships.

The commander of Task Force 58, Vice Admiral Mitscher, paced the toughness of the men fighting under him. On 11 May, Vice Admiral Mitscher's flagship, the Bunker Hill, was rocked by two direct suicide hits. As soon as the boiling gasoline fire was under control, Vice Admiral Mitscher shifted his flag to the Enterprise and carried the fight to the enemy. Three days later, a suicide plane dived into the deck of the Enterprise. Vice Admiral Mitscher and his decimated staff moved over to the Randolph and went on fighting.

When the fleet came to Okinawa, Japanese propagandists blatantly boasted this was the opportunity they had been waiting for. With the entire fleet within range of the Special Attack Corps, it would be an easy matter to wipe United States naval power from the Pacific seas. The Kamikaze pilots would "trade a plane for a ship" until the U. S. Navy was smashed.

It cost the Japanese more than 3,500 planes to learn that the terms of the trade would not be that easy. It cost them nearly the entire effective operational strength of their air force to discover that the Navy had come to stay at Okinawa.

IN HIS OPERATION PLAN FOR THE OKINAWA CAMPAIGN, Admiral Spruance, Commander of the Fifth Fleet, expressed concern that our Okinawa forces might be raided in early morning light by fast enemy surface ships starting out from Inland Sea bases the previous dusk. Against this possibility, a careful network of day and night searches covered all the approaches of Okinawa. As a further precaution, submarines were placed on station at all the exits from the Inland Sea.

The enemy tried a surface raid once — only once! The result was one of the most decisive victories ever achieved by naval air power.

On the night of 6 April, one of the submarines on dreary station patrol off Bungo Suido — the channel leading into the Inland Sea between Kyushu and Shikoku — struck pay dirt. Radar contact was made with a large enemy force heading south. The submarine's radio ticked out the coded report, "At least one battleship . . . supporting units . . . course one nine zero . . ."

The intent of the enemy seemed plain. During daylight on the sixth of April, he had launched his heaviest Kamikaze attack of all time. Apparently banking on the success of that attack, this surface force was to proceed to the Okinawa area and dispose of the remnants of our fleet. To meet the enemy threat, Vice Admiral Mitscher immediately ordered all three task groups of Task Force 58 to concentrate northeast of Okinawa.

At dawn on 7 April, forty fighters in units of four were launched to search a wide, pie-slice quadrant which covered every area of the sea to which the enemy task force might have proceeded during the night. Planes from the Essex made the first contact at 0822.

The terse radio message reported a Yamato-class battleship, one or two cruisers, and probably eight destroyers on a westerly course. The force had cut across the southern tip of Kyushu through Osumi Kaikyo and now was headed out into the East China Sea.

The change in course was puzzling. Was the enemy force not to attack at all but simply head for the west coast Kyushu naval base at Sasebo? Or was the force intended to come down the western side of the Ryukyus and smash into the major concentration of our invasion forces gathered to the west of Okinawa? We did not wait for an answer.

On the carriers the word was passed quickly. Planes were loaded, the Avengers with torpedoes, the Helldivers with semi-armor-piercing bombs, and the fighters with 500-pound general purpose bombs. Here was an opportunity our naval aviators had not experienced since the Battle for Leyte Gulf. An enemy task force was on the loose, the first to poke its nose out in many a lean month.

By 1000, the planes of Task Group 58.1 and 58.3 were ready to launch. Loaded to maximum capacity with bombs and gas, the planes strained to take to the air. West wind half an hour 132 fighters, 50 dive bombers, and 98 torpedo planes were setting noses toward the estimated position of the enemy task force some 240 miles to the northwest.

After the strikes were off, further reports came from the shadowing planes. The size of the force was established as 1 battleship, 1 light cruiser (subsequently identified as the Yamato and Yahagi), and 8 destroyers. Weather at the target was unfavorable. Cloud ceiling was 3,000 feet; visibility five to eight miles with occasional rain squalls. Inexplicably there was no cover of enemy fighters to protect the task force.

Planes of Task Group 58.1 were the first to locate the enemy fleet almost directly west and 110 miles distant from the southern tip of Kyushu. Weather in the general area was so bad that the Hancock strike group never reached the target. The low ceiling and great numbers of planes which soon arrived at the target made coordinated attacks difficult. Antiaircraft fire was intense, but inaccurate; Our planes pressed home their attacks. Exact assessment of results was impossible, but the battleship and the cruiser were heavily hit; two destroyers were sunk.

Forty-eight Hellcats, 25 Helldivers, and 33 Avengers of Task Group 58.4 were launched an hour after the strikes from the other two task groups. When they arrived over the enemy fleet, the battleship Yamato had a port list, but was still making good speed and still shooting. The light cruiser was dead in the water, spewing oil. One of the remaining destroyers was burning and trailing an oil slick.

Futile antiaircraft fire, by now reduced in volume, met the attack. The planes split effectively for the clean-up. The badly damaged Yahagi was literally pulverized. A dozen bombs and eight torpedoes were smashed into the ship which broke into a churning mass of twisted steel and struggling crewmen. Two more destroyers were sunk, another badly damaged and a fourth left burning.

The big prize, the Yamato, pride of the entire Jap battle fleet, took a wild zigzag course, her foaming wake cutting crazy figures in the oily waters. Relentlessly the planes closed in. The USS Intrepid air groups swarmed in first, claimed one torpedo and eight bomb hits. The coup de grâce was administered by six Yorktown torpedo planes.

The young skipper of the torpedo squadron and leader of the six attacking planes sized up his lumbering target with split-second wisdom and acted with typical initiative. The Yamato was listing to port. Her protective band of anti-torpedo armor on the starboard side was lifted high out of the water, exposing the thinner plates of her underbelly.

The torpedo squadron leader called his planes and instructed all crews to change the depth setting on the torpedoes from 10 to 20 feet. Then he brought his planes around to the Yamato's starboard side for the attack. The six planes made perfect runs. Six torpedoes dropped. Five were observed to hit, and before it was lost to sight the sixth was running hot, straight, and true.

These torpedoes tore the bottom out of the Yamato. The huge ship, pushed by the tremendous force of the underwater explosions, slowly rolled over and exploded in a great geyser of flame and smoke. It is doubtful if a single member of the Yamato's crew survived.

Triumphantly, our strike groups returned to their ships. At a cost of 4 dive bombers, 3 torpedo planes, and 3 fighters from which all but 4 pilots and 8 air crewmen were rescued, the supposedly invincible 42,000-ton dreadnaught Yamato, the light cruiser Yahagi, and four destroyers had been wiped out. No units of the enemy fleet ever again sortied from their Inland Sea haven. Our pilots had scored another signal victory.

A PANORAMIC VIEW OF THE VAST OKINAWA CAMPAIGN — the largest amphibious operation ever undertaken in the Pacific — tends to slight some of the most vital components of its success. The swirling Kamikaze attacks, the great air battles, the heroic stand of the radar picket destroyers, the sinking of the Yamato, overshadow the contribution of some of the integral units of the aviation team that helped smash the enemy on the threshold of his homeland.

For the infantryman slogging through the mud and battling up bloodied escarpments, the air attacks on the ships were of remote importance. For this man one of the most welcome features in his existence was the support his desperate forces received from Helldivers and Avengers blasting tenaciously held enemy positions.

Naval and marine air support of ground troops in the Okinawa campaign was used on a larger, more intensive scale than ever before. Ground troops confronting an especially stubborn enemy position needed only to get the word to their air support liaison team and within minutes a squadron of Helldivers would be screaming for a pin-point attack. Repeatedly, aircraft successfully smashed targets that could not be reached by artillery or naval bombardment, or which needed an especially heavy explosive load.

By 17 May, 5,800 tons of bombs had been dropped on the enemy by 1,388 missions involving a total of 13,950 individual plane trips. Precise, standardized techniques, developed and refined since the days of the Gilberts operation when air support was first used, contributed to the success of the missions. To the aviators, an air support mission was a "milk run," a dull routine flight, but to the man on the ground it was often the margin of victory over a fanatical enemy.

The real work horses in air support in the Okinawa campaign were not the fast carriers, but the escort carriers of Task Group 52.1. The jeep carriers, despite their limited facilities as compared with a large carrier, matched the CV's sortie for sortie in sending support missions over Okinawa front lines.

On the job from 24 March to 24 June, day after day from twelve to seventeen of the "jeeps" were steaming in the waters southeast of Okinawa. In addition to their daily quota of air support missions and combat air patrols, the CVE's neutralized airfields in the Sakishima Gunto when the British Task Force 57 was refueling and replenishing. And when the fast carriers retired to base, the CVE's sent long-range sweeps and bombing missions as far as Kyushu.

Another extremely important phase of our air activities at Okinawa was the work of the land-based planes of the Tactical Air Force commanded by Major General P. F. Mulcahy, USMC. Marine Air Group 31 was the first unit to garrison Yontan airfield. Arriving on 7 April with 90 Corsairs and 15 night-fighter Hellcats, the Marines sent up combat air patrol missions the same day. Marine Air Group 33 arrived two days later. Within six days, the Marines were flying air support missions and a growing schedule of combat air patrols.

For the first month of the campaign, Yontan airfield was our major land base, supplemented by Kadena airfield, located a few miles to the south. Yontan was all-weather operational by 8 April, but Kadena was rained out frequently and for several weeks was within range of enemy artillery fire. The only other important field developed was Ie Shima. Although this airfield was secured by mid-April, it was heavily mined and crisscrossed with trenches, and did not receive any garrison aircraft until 13 May when P-47's flying in from Saipan became the first welcome addition of army fighters to the strength of the Tactical Air Force.

While battling the Japanese in the air was the primary job of the Tactical Air Force, it was not necessarily its most difficult task. The

Tactical Air Force was confronted with endless difficulties in establishing its bases. The enemy left his fields heavily mined. Frequent rains made the runways mushy. It was difficult to get gasoline ashore; in fact, not until the twenty-fourth of April were pipe lines laid ashore to insure Yontan Field, our most important base, an adequate gasoline supply. Air attack always was imminent and frequently sneak raids were carried out at night. Air attack even posed a serious problem indirectly. Our ships in the anchorage threw up so much flak during air attacks that the pieces of spent shrapnel raining down on the fields frequently damaged our land-based planes beyond repair.

The Tactical Air Force overcame these difficulties, and by mid-May its planes had flown nearly 5,000 sorties on combat air patrol alone and had destroyed more than 250 enemy planes. By 16 May, the Tactical Air Force was ready for its first distant support mission, and thereafter sweeps were carried out ranging the 700 miles from Kyushu to the Sakishima Gunto.

Vital to all airmen, land- or ship-based, were the spectacularly successful rescue operations for pilots and crews whose planes were forced to make water landings. From the beginning of the Okinawa campaign through its critical months of April and May, 132 pilots and air crewmen were rescued out of a total of 186 who conceivably could have survived from planes that were downed. It was a great boost in morale for pilots weary to the point of exhaustion with the heavy schedule demanded of them to know that even if shot up and forced down their chances of survival were better than 70 percent.

Every known rescue facility from life rafts to light cruisers was used in the Okinawa campaign. In the waters close to our shipping concentration, the bulk of rescues were performed by destroyers and smaller craft for it was all but impossible for any plane to go down in this area without being sighted by at least one ship. Kingfisher seaplanes catapulted from battleships and cruisers were used with good results. With customary efficiency, submarines operating off Amami O Shima and the Sakishima Gunto fished out aviators downed in these waters.

Most successful of all, however, were six twin-engined flying boats of Rescue Squadron 3. These big Mariners with the heavy belly and graceful gull wing (like their predecessors, the Catalinas) also known as the Dumbos, rescued more aviators than any other agency. In addition, they operated over the greatest distances and performed their rescues under the most difficult and hazardous conditions.

In the first two months of the campaign, the Dumbos saved 63 aviators and air crewmen. To affect these rescues, they flew missions ranging from Kyushu to Formosa. They flew in impossible weather and made 33 landings in rough, open seas. They made 21 rescues within ten miles of enemy-held islands and 7 while under fire from shore batteries.

Typical of the hazards and difficulties the Dumbos faced was one rescue insolently perpetrated within a few miles of the enemy mainland in Kagoshima Bay, a long narrow strip of water on the southern tip of Kyushu. Two carrier pilots were down in the bay. A search seaplane happened along, tried to land for the rescue, cracked up and added ten more survivors in the water. Already on the way, the Dumbo went in under Jap coastal guns, took aboard twelve survivors, and although heavily overloaded made a precarious take-off with the assistance of special jet propulsion units. A few hours later the twelve survivors were delivered safely to base.

The Dumbos were not the only seaplanes to share in the aviation team that helped take Okinawa. For the fullest protection of our forces, a vast network of air searches was necessary. Fanning out from the Kerama Retto like a giant spider web, the searches stretched down to Formosa, skirted along the China coast as far north as Korea, then across Tsushima Straits and along both the west and east coasts of Kyushu.

For the first three weeks, Mariners and Coronados carried the entire search load. More than 8,000 flying hours were logged by the big twin-engined and four-engined boats as they covered the sea approaches to our beachhead. Finally, on the twenty-second of April, six PB4Y-2's, the new Privateer model of the sturdy Liberator, roared into Yontan Field to reinforce the efforts of the seaplanes.

Throughout the month of April, the search effort was almost wholly defensive. The search planes were the far-seeing eyes of the fleet. In this role, Mariner search planes figured in the sinking of the Yamato. Mariners spotted the force in the morning, stayed with it all day vectoring carrier planes to the target, and finally finished off their work by landing on the open sea to rescue several aviators downed by the antiaircraft fire from the enemy task force.

By May, the search planes were ready to take on new duties. Loaded with bombs, they harassed enemy shipping plying between China and Korea and the Japanese mainland. Aggressive pilots took their planes more than seven hundred miles north of Okinawa to smash into this last slender supply line of the enemy. So devastating and thorough was the attack of the search planes that ultimately the enemy was forced to limit his trips to slinking night runs across the

East China and Yellow seas. But even at night he found no refuge from our radar-equipped search planes. The lumbering Coronados and the leaner, high-tailed Privateers searched him out on the seas and even followed his ships into their "safe" anchorages along the Korean coast.

The heavily armed, rugged Privateers were tailor-made for the anti-shipping attacks. In a little more than two months, the Privateers sank 78 enemy ships and damaged 105. The flying boats added 61 more sunk and 89 damaged to this score. By the end of the Okinawa campaign, the search planes had imposed an effective blockade across every sea area within range.

Special radar-equipped night fighter planes also assumed a vast importance in the operation. Night after night these planes orbited out ahead of Task Force 58 or north of Okinawa in constant readiness to intercept any enemy aircraft that might try to sneak down under cover of night for attack or search. Operating under the control of a fighter director officer who guided every movement of the pilot from his post in the Combat Information aboard ship, the night fighters of Task Force 58 alone in the first two months destroyed 47 enemy planes around the task force and 43 more at Okinawa. With this score which far exceeded all previous night-flying efforts, the skepticism with which the night fighter had long been regarded was dispelled.

The contribution of the night fighters cannot be measured merely in terms of the number of planes shot down which, compared to the day actions, was relatively insignificant. The night fighter was most valuable in a preventive sense. The vigilance of the night fighter permitted sleep for weary ship personnel who, after weeks of nerve-trying attacks, were at the point of physical exhaustion. Entirely aside from the damage that might have resulted from night attacks, had the Japanese been able to come close enough to keep our ships at battle stations throughout every night, the resultant physical exhaustion might well have been a marginal factor in the victory that was ultimately achieved.

Rounding out the aviation forces at Okinawa, mention must be made of the full-fledged British carrier fleet which joined in that battle. Designated Task Force 57, the fleet included 4 carriers, 2 battleships, 4 light cruisers and 11 destroyers. Between 26 March and 25 May, except for a brief period of replenishment at Leyte, Task Force 57 plied the dangerous waters to the east of northern Formosa and conducted neutralizing attacks on the airfields of Sakishima Gunto.

All the carriers present — the Indefatigable, Formidable, Indomitable, and Victorious — took suicide hits in the course of the operation. Their armored flight decks reduced the damage and all carriers remained operational. Although the British carriers carried fewer planes and inherently presented more operational difficulties than our CV's, Task Force 57 proved its merit against the Formosa-based enemy air force and earned a full partnership in Pacific operations with the U. S. carrier forces.

IN A DISMAL, DRIVING RAIN ON 28 MAY, the Japanese staged their last mass aerial suicide assault on the forces at Okinawa. Coincidentally, this was the day that Admiral William F. Halsey, Jr., Commander of the Third Fleet, relieved Admiral Spruance of the Fifth Fleet as overall commander of the Okinawa campaign.

As the enemy's suicide attacks tapered off, the entire campaign was accelerated. The ground forces finally wore down resistance in the grinding bottleneck around Shuri Castle and were ready to fan out in the wide southern end of the island to cut the enemy to ribbons piecemeal. The Tactical Air Force, with its bases at last firmly established and adequately supplied, assumed a rapidly increasing share of the responsibility for air defense of Okinawa. Long-sought relief for many of the destroyers and smaller ships on radar picket duty was provided as small expeditionary forces reached out to take the outlying islands of Tori Shima, Aguni Shima, Kume Shima, and Iheya Shima for land-based radar stations.

Suicide planes still came down the Ryukyus; our ships still absorbed punishment. But after the twenty-eighth of May there was no day on which more than fifty enemy planes attacked, and before that time a fifty-plane day was virtually a vacation. Before the twenty-eighth of May, destructive or damaging aerial attacks had been scored on 216 of our ships. After that date, only 22 ships were hit.

The waning of the enemy's aerial offensive freed our forces for other activities. Task Force 58 under Admiral Mitscher had taken a last poke at Kyushu air bases on the thirteenth and fourteenth of May, and had destroyed 72 airborne enemy planes and 73 more on the ground. As Task Force 38, under the command of Vice Admiral John S. McCain, the fast carriers made sweeps over Kyushu bases on 2 and 3 June, but found few enemy planes either in the air or on the ground.

On the eighth of June, a concentrated blow by strong sweeps from the two remaining task groups hit Kanoya airfield, the enemy's principal Kyushu base. A scant handful of enemy planes — possibly a dozen — rose to contest our thundering air armada but were soon shot down or chased. It was apparent the enemy's ability to strike from his Kyushu bases was irretrievably smashed.

Its task against enemy air power finished, Task Force 38 retired to San Pedro anchorage in Leyte Gulf for a deserved and long needed replenishment and rest period. The escort carriers stayed on the job as did the now powerful Tactical Air Force. Both escort carriers and the Tactical Air Force extended increasing efforts against the battered Kyushu bases. Search planes, too, stepped up the offensive against enemy shipping supply lines. Our troops blasted the remaining Japs so mercilessly that for the first time in Pacific history, enemy troops began to surrender by the hundreds instead of singly or in isolated little groups.

At 1305 on the twenty-first of June, effective enemy resistance was declared at an end. Okinawa Gunto — "a prefecture of Japan . . . a densely populated chain of rugged islands" — was secured.

Many of our military techniques applied in the long march across the Pacific met their toughest trials in the battle for Okinawa, the most extensive amphibious operation of the Pacific war. None was more extensively tested than our concept of the carrier task force.

At Okinawa, Task Force 58 took on the principal land-based air force of a major military power. For 84 days our carriers were within easy range of a potential force of some 5,000 Japanese planes. Nine days out of ten, Task Force 58 tied down by its obligation to the desperately fighting forces at Okinawa, operated within an area hardly a hundred miles square. And by virtue of the Kamikaze crash to which all Japanese pilots were committed, the enemy air force became a threat entirely out of proportion to its size and efficiency.

The cards were stacked in the enemy's favor. No better opportunity could have been afforded the enemy to carry out the threat of his propagandists to wipe out the United States Fleet. By the same token no more effective proving ground could have been provided for a long-standing question mark in naval warfare: Could a carrier force stand up against a strong land-based air force?

The answer of the navy air force at Okinawa was an emphatic affirmative, abundantly attested by 3,863 destroyed enemy aircraft, by smoking, smashed airfields everywhere within the carrier planes' range, and by an enemy fleet, largely sunk, with the remaining units rendered hopelessly impotent by the mere threat of carrier-based air attack.

27 – INTO THE HEART OF JAPAN

WITH THE RETAKING OF THE PHILIPPINES and the conquest of Iwo Jima and Okinawa, the way was cleared for a series of blows at the heart of Japan. The capture of Okinawa had removed the last defensive ring of islands protecting Japan. Life for the Japanese during the ensuing months was to be a continuing nightmare, for from the newly acquired and quickly developed airfields on Iwo Jima and Okinawa, and from the carriers of the United States and British fleets, swarms of planes took to the air.

Penetrating the core of the home islands, these planes challenged Japanese air power to defend itself, and if the challenge was refused, sought out enemy planes and destroyed them on the ground. Not content with attacking simply Japanese air power, our planes also struck at industrial plants, navy yards and other military installations in an all-out effort to destroy the enemy's power to resist.

The Navy's greatest striking air unit during these last months of war was Task Force 38, now organized into three task groups with a total strength of 9 large and 6 light carriers, 9 battleships, 3 heavy and 16 light cruisers, and 62 destroyers. On 1 July, this force, under the command of Vice Admiral J. S. McCain, sortied Leyte Gulf bound for Japan.

Three days later units of the British Pacific Fleet, under Vice Admiral Sir Bernard Rawlings, totaling 4 large carriers, 1 battleship, 6 light cruisers and 18 destroyers, reported for duty with the Third Fleet as Task Force 37. Another extremely important but little publicized group joined the Third Fleet on 8 July. This was a fueling group consisting of 26 fleet oilers, 1 store ship, 4 tugs, with 6 escort carriers, 13 destroyers, and 19 destroyer escorts. The presence of this group made possible a more sustained period of attack than would otherwise have been the case.

While this fast carrier Task Force 38 was en route to the scene of action, army, navy, and marine air forces based on the islands surrounding the empire continued the air war against the home islands of Japan. B-29's of the Twentieth Army Air Force, operating from bases in the Marianas and escorted by fighters based on Iwo Jima, fired the industrial cities of Japan with area-bombing incendiary attacks. Army Thunderbolts (P-47's), Black Widows (P-61's), Mitchells (B-25's), and Marine Corsairs (F4U's) from Okinawa, attacked and heckled the airfields on Kyushu, the southern island of Japan.

Search planes of Fleet Air Wing 18 continued operations from Iwo Jima against merchant and other vessels off the southern and eastern coast of Honshu. Fleet Air Wing 1 Privateers (PB4Y2's) from Okinawa continued to harass and destroy merchant and naval shipping in the East China Sea and extended their rovings to the waters of the Yellow Sea around the Korean peninsula.

After a speed run to the launching point, the carriers of Task Force 38, early on the morning of 10 July, were ready to make their first strike of an intensified campaign against Japanese air power. Aircraft, airfields, gun positions, fuel storage areas, repair facilities, and industrial installations in a broad arc around Tokyo were the targets for the day. Complete tactical surprise was achieved on the first strike, and the attacks continued throughout the day under ideal weather conditions against little air opposition.

On the following day the weather changed, and remained bad for flying for four days. During this time the task force moved north, and as soon as the weather lifted slightly, made a two-day attack on the previously untouched enemy territory of northern Honshu and southern Hokkaido. Weather over the airfields made visibility poor on the first day, and the strikes were diverted to shipping targets.

The results were gratifying from the American standpoint, and targets were found principally in the Tsugaru Strait between Honshu and Hokkaido, through which vessels were found trying to escape into the Japan Sea. Five of the estimated nine train ferries operating across the strait were sunk and a sixth was beached. In the two-day attack, 140 ships totaling 71,000 tons were sunk, and 234 ships and small craft damaged. The enemy sent up no air opposition, but had 37 planes destroyed and 45 damaged on the ground.

On the fourteenth of July, as an illustration of the versatility of the task force, a bombardment group was detached and moved into gun range off the coast and for an hour and a half with telling effect shelled industrial targets at Kamaishi. On the following night, the task force moved south and fueled on the sixteenth. As it made this movement, this task force that had been blasting the Japanese back across the Pacific and would have been counted pretty fair fighters in any league, conducted training exercises! Naval aviation was, in addition, still trying out new techniques of warfare. In the first operation of its kind, a night fighter combat air patrol from the Bon Homme Richard, on the sixteenth, covered a night bombardment of the coast by the cruisers and destroyers of the task force.

Having fueled, Task Force 38 rendezvoused with the British Task Force 37, and on 18 July the combined forces launched an air attack on the Yokosuka naval base at Kure. In spite of the heaviest antiaircraft fire encountered since the last attack on this base in March, the planes inflicted serious damage on the base, bombed the battleship Nagato, sank an old heavy cruiser, two destroyers, and several craft. Unfavorable weather reduced the effectiveness of the attack, however, and it was called off after three strikes had been made.

The forces next retired to sea to conduct, within the succeeding three days, the largest refueling, replenishing, and rearming operation ever undertaken at sea. The essential co-operation between naval and air forces was never more thoroughly demonstrated. On 23 July the force proceeded southward toward the next launching point and carried out a two-day continuous attack on lower Honshu and the southern islands of Japan, Shikoku and Kyushu. Day strikes launched on the twenty-fourth were continued through the night by night hecklers and intruder missions over the airfields. Once again bad weather came to the aid of the enemy, and the attack had to be called off toward midafternoon on the twenty-fifth.

Before bad weather intervened, however, naval air forces had hit the jackpot at the naval base at Kure, in the Inland Sea. In addition to other heavy devastation, 22 warships (258,000 tons) were damaged. Notable among the casualties were 1 battleship (Hyuga) sunk and 2 (Ise and Haruna) damaged, as were 3 cruisers. The airfields at Nagoya-Osaka and Miho were well worked over, and fortunately so, for it was found that the enemy planes were fueled and apparently ready for an attack on the fleet. The first concerted air opposition since 10 July was encountered in this attack. There was, however, no damage to the ships of the force, and 18 enemy planes were shot down during the day.

The lack of air opposition and the wide dispersal of grounded aircraft which characterized the enemy's reaction to attacks during this period made it apparent that the Japanese had decided that they were no longer able to oppose our carrier aircraft except under the most favorable conditions. The enemy's air power, which had been so lavishly expended during the Okinawa campaign, was now being hoarded for a final assault on our forces when they invaded Japan.

The wide dispersal of Japanese aircraft, some planes being found as far as five miles from an airfield, and the excellent camouflage used made the destruction of these planes quite difficult and required the extensive use of photographs and photo interpretation.

By the use of these, however, pilots successfully located and destroyed enemy aircraft, thus reducing the strength that the enemy was trying to desperately to conserve.

The next objective was the Inland Sea, the inner hiding place of the Japanese Fleet, that was to be blasted by air attacks. It had been hit before, at Kure, and elsewhere, but this raid was designed to launch continuous waves of attacks from Nagoya to North Kyushu and Miho. Under favorable weather conditions, operating from the task force located about a hundred miles off the coast of Shikoku, on 28 July, our planes scored heavy damage to naval and merchant shipping, transportation facilities, aircraft factories and to the Japanese air arm.

Noteworthy victims sunk during this attack included the Haruna, which had been frequently reported sunk, the battleship Ise (fitted with a half flight deck), the heavy cruiser Aoba, and the light cruiser Oyodo. Some air opposition was encountered, and 21 enemy planes were destroyed in the air.

In spite of the wide dispersal of aircraft, 123 were destroyed on the ground. In addition, there was serious damage to ground installations and equipment, including destruction of 14 locomotives, 1 hangar, 3 warehouses, 1 transformer station, 3 oil tanks, and 3 roundhouses.

The flexible operation of naval aviation was well demonstrated on the thirtieth. Strikes were launched against airfields in the Tokyo area, but weather being unfavorable in that area, the planes were diverted to the Maizuru area on the coast of the Japan Sea. More than 1,200 sorties were flown, during which the pilots continued to locate well-hidden Japanese aircraft and destroyed 144 on the ground. Merchant shipping was hard hit, with 24 ships sunk and 133 damaged. Industrial targets, also, were not spared, and among the chief victims were three aircraft plants, one steel mill, the arsenal at Nagoya, and naval docks at Maizuru.

On the following day, the combined forces retired toward a refueling rendezvous and were forced to head south to avoid the path of a typhoon. The stormy weather caused the cancellation of strikes scheduled for 4 and 8 August, and it was not until the force moved northward on 9 August that the attack on Japan was resumed.

It was during these months of attack on the home waters of the Japanese Empire, and the enemy-controlled Chinese coast, that one of the finest examples of naval air-sea co-operation was demonstrated time and time again. This was the liaison between our long-range navy patrol planes and submarines in the area.

Two main benefits resulted from this co-operation. In the first place, the patrol planes spotted potential targets for the submarines, which went in to make the kill. The combination proved to be a potent one, and as a result Japanese shipping losses mounted higher and higher. In the second place, during the last six weeks of the war, the American submarine proved to be the greatest "Dumbo" of them all.

Outstanding work had been done in air-sea rescue by Catalinas, Mariners, and scouting craft. As the pressure on offensive was relaxed somewhat for the submarines, however, they turned to the task of rescuing downed Allied airmen, army, navy, and marine. Operating under the noses of the Japanese, the submarines were responsible for the rescue of over 80 percent of the total personnel saved. Their presence gave a morale lift to pilots and air crewmen that can hardly be exaggerated, and naval aviation feels a deep sense of gratitude to the "Silent Service" for its performance on behalf of aviation's fighting men.

Meanwhile, as this fast carrier task force had been opening with such devastating results off the coast of Japan, the strategic and tactical air arms of the AAF, the marine fighter-bombers, and the search planes of the Navy added to the pressure of our air offensive by continuing the delivery of impressive evidence of what the continuation of the war would mean to the people of Japan. Large groups of B-29's literally rained incendiary bombs on the industrial cities, and shipping, airfields, aircraft, transportation and industrial facilities were fired and destroyed over such a large area that all Japan felt the impact of our power.

This mounting offensive reached its climax on the morning of 6 August 1945, when a lone B-29 dropped a single bomb which hit Hiroshima with earth-shattering force and rudely awakened Japan and the rest of the world to the awesome power of the disintegrating atom. Reconnaissance photos showed that 4.1 square miles, or 60 percent of the city's built-up area, had been totally destroyed. Two days later, Soviet Russia entered the lists against Japan, and on the following day a second atomic bomb was dropped, this time on Nagasaki.

Meanwhile Task Force 38 continued pummeling Japan. In a two-day attack, beginning on the ninth, over 2800 sorties were flown against targets on Honshu north of Tokyo. The primary objectives were airfields and aircraft, but in addition 44 small vessels were sunk, and 38 damaged.

The attacks against the primary targets were even more satisfactory; 412 planes were destroyed on the ground. A few enemy planes from fields bordering the area being attacked succeeded in reaching the ships of the force and damaged the destroyer Borie and narrowly missed the Wasp in suicide attacks. No enemy planes approached the ships of the force on the second day, but word of Japanese surrender moves resulted in special alertness against surprise attacks.

The task force returned to the Tokyo area on the thirteenth and continued attacks throughout the day in spite of bad weather. There was no air opposition over the target, but unsuccessful air attacks were launched against the ships of the force at various times during the day.

After fueling on the fourteenth, strikes were launched early on the morning of the fifteenth. The first strike was met over the Tokyo area, which was the target, by 45 planes, 26 of which were shot down by our aircraft. Before the second strike reached the target, the news everyone had been waiting for came through. This was a dispatch from Admiral Nimitz announcing the end of the war. The planes returned to their carriers, and aside from the small task of shooting down eight attacking Jap pilots who hadn't heard that the war was over, the shooting was finished. At 1600 the task force retired from the area to await further orders.

During its period of operation off Japan, from 10 July to 15 August, in which strikes had been launched on thirteen days, the task force had caused heavy damage to the enemy. More than 10,000 sorties had been flown against land and shipping targets, 4,619 tons of bombs had been dropped, and 22,036 rockets expended on the enemy.

A total of 1,232 enemy aircraft had been destroyed and 1,181 damaged; 86 ships totaling 231,000 tons had been sunk, 118 ships totaling 568,000 tons had been damaged, and 618 small craft had been sunk or damaged. In contrast, our losses amounted to 307 planes, of which only 174 were actually lost in combat, and 1 vessel, the Borie, damaged. The Navy's air war had come to a triumphant conclusion.

PART V: FORCES BEHIND THE NAVY'S AIR WAR

28 – Training Naval Aviation's Manpower

THE FORCES THAT HAD BEEN HOLDING BACK TRAINING, and as a matter of fact, all war preparations, were swept away by the attack on Pearl Harbor. The first year of the war saw tremendous strides made in training manpower for naval aviation. In the field of pilot training, perhaps the outstanding development making the huge output possible was the standardization of the program. The main framework of pilot training was to be modified to fit changing conditions as the war developed, but in essence it was welded during the first months of conflict.

The supply of pilot material was not limitless, even in a nation the size of the United States. The consensus of opinion was that the aviator had to be young. As science improved the ability of airplanes to withstand the strain of pulling out of a dive, executing sharp maneuvers, operating at ever higher altitudes, it became imperative that those flying the planes be able to stand up under the punishment that such operations entailed. It came to be believed that youths between the ages of eighteen and twenty-six could best stand the life, and the emphasis was on the lower age groups within this bracket.

Only a relatively small percentage of those within these age limits could meet both the physical and mental qualifications that indicated ability to pass the training period successfully. Young men who thought that they were the acme of physical perfection found that they possessed some hitherto unsuspected defect that disqualified them.

Perhaps the greatest obstacle was the eye test. It was essential that a pilot be able to distinguish colors, because of their frequent use in landing and other operations. Many a would-be pilot was stalled on the dot type of color chart that enabled only those with correct color perception to see the numbers on the page. Ironically enough, this test, which helped the United States pick the cream of future aviators, was devised by a Japanese.

Because there was a possibility of flying in high altitudes in which oxygen apparatus had to be gripped firmly between the teeth, there was a rule that the prospective pilot had to have the proper number of teeth in the right places. Some other defects were remediable, and many young men voluntarily submitted to operations to correct a defect and make them eligible. Football and other athletic injuries, especially to knee cartilages, ranked high in this category.

The supply of potential pilots was limited. On the other hand, it was larger than could be handled at one time by either the Navy or the Army. As a result, both forces developed systems to "earmark" young men for future training. The Navy's deferred enlistment programs served a number of very useful purposes. In the first place, it made readily available the right type of manpower to be trained at the Navy's convenience, and removed the possibility of these men being drafted not only out of civilian life but away from the possibility of naval aviation training. In the second place, it was a definite morale booster for the individual concerned.

It removed him from the category of "draft dodger" or "slacker," for the responsibility had been shifted to the Navy. He had volunteered, and he was ready to go whenever his country needed him. Furthermore, as the programs developed, he was given a fairly definite idea as to when he would be called, and could make his plans accordingly. In the spring of 1942, the Navy offered two programs to young men, both of which could lead to pilot training.

The V-5 program, as it was called, led definitely to this training, with the assurance that if the young man could meet the standards set up by the training program, he could become a commissioned officer in the Naval Reserve. As the program was expanded, he could be called to active duty in fairly short order; or, if he were a college student, he would not be called until the conclusion of the school year. This system permitted young men to complete a year of college with ease of mind.

The other program, launched a little later, was called the V-1 program. This program was available to college freshmen or sophomores and enabled them to finish the first two years of college before being called to active duty. Not all the V-1 students were tabbed for aviation.

At the conclusion of their sophomore year, the students could, if they had met certain liberal scholastic requirements, decide where they wished to transfer to the V-7 program, permitting two more years of college with ultimate training as a deck officer, or, shifting to the V-5 program and as soon as they were needed, enter aviation training. Toward the end of the year, these programs were consolidated into the so-called V-12 program which continued through the end of the war.

In the V-1 and V-5 programs the policy was to permit the college student to remain a civilian to all intents and purposes until he was called for training. He was sworn into the Naval Reserve when he signed up for the program, but he remained without pay, on inactive duty, and in civilian attire until he went into training.

The V-12, on the other hand, took the individual, put him in uniform and then sent him to college. He was no longer a typical college student in sweater and slacks, free and easy as the air; he was an apprentice seaman and sported the bell-bottomed trousers and coat of navy blue — and had his way paid. The V-12 program, comparable to the Army's Specialized Training Program, was designed to prepare men not only for flight training but as officer material for other billets in the Navy, from chaplains to engineers.

As time went on, more and more V-12 candidates were selected from the enlisted ranks themselves. Radiomen in the Caribbean, ordnance men in the Pacific, when given the opportunity, willingly gave up their petty officer's stripes for a crack at a college education and a chance to better both themselves and the Navy.

During the first year of war, the Navy profited by the Civilian Pilot Training Program. This government activity, which had been so helpful in the year prior to our entrance into the war in preparing men for flight training, continued its work. Changing its name to War Training Service, it accelerated its program to the preliminary training of 20,000 pilots a year in 92 schools and colleges beginning in July 1942. Since Reservists in both the V-1 and V-5 programs were trained by WTS, considerable time was saved for the Navy, and at the same time, obviously unfit men were eliminated from future training.

These were the preliminary steps — the recruiting and testing of young men by aviation cadet selection boards throughout the country, or by traveling committees sent out from these boards, and a certain amount of training by the War Training Service. In 1942, naval aviation undertook a standardization of its own training program. Out of this emerged five main stages of training.

The first stage was the establishment of nationally known "pre-flight schools." These were the outcome of a definite train of thought. In the first place, as has been pointed out, it was felt that the pilot should be not only a good physical specimen but should be in top physical condition.

Secondly, in view of the terrain over which much of the fighting was to be carried on and with the prospect of possible hand-to-hand struggle with the enemy in case of forced landings, the pilot should know the rudiments of the art of self-defense against a foe that paid no respect to the rules of the Marquis of Queensberry. We were fighting a vicious foe; pre-flight was to prepare pilots for this fact, in mind as well as body.

In the third place, air combat was no romantic and aerial counterpart of the medieval tilting contest in which two opponents made passes at each other and then waved silken handkerchiefs at the conclusion of the fray. Air combat was a matter that required teamwork. Your wingman was just as important to you as you were to him. Success would lie in a unified attack. Consequently, one of the aims of pre-flight schools was to teach co-operation. It was felt by those in control, whether rightly or wrongly, that football was the best sport to bring out this aspect of teamwork, and this sport was given wide emphasis in the training program.

Four pre-flight schools were announced in February 1942, for opening in May or June of that year. They were located at Athens, Georgia, Chapel Hill, North Carolina, Iowa City, Iowa, and St. Mary's, California. Toward the end of the year a fifth school was established at Del Monte, California, raising the capacity from 7,500 to 9,350. Leading coaches from colleges and schools all over the country were commissioned to take charge of the course.

These men were sent to Annapolis for a one month's indoctrination training at a pace so rapid that, as one coach put it, he didn't have time to smoke his pipe for the whole period, and couldn't have stood it if he had. Three other courses of similar length followed, until on 18 July the "school" closed, having graduated a total of eight hundred officers.

In response to requests from all over the country, a two weeks' exposure course or clinic was offered college and high school coaches in an attempt to co-ordinate physical training in private and public schools of the country with the Navy's physical training program.

These coaches were welcomed at the pre-flight school, put through the course for two weeks, and sent home with a new enthusiasm for the rugged techniques of Commander Tom Hamilton, who was the guiding light in the program. Almost a thousand coaches received this training.

The football teams turned out by the pre-flight schools created a great deal of publicity for the naval aviation training program and may have been an important factor in creating teamwork and putting men into top physical condition. Less spectacular, but possibly more important, was the swimming program. Nearly 30 percent of the entering students could not swim. The leading swimming coaches in the country contributed their talents, and the result was that when the classes finished not only could everyone swim, but he could also support himself in the water "for hours." The emphasis was not on speed but on endurance.

It did not matter how fast a man could swim a hundred yards; the important thing was to give him enough swimming ability to stay afloat until he could be rescued. The slow breast stroke was given precedence over the Australian crawl. Versatility was taught, and a man had to gain reasonable mastery over four different strokes, and to splash his way through burning oil, or learn to swim under it. The cadets were taught the basic elements of lifesaving, both of their own lives and those of others through such devices as filling their pants with air as a crude float.

Football and swimming were not the only types of athletics. The cadet was exposed to many sports, and specialized in one. The responses were varied. Some swore that pre-flight killed their interest in any organized form of sport, others enjoyed the life. One fact, however, stands out. Less than a year after the inception of the program, its director reported a 22.7 percent increase in the physical fitness of the average cadet from the time of entrance to his graduation from the school.

As the name implies, the cadet was not given flight training at pre-flight school. The period was one of indoctrination and seasoning. In addition to the rigorous physical program, the student was introduced to the main features of naval aviation, took courses in seamanship, recognition, communications, and learned the rudiments of the manual of arms.

The next stage in the training program was called primary flight training. Formerly conducted at Pensacola and a few new fields, this training was transferred to the Reserve aviation bases. Officers and men of the Regular Navy reporting for aviation training were inducted into the program at this point, escaping the indoctrination and physical conditioning of the pre-flight schools.

Here a three months' course was given, divided approximately evenly between ground school and flight training. It was estimated that the program at this stage would require 3,000 training planes, 1,900 flight instructors, 300 ground instructors, barracks and ground facilities for 7,500 students, between 95 and 100 outlying fields, and an increase of assigned enlisted personnel from 2,500 to 17,000.

Having successfully completed primary training, the cadet passed to intermediate training. This was a fourteen-week course, concentrated at Pensacola and the growing facilities at Corpus Christi. In order to handle this phase of the program additional fields and seaplane facilities were required and constructed. Two thousand intermediate trainer planes were needed, each station to have 400 scout and patrol bombers, 300 fighters, and 200 torpedo bombers.

An attrition of approximately 30 percent in the original candidates was expected by the end of intermediate training. At the end of this period, the successful students were designated naval aviators (officers) or naval aviation pilots (enlisted men).

With the completion of intermediate training, the cadet won his wings and his commission. The group, which had been kept more or less together (on paper at least) until this time was now broken up. The pilots moved to one of the various phases of the next step in training, called operational training.

The marine aviators, who until this time had trained with navy cadets, went to their own operational training course at Cherry Point, North Carolina, or to stations on the west coast, where they learned tactics most likely to be useful to them, such as close ground support in amphibious operations, ground strafing, and close support bombing.

Ferry pilots and NATS (naval air transport pilots) were separated from the program at this point, and still others were plowed back into primary and intermediate training as instructors. Foreign students also left the program at this stage.

For the rest, operational training concentrated on the first major service type planes. Observation-scouting pilots received a two months' course preparatory to being assigned to air units aboard battleships and cruisers, or those destined for inshore patrol received an abbreviated course.

Carrier pilots and patrol plane pilots of various types also got roughly a two months' course, and an innovation was made in the training techniques at this stage. This was a further recognition of the principle that it takes teamwork to win a war, and consisted of a joint training of the pilots and the air crewmen. This type of training was improved upon as the program continued and proved to be of great value.

In order to take care of this expanded operational training, new fields were necessary. A nucleus was already available in fields at Jacksonville, Miami, Key West, and Banana River, and in due course additional fields sprang up all over east Florida. In the early period of this program, it was planned to have simulated carrier landings at Florida bases, with actual qualification landings on near-by carriers.

This plan was expanded by the introduction of carrier training in the Middle West. Two Great Lakes steamers, formerly coal-burning, side-wheeling excursion vessels, were converted to carriers, the USS Wolverine and the USS Sable, and aboard these vessels much of the later carrier qualification training was conducted. The Charger fulfilled a similar role as a practice ship off the east coast. By the first of December 1942, carrier pilots alone were being trained at the rate of three hundred a month.

Gunnery training became an important part of operational training under this program. Depending upon the type of craft they flew, pilots received training in fixed gunnery, dive and torpedo bombing, and gunners became experienced in aerial free gunnery. Except for additional facilities for torpedo training, new facilities were not needed, and the main emphasis on training for enlisted men was increased specialization on specific types of gunnery. The whole program was speeded up by the introduction of synthetic training devices, which both made up for a temporary lack of sufficient ordnance equipment and made mass training more easily possible.

Navigation training, so thorny a problem in the First World War, continued to be a mildly controversial subject in the second. Obviously essential for carrier and long-range patrol operations, European experience proved the necessity of top-notch navigation training for all pilots. In the prewar and early war years, however, the program in the United States was greatly handicapped by a lack of competent instructors and, on the part of the pilots, a dearth of interest and appreciation of so complex and technical a subject. To counteract these factors, navigational schools for instructors were established at Atlanta, and for students at the University of Miami.

Formerly, each large American patrol bomber had carried three pilots, one of whom served as navigator. With the expansion of the patrol plane program, however, a temporary shortage of skilled and experienced pilots for this type of craft developed. The result was a resort to the expedient of training a certain number of non-pilot navigators. These were in addition to the regular pilots, who continued to receive navigation training. Arguments pro and con non-pilot navigators were to receive ardent supporters among high naval quarters as late as the fall of 1944.

Fully as important as this aspect of navigation were the changes in technique of training. At first, the Navy wisely sought the assistance of men who were skilled in navigation. These were the pilots of commercial airlines, especially Pan-American Airways. In 1942, the Navy began the use of a valuable synthetic training device known as the Link Trainer. This device, which underwent many improvements, was an instrument that enabled the pilot to practice navigation under conditions that closely resembled actual flight. The Trainer was a piece of equipment resembling the cockpit of a plane with the necessary controls that enabled the pilot to simulate flight. While he never actually left the ground, his course could be mapped out by the Link-Trainer operator.

It was less expensive than actual navigational hops and had the add advantage that the pilot could fly upside down ten feet below the friend and land at the wrong airport with no greater injury than a red face. As the war progressed, Wavesa were introduced as Link-Trainer operators and performed excellent service in this connection.

Flight without wings. A Link trainer used for instrument flight instruction. The Wave operator, like thousands of her sisters, did a man's job in naval aviation.

Another important aspect of training, and one that was continued throughout the pilot's career even in the combat area, was recognition training. The best training in the world was of little use if the pilot used this skill to knock down our own planes, bomb our own submarines, or attack our own vessels. It was equally of little avail to train men to the peak of perfection in combat technique, if they sat back to watch casually the approach of a friendly plane, only to find out too late that they were welcoming a Jap Zero or Betty.

To combat these possibilities, recognition courses of limited scope had been given as early as the First World War. A smattering of such training had been given in American models prior to the Second World War. The first wartime innovation was the launching of the Aircraft Model Program, in which the Navy in January 1942, asked the youth of the nation to build 500,000 models of standard combat aircraft.

Plans and standard templates were furnished through the Office of Education, which co-sponsored the project, to the school systems of the nation. Soon small-scale models, carefully drawn to such a scale that at 35 feet the model would appear as a plane would at half a mile, were flying from the ceilings of recognition classrooms, pilot ready rooms, and barracks reading rooms.

So hearty was the initial response of high school boys and girls all over the nation, that an additional 300,000 models were requested before the year was out. Certificates were given students after they had completed a certain number of models. As they passed each production milestone, they were advanced in aircraftsman rating, the highest rating being admiral aircraftsman, a title awarded for the completion of fifty model planes of different types representing the four belligerents covered by the program, the United States, Great Britain, Germany and Japan. The Navy and the nation owe a debt of gratitude to this unselfish service of its youth.

The early stages of recognition training were based on the so-called WEFT (Wings, Engine, Fuselage, Tail) system. It was found, however, that in actual combat there was barely time to count to four, much less attempt to analyze four different characteristics of a rapidly approaching plane. Since recognition so often had to be practically instantaneous, a new type of visual recognition was perfected by Dr. Samuel Renshaw, of the Ohio State Research Foundation.

The basic principle of this system was to learn the appearance of the plane or ship as a whole from any angle. A crude analogy of this theory was often given by recognition instructors. They pointed out that one could recognize a friend walking down the street, without knowing, or at least without having to call to mind, before recognition, the color of his eyes, the shape of his head, or his waist measure.

The Ohio State system was adopted by the Navy in May 1942, and a special school was opened at Ohio State University, 20 July 1942, to train twenty-five recognition instructors every two weeks, a course which was quickly extended to two months. The system featured the use of slides, flashed on a screen at rates varying from an initial $1/10$ of a second working up for a while to $1/75$ of a second.

This put a premium on the flash or instantaneous recognition of friendly and enemy types so essential under combat conditions. The Ohio State system became universally adopted and was taught in the fleet, in shore stations, and in practically all training schools other than the technical training maintenance schools.

An important aspect of flight training from the diplomatic standpoint was the training of foreign pilots. Under lend-lease agreements, training of Latin-American pilots and a few Free French pilots was begun in 1942. Occasional representatives of Latin-American countries had been trained in the past, but the program had not been on a firm basis.

In April 1942, however, invitations were issued to most of the Latin-American countries through the State Department, and favorable responses were received immediately from seven countries and later from others.

The first men arrived from Peru in May 1942, and a small but steady stream followed from Argentina, Brazil, Chile, and other countries to the south. The largest number trained were from Brazil. In all more than three hundred received this training. The training program for these individuals was modified and in general, for purposes of security, excluded operational training. Toward the close of the year training was inaugurated for the Free French, and since these men were at war, they received more thorough training.

In 1944, when Catalinas were transferred to the Soviet government, Russian flight crews were sent to this country and given training to acquaint them with this type of plane and its equipment.

THE TRAINING OF GROUND OFFICERS, newly commissioned from civilian life, tentatively begun in the field of aeronautical engineer officers in February 1941, swelled immediately after Pearl Harbor. Following British precedent, a school for training officers in photographic interpretation was convened at Anacostia, District of Columbia, 5 January 1942.

An outstanding school for developing ground officers was that conducted at Quonset Point, Rhode Island. The original plans to have all Reserve officers for the aeronautical organization trained at Anacostia, under discussion in December 1941, were quickly dropped, and in February 1942, the Naval Training School (Indoctrination) was established at the huge and relatively new air station at Quonset Point, Rhode Island.

The underlying theory of this school was that men who were successful lawyers, businessmen, teachers, and newspapermen could with a modicum of indoctrination become very useful in different administrative capacities to naval aviation, and could relieve pilots for combat or instructional purposes. As a result, men from all walks of life, from their late twenties to late forties were commissioned and poured into Quonset. Here they were subjected to a two months' period of military drill, courses in the fundamentals of naval service, recognition, naval aviation, naval regulations, naval courts and boards, and seamanship.

Those who lived through this period will always have certain ineradicable memories: the Iowa merchant, who a few weeks before had been bossing clerks around in a dry-goods store, painfully trying to unravel the intricacies of an official naval letter; the North Carolina realtor having trouble differentiating between an F4U and a navy blimp flashed on the screen at 1/25 of a second; the Maine hotel man vainly trying to recall which was his right shoulder on the drill grounds; the California professor shining the shoes of an Ohio newspaperman preparatory to inspection; the New York stockbroker fussing over a wrinkle in his bunk sheet with all the meticulousness of an elderly spinster.

Upon the completion of the course, which handled as many as 750 officers at one time, the graduates were parceled out among the various establishments of naval aviation. They went to sea frontiers, air stations, further training for specialized tasks, to become photographic officers, fighter directors, and engineering officers — to mention only a few of the many tasks performed by this group.

An important outlet for the younger Quonset men especially was the Air Combat Information School, which was also established at Quonset. This was a two months' course that trained ground officers for necessary duties in combat areas or at coastal operational fields. These were the men who briefed the pilots before and after flight, kept them abreast of operational data, and collected intelligence information both for the pilots and for higher echelons.

Hardly a history of a naval squadron has been written without tribute being paid the willing and effective contribution of these men. Much of their entrenchment cannot be measured by any tangible gauges, but by the intangible factor of personal relationship. The ACI officer, though often some years older than the pilots, ate with them, and when occasions permitted, drank with them, worked with them, counseled them, and all in all acted as a leveling influence on young men living under the terrible uncertainty of modern war.

In the first year of war, the technical training program more than quadrupled. On 1 December 1941 there were 7,905 men in training; one year later there were 31,529 men in the primary schools alone. This phenomenal progress was not accomplished without effort and occasional difficulties.

The increase was almost immediate. Shortly after the outbreak of war, Jacksonville's service schools went on a double shift, coincident with the request of the Marine Corps for the training of a large number of technicians, and construction was started immediately on still further facilities.

A decision to shift much of the training away from the increasing congestion in the seaboard stations to the Middle West was approved by the Secretary of the Navy, 24 April 1942. Two large centers were to be created which were to have an initial capacity of 10,000 men each. By 15 September 1942, this total had been increased to 15,000, and in the fall shifting of training schools to the Middle West was begun.

This great expansion brought into being a problem that required solution. This was the necessity of securing enough instructors to handle the influx of students. Teachers were obtained for the technical training program in a variety of ways, which generally speaking can be boiled down to two. In the first place, the Navy attempted to make use of civilian instructors and then tended to rely more and more on uniformed personnel.

A recommendation of the Training Division of the Bureau of Aeronautics as early as August 1941, had suggested civilian instructors. Consequently, early in 1942 a teacher training school was established in Chicago.

The students who were to become teachers were instructed by men brought from the Navy's schools at Jacksonville. The graduates retained their civilian status and gained further training in practice schools, which were started in May under the supervision of more experienced navy instructors, for aviation machinists' mates, aviation metalsmiths, and aviation ordnance men.

In addition, civil service positions for instructors had been set up as early as the fall of 1941, and a special Army-Navy board at Chanute Field, Rantoul, Illinois, processed all civil service applications of possible instructors for the naval aviation schools and the Army Air Force. Civilians selected by the Navy were sent to the Teachers' Training School at Chicago. Other civilian instructors were obtained until the end of 1942 through the Office of Education. Some of the civilian teachers were not as effective as was desired, and this, coupled with the dissatisfaction among enlisted instructors with the fact that civilians received more pay for the same duties, made for friction on teaching staffs.

These factors, plus continued inroads on civilian instructors by the draft, led to a decision to return to the use of military instructors only. A classification of Specialist (T) was created for those civilian instructors who were willing to accept an enlisted status. With these and with the retention of the best qualified graduates of the course a new faculty was quickly built up.

Another training obstacle was the problem of obtaining modern equipment. As in the First World War, priority on all new operational equipment was, at first, given to the fleet. The training program was forced, therefore, to continue to use obsolete material, despite the fact that this meant training men in the use and maintenance of aircraft no longer found in the fleet, when they were supposed to be fully qualified to handle the latest equipment. This situation became especially critical in the aircraft used for technical training. The older the aircraft, the more maintenance they needed and the more spares they consumed. Training, therefore, was at times held up by the lack of spares or equipment.

As early as January 1942, an attempt was made to counteract this difficulty by the assignment of a certain percentage of all new aircraft to training, but this quota underwent constant revision as the needs of the fleet increased, and technical training's needs were subordinated in some degree to the needs of other training.

Another type of technical training met with mixed success. This was the effort to train men in the factories that produced the equipment. Difficulties of administration, overspecialization, lack of living facilities, all hampered this type of training. The result was that some ventures were abandoned, while others, weathering these hardships, became fairly well organized. The following year, however, was to see a shift to navy training schools.

The training of radiomen, which was critical in the year before the war, continued to present difficulties. These were ironed out to a considerable degree by the establishment of new schools. With the increased use of radar, additional training was provided in this field.

One of the most important developments in training during the first year of war was the change in administration that was affected in the fall of 1942. The great expansion taxed the facilities of the small staff at the disposal of the Training Division of the Bureau of Aeronautics and those of the commandants of the naval districts under whom the schools normally functioned. To ease this burden so-called functional commands were set up.

The first of these was the Air Operational Training Command which foreshadowed the trend by coming into existence in April 1942. This was followed by the inauguration of the Air Primary Training Command, the Air Intermediate Training Command, and the Air Technical Training Command. With this organization naval aviation was ready to proceed to even greater expansion.

The second year of war saw aviation training swing into its full stride. The main groundwork of the program had been laid, though the remaining years of the war were to see modifications and elaborations to meet changing needs. The building of the program went on apace, each step marking a fuller realization of the problems involved and a further move toward the ultimate goal.

There were several important organizational changes during 1943. One of these was the establishment of a separate command for the training of lighter-than-air pilots. This was a logical step, and the new direction of this training from Lakehurst ended the stepchild position of LTA work under the previously established commands.

The second important change was the transfer of the Training Division from the Bureau of Aeronautics to the newly created Deputy Chief of Naval Operations for Air (DCNO) in August 1943. Since the main reason for the establishment of this office had been to create an aviation chief who would have authority to co-ordinate the work of the various bureaus on aeronautical matters, it was logical to put the Training Division under it so that it would have added authority to carry out its programs, particularly through the Bureau of Naval Personnel. Since steps had already been taken in this direction, this shift was perhaps more important in what it portended for the future than for any change in the actual situation in 1943.

The third important organizational change was the co-ordination of all primary, intermediate, and operational training under a new officer, the Chief of Naval Air Training. This move, while it took over to a certain degree the coordinating functions of the Training Division, proved to be an effective device for further integrating the entire program.

From the beginning there was an informal type of co-operation between the Navy Department and our allies of the British Commonwealth, especially Canada. In 1943 this interchange of information was placed on a more formal basis through the establishment of the Combined Committee on Air Training in North America. In a variety of ways, particularly in the fields of recognition, navigation, and aerial gunnery, this committee assisted materially the training of both us and our allies.

THE SIZE OF THE TRAINING PROGRAM WAS AGAIN increased by the fixing of a new upper limit of planes for the Navy at 31,447 on 15 June 1943. Jacksonville and Miami were diverted from intermediate to operational training, and another phase was added to the training program early in 1943.

This was the introduction of the so-called Flight Preparatory Program, which was designed to give college training to future cadets, and to improve morale in the large backlog of men who had volunteered for aviation training but who had not yet been absorbed in the program. Together with the V-5 and V-12 programs, it had the additional merit of helping to keep alive some of our institutions of higher learning in the lean years when the bulk of the regular student body was in the armed forces.

Twenty colleges throughout the country were selected as flight preparatory schools. Their wide geographical separation is indicated by the listing of a few: Williams College, Williamstown, Massachusetts; University of Virginia, Charlottesville, Virginia; Monmouth College, Monmouth, Illinois; University of Washington, Seattle, Washington. In these schools a twelve weeks' course was given in subjects including mathematics, physics, principles of flying, navigation, military drill, and physical education. The academic courses were taught by civilian professors of the institutions participating, while discipline, physical education, drill, and administration were under naval officers.

By March 1943, over 12,000 men were enrolled in this program. This training had a number of advantages, since it lessened the amount of work to be covered in ground school in later phases of the program, strengthened the future AvCads (aviation cadets) in subjects that would be used later in their training, and allowed the Navy to do at least some weeding out of men who were obviously incapable of absorbing their future ground school work.

There was also a backlog of enlisted men who were destined for aviation duties. In order to handle this group more effectively and equitably, the so-called Tarmac program was instituted. This term, unlike many other war-coined names, was not an abbreviation of a lengthy title, but came from an English nickname for men who worked about the early airstrips of tar or macadam. The Tarmacs were given preliminary, if manual, work around air stations and seaplane facilities as a type of indoctrination for their later duties in connection with aviation.

Another evidence that training was in full swing was the fact that carrier training was improved. More and more carriers were coming off the ways, with the result that it was no longer necessary to rush them immediately to the scene of battle. Instead, the carriers, with their new squadrons and other complements could take "shakedown" cruises that enabled all hands of the accustom themselves to carrier routine.

The schools for ground officers and enlisted personnel continued through 1943, but there was an increased demand for persons to fill nonflying billets. An important phase of the solution of this problem was the introduction of women into the Naval Reserve. The Waves, or Women Accepted for Volunteer Emergency Service, proved to be a valuable asset in many fields of naval activity. Their contribution to naval aviation was outstanding. They were first used in clerical and professional capacities in various divisions of the Bureau of Aeronautics, such as Training, Planning, and Personnel. More important than this, perhaps, they came to be employed in different specialized ratings in the field.

Many were trained to work in ground crews as mechanics; others did stellar service in control towers; still others were trained in aerology, photography, radio, and radar work. Some served as gunnery instructors, others as recognition instructors; some, as already noted, became Link Trainer operators, and others parachute riggers. By 1 March 1943, 1,050 were under technical instruction.

We have already noted that during the early period of the war, maintenance schools had been established in various factories, and we have seen that there were certain weaknesses in this type of training. In order to counteract these difficulties, in 1943 the Navy inaugurated line maintenance schools of its own. Located in the Midwest, with the center at Chicago, these schools gave brief courses in such maintenance subjects as magnetos, superchargers, turrets, and carburetors.

The next development was the creation of courses to familiarize mechanics with certain types of planes with which they would deal most frequently. By the end of the year, four weeks' courses were being given on the FM's and the F4U's, and the corresponding courses at the factories manufacturing these planes were closed.

An important development in maintenance training grew out of the need to keep men in the field informed of the many changes that were being made in engineering. It was obviously impracticable to send a machinist's mate back from the South Pacific for a few days' course on carburetor changes, much as he would have appreciated the assignment. Instead, the schools were taken to the field.

By the end of the year, the Mobile Training Unit was in operation. Huge trailers, equipped with cut-away models, charts, and other materials necessary to give a short course on recent changes, were sent to continental and outlying bases. This program proved to be one of the most successful innovations made in the field of training.

With the great increase in output of enlisted men and with growing specialization, it was found that the existing ratings were too general and as a result a man who had been highly trained in one phase of maintenance might not have a chance to put into practice what he had learned.

Consequently, more information concerning his navy education was placed in the enlisted man's record, and in many instances, he was limited to specialized work. In addition, at the suggestion of the Bureau of Aeronautics, the Bureau of Naval Personnel expanded the ratings to give a clearer indication of the abilities of the man bearing the rate.

For example, a man was no longer a mere AMM, or aviation machinist's mate to be thrown into any number of jobs. He was, instead, an AMMC (aviation carburetor mechanic), or an AMMH (aviation hydraulic mechanic), or, perhaps, an AMMP (aviation propeller mechanic).

The outstanding new development in flight training was that of night fighters. In the account of combat operations, the significance of this type of fighting is pointed out, but it is obvious that careful training was an absolute essential to the success of the program. The training was carried on in Florida and in Texas. Made valuable as a weapon by the development of radar, night fighting necessitated extensive training in navigation, instruments, and radar.

Under the rather noncommittal title, Special Devices, there lies one of the most fascinating technical stories of the war. There were several factors that brought these devices into being and made them successful. As we have already seen, there was continued difficulty in securing sufficient operational material for training purposes, since these planes and other equipment were sorely needed in the combat areas.

Consequently, synthetic equipment was produced. It could be built more rapidly and less expensively. The use of this equipment reduced the number of casualties that would have resulted from the use of actual flight equipment in the early stages of training.

The devices could be used twenty-four hours a day, were not affected by adverse weather conditions, and thereby speeded up the training process. There were other features that gave special devices an added educational advantage. More persons could be trained at one time by a single instructor, and in many cases, they could be trained more effectively, as in the case of free gunnery training. Many of the devices partook of the slot machine or penny arcade character, with the result that training became a game that held the student's interest.

What were these special devices? A mere catalogue of them fills an entire volume, and they range from specially contrived rulers to mechanisms filling a room and weighing thousands of pounds and costing sizable sums.

They are not makeshift affairs, but represented the achievement of the finest technical skill and engineering knowledge. They taught the pilot not only to fly, but to navigate, handle various types of armament, recognize a submarine, operate radar, use the radio and other devices for communications. They taught men how to repair engines, patch surface structures, install electrical gear. They helped pilot and crew to fly as a unit. They trained men in co-operation under simulated flight and combat conditions. They taught pilots how to get into their anti-blackout suits and how to get out of a submerged cockpit. They taught men how to drop equipment for air-sea rescue and how to pick up a periscope on the radar screen.

The synthetic training program began before we entered the war. The early period was devoted primarily to experimentation and analysis. The question of procurement was a difficult one, as in other fields of aviation, but was gradually solved, and by 1943 special devices formed an important part of the training program. It was believed that about 25 percent of the training program for aviation personnel could be covered effectively by synthetic devices, and that the cost of this training would be but 2 percent of the entire training cost.

One of the most important training devices was the movie film. This, together with the other photographic materials, performed a significant role in the training program. Here again, a mere listing of the titles of films would fill a volume. Between March 1941, and March 1945, well over 2,000 different films were completed by the Training Division for use in aviation training. These have included everything from a film on the care of a machine gun to such a documentary epic as "The Fighting Lady."

One film showed the use of the lathe, another gave a picture of the duties and functions of the plane captain in caring for an airplane. Slide films were developed for the use of an instructor, so that he could stop to explain the subject matter as the films were projected on the screen. Moving pictures of simulated operations were made that utilized the "stop and go" technique. These enabled members of a crew to work out actual combat problems including navigation, communication, and recognition during the showing of the picture.

In the production of movie film, the Navy showed true professional skill and produced results that were in the realm of "big business." Their photographers were able; some, for example, had been specially trained by Fox Movietone operators, and their equipment was of the best. Their contributions to the war effort lay not only in the fact that they speeded up training, but that they standardized training on a high level.

The Bureau of Aeronautics realized that training did not stop with the conclusion of formal training. The pilot and other personnel of naval aviation found that they were exposed to various types of training for the remainder of their naval careers. The bureau also realized that constant repetition can lead to boredom which may defeat the purpose of continued training. In order to counteract this effect, a section of BuAer was devoted to the task of producing pamphlets and other publications that would have training value and at the same time be palatable.

Perhaps the most effective writing of this sort was to be found in the so-called sense manuals. By ridiculing the stupid in pungent language and with amusing cartoons, these manuals brought out important points that personnel could not afford to forget. There was, for example, "Gunnery Sense" which gave detailed instructions on "How to Be the Oldest Living Gunner," and had some well-chosen sentences on the "Care and Feeding of Machine Guns."

Some of the pamphlets were morale builders. One on "Dunking Sense," for example, gave advice on survival at sea after being forced down and asked the reader, "Are you the sort of fellow who dives merrily into an empty swimming pool, or who bails out at ten thousand feet only to have to go back for his parachute? If you are, don't read this little pamphlet."

In a more sober vein, the Bureau of Aeronautics published a semi-monthly magazine called Naval Aviation News. Well and profusely illustrated, this journal covered all phases of aviation from operations to maintenance. One of the most popular features consisted of the comments of a character called Grampaw Pettibone.

In reality a naval aviator of long experience, Grampaw, illustrated as an irascible old codger, gave his pointed comments on a variety of accident reports. His main thesis was well expressed in one of his early remarks, "It is better to be a live pigeon than a statistic." He was devoted to the cause of saving pilots' lives by bringing before them accounts of accidents brought about by carelessness or recklessness. Pilots as a rule do not like to be preached to; Grampaw Pettibone's style of writing was such that he could express his opinions in the most forcible manner and at the same time avoid causing irritation.

One of the most popular official publications was the Recognition Journal. Conceived in the Bureau of Aeronautics, this magazine was a joint effort of Navy and the Army Air Force, and it attained an average monthly circulation of over 400,000. Published by the editors of Time, the Recognition Journal possessed some of the characteristics of Life magazine. In addition, it was timely, with pictures of planes, ships, and, during the push in Europe, of tanks. It had several types of quizzes that tested the recognition skill of the reader.

Generally, quite factual, the Journal did not hesitate to take a dig at those who failed in recognition, as, for example, the pilot who bombed a denizen of the deep, by writing "Sighted sub, sank whale."

The comic poster was also utilized to fight carelessness. The inimitable Dilbert was invented, a slap-happy, stupid oaf, who inevitably did the wrong thing, who entered the landing circle the wrong way, landed with the wind, taxied into other planes, left his monkey wrench in the motor of the plane he was repairing. Dilbert was immortalized on posters placarded on barracks and shop walls as a gentle reminder of things not to be done.

The training program achieved its maximum development and pretty much its final form in 1943. Thereafter the story was one of modifications. In 1944 certain cutbacks were made in the number of persons trained, and resulted in the elimination of some facilities and the improvement of the ones that remained.

In March the first cutback was announced; the Navy was to train 20,000 pilots instead of 25,000. Further cuts were made in the planned output, and such reductions lowered the need for training facilities. Six primary air stations, one intermediate training station, and one pre-flight (Del Monte) were eliminated from the training program.

This sizable cutback naturally affected many cadets who had already begun their training. As will be seen, the Navy pursued a policy toward these persons that made it possible to take them back into training program when the need arose. The "De-selectees" were given a number of choices: (1) returning to civilian life, where the draft would probably put most of them back in service in short order; (2) transferring to the air-crewman program, where a shortage of substantial proportions had existed for some time; (3) transferring to other sections of the naval program, including for some a chance to try out for commissions as deck or line officers.

The Navy made every effort to make a sensible and fair provision for the disappointed would-be fliers, all of whom had volunteered for flight training, and the adjustment was made with a minimum of discontent and disturbance.

We have seen that, in 1943, to handle the greatly increased training program, flight preparatory schools had been established in twenty colleges and universities throughout the country. With the reduction in the volume of the program, these schools were dropped, and the work that they had handled was shifted to the remaining schools. The pre-flight schools reduced some of their emphasis on physical education and absorbed much of the academic work of the flight preparatory school, the total course being extended from eleven to twenty-six weeks to accomplish this task. The total training program was increased to approximately seventy weeks, giving more conditioning, more time for ground school, and more opportunity for flying.

One of the most unusual types of training was a project introduced in pre-flight schools with the anomalous title, "An Intensive Course in Relaxation." This course was based on an appreciation of the fact that in the combat area, pilots often had to go without sleep for extended periods, and then had to snatch rest in odd moments and under trying circumstances. Under such circumstances, the pilot could not afford to toss about in his bunk or sprawl in a chair, plagued by disconnected and often disconcerting thoughts about the girl he had left behind, the flight just completed, or the mission that was scheduled for the next morning. His job was to relax and, if possible, sleep. The course attempted to solve this problem by showing the student how to relax, by teaching him where and how to relax the various points of tension throughout the body.

With the pressure on output lessened, it was possible also to extend the V-12 program. Students now found that they were to have three semesters of college instead of two and by the close of the year, every cadet then in training had had at least some academic work in the V-12 program. The significance of these changes is obvious: The Navy took advantage of a reduction in the quantity of training to improve quality.

The joint training of pilots and air crewmen received improvements and refinements during 1944. The teams were given more training before being sent to the fleet, and multi-engined bomber operations were thereby improved.

The new technical advancements in warfare issued a challenge to training that was met by proper expansion. Radar training underwent marked improvement. The introduction of rockets in aerial warfare necessitated a great deal of training, not only in firing but in maintenance. Special Devices made one of its many contributions by devising a cheap tow-kite which proved to be the most successful and maneuverable towed target naval aviation was to have during the war. The rapid development of target drones, small-scale planes which were radio-controlled and carried their own power plants, reached quantity production by the beginning of the year and proved to be another innovation, invaluable in improving both aerial gunnery and particularly the fire of shore and ship-based antiaircraft batteries.

IN THE LAST YEAR OF THE WAR, aviation training demonstrated its ability to meet changing conditions with a minimum of disorder. By the end of 1944, it was found that earlier cutbacks in the number of cadets had been too drastic. The need for additional pilots came about because experience in the Philippines showed that, when approaching large land masses under enemy control, the tempo of air operations rapidly increased with a resultant growth both in casualties and pilot fatigue. The goal for future training, accordingly, was raised.

To meet this goal, the "de-selectees" of 1944 were given the opportunity to apply again for flight training. Of the seven thousand men who had been removed from the flight training program as a result of the cutbacks, about 60 percent volunteered for a return to the program. The decision to raise the quantity of training was based in part on the increased attrition rate as our planes approached nearer the heart of the Japanese Empire, and partly on the recognition of the fact that two tours of duty in combat areas were all that the pilot could be expected to stand.

A further calculation that changed the picture was that fleet pilots should be sent out for shorter periods of active combat duty than had obtained in the earlier days of the war. As a result, the ratio of carrier groups to carriers was increased, and a decision was made to put marine squadrons on CVE's.

During the year there was a tendency to increase the number of fighter pilots and to give them additional bombing training. This shift in emphasis threw a burden on training that was quickly met.

An important training development arose to meet an improvement in the technique of air warfare. Fighter Direction was succeeded by what was known as Combat Information Centers (CIC), which tied more effectively together and improved co-ordination between aircraft operations and carrier or ground activity. An extensive use of radar was involved in this process. Once again, training met the demand by improvement and further specialization.

On the administrative side, a final move was made toward co-ordination. The Naval Air Technical Training Command was placed under the command of the Chief of Naval Air Training. This completed the integration of the administration of the training program begun with the establishment of the Air Operational Training Command in April 1942, and brought the last of the functional training commands under the Chief of Naval Air Training.

The main trend in the training of enlisted men was a natural one and once again demonstrated the determination of the Navy to improve quality wherever possible. There was a decline in the number of primary schools, but at the same time there was a corresponding increase in the enrollment in the advanced technical schools.

As hostilities drew to a close, the Navy began to make plans for the future. Training had been welded by the hard flame of war. Non-essentials had been done away with, and the Navy wanted to consolidate its gains. A committee was drawn together in the summer of 1945 to help draw the complete picture of training.

The main achievement was constructive work on a master collection of syllabi on training, given the expressive title, "Embryo to Tokyo." The primary aim was to eliminate duplication of training as much as possible, and though the task was far from complete by the end of the war, important strides had been taken.

The Navy can well be proud of its aviation training program. Taking the finest raw material in the world, American youth, the Navy, with a flexible and expanding program, turned these young men into the best pilots and aviation experts that any nation could require.

29 – PRODUCTION TO MEET CHANGING WAR NEEDS

ORGANIZING THE MASS PRODUCTION of airplanes for naval use acquired a new meaning for the Bureau of Aeronautics as a result of the attack on Pearl Harbor, and work already under way was driven ahead with increased intensity. Thanks to Allied orders, the aircraft plants had expanded. The question of the moment was whether or not these companies could meet both Allied needs and our own.

Almost immediately the challenge was thrown down to industry in the ringing tones of the President's message: The United States would make 60,000 airplanes in 1942, and it would more than double that number by producing 125,000 the following year. Engineers whistled on hearing these figures, and sat down with sharpened pencils to figure out just how much floor space and how many machine tools would be needed to complete the job.

The task presented was not an easy one. Airplanes were complex machines with a very short life, whose manufacture on a great scale was being planned for the first time in the United States. During the First World War, American industry had produced only some 15,000 planes of all types, while France, with the best record for aircraft production, had made slightly less than 68,000 planes during the period from 1914 to the signing of the Armistice.

As we have seen, the Navy and the Army at the time of the Pearl Harbor attack had had only modest experience in ordering large quantities of aircraft. In December 1941, the Navy had on hand about 5,000 useful aircraft and by July 1941, the beginning of that fiscal year, orders had been placed for only some 7,000 planes.

Now in the short period of two years, the Navy was to obtain almost 32,000 planes under the President's program. To be successful, the program had to overcome the scarcities of machine tools, materials, skilled labor, and the lack of experienced management for all the new plants.

Two important problems affecting this contemplated production had been in evidence months before the break with the Axis. In the first place, both the Army and the Navy were in the market for planes. While both armed forces required planes in numbers that were astronomical in comparison to prewar orders, the Army had a greater demand than the Navy.

It was evident that there must be co-ordination and not competition between these two purchasers. In scientific and engineering work the precedent had been set before 1917 by the establishment of the National Advisory Committee for Aeronautics and the Aeronautical Board, and when America entered the war other agencies had been set up to deal with the production of airplanes.

The other problem, and one that affected both the Army and the Navy, was the question of the share that the Allied nations were to have in the total output of the American aviation industry. Both British and French cash had built our aircraft factories to new size and capacity, and it was our policy to help these countries in their struggle against the Nazis. This policy was necessary to give us time to build our own defenses.

The President, therefore, declined to stop the flow of arms and planes to France and Britain. When France fell, it became our need to arm more rapidly than before and at the same time help save Britain from invasion. To meet this peril, England took over all the planes marked for foreign use.

In September 1940, all airplane production was placed under the Joint Army-Navy-British Purchasing Committee. At first this committee considered only airframes, engines, and propellers, but by the winter of 1940-1941 it was clear that the scope of its authority would bear closer scrutiny because other kinds of aviation material were not being manufactured rapidly enough. The committee, therefore, was cloaked with more power and on 22 April 1941 became known as the Joint Aircraft Committee (JAC).

To the end of the war, this remained the principal agency for scheduling Army-Navy purchases of airframes, engines, radios, radars, and aviation equipment. Both the Army Air Force and naval aviation were represented on the JAC by their senior members, who were authorized to commit their own departments.

One of the first accomplishments of the JAC was to complete the division of aircraft plants between the armed forces according to plans that had been drawn up in 1940. This was a measure that had been used during World War I to minimize competition between the armed forces, and once again experience in our first air war stood us in good stead.

In 1941 there was special urgency for such a move, since the Army had already launched upon its expansion, whereas the Navy had just begun.

Both forces had great need for trainers, and it was essential to agree at once on similar types. Scheduling and standardization became the real work of JAC. Through twenty technical subcommittees, JAC hammered down the differences in design of armament, communications, and navigation equipment used in American planes. The committee also created a master plan of aircraft production that set forth the construction of every type, class, and model called for in the President's program and drew together our own and British requirements so that similar equipment could be used by all.

After the United States entered the war, the actual assignment of finished aircraft was entrusted to the Munitions Assignment Committee (Air), which functioned under the Combined Chiefs of Staff, to distribute among our own and Allied services the weapons being forged in a now overtaxed arsenal of democracy. It was not until after the first year of our war had been fought that further changes were made in this organization.

Meanwhile, in order to solve the manifold problems of expanding plant capacity, speeding production, and sharing output with the Army and England, the Bureau of Aeronautics needed an ever-increasing staff. This requirement was met by commissioning additional officers, many of whom were experienced engineers, businessmen, or lawyers.

These men were assigned to procurement duties in Washington or in the field. With the officers, an increasing number of Waves, enlisted personnel, and civilian employees, the administrative aspects of aircraft procurement and production were soon well taken care of.

Throughout the first year of war, the old system of having all final contracts made by the Bureau of Supplies and Accounts remained in effect. While laying the groundwork for the growth of the aircraft industry and for placing immediate orders for thousands of additional aircraft, it was prudent to use the established procurement methods.

In the second year of war, after the initial shock of the transition from peace to war had been absorbed, and the regular production and delivery of aircraft assured, we shall see that it was to become expedient to develop administrative methods more suitable to the requirements of the vast new industry that was coming into existence.

With these readjustments in organization, the Navy pushed ahead to meet the President's program for production in 1942. The job at hand was simplified by the groundwork that had already been laid. It was decided that planes that were in major production, the Brewster Buffalo and the OS2U (Kingfisher), were to be pushed with new speed.

Another fighter, the F4F (Wildcat) and the SBD (Dauntless), a dive bomber, were coming off the lines in small numbers, and strong efforts were made immediately to increase the volume of production.

The design problems that had been so troublesome in getting the TBF (Avenger) into full production were tackled by engineering teams in order to get improved torpedo planes into action. On the whole, production of training planes was more satisfactory than the output of combat aircraft.

Until the summer of 1942, the bulk of the Navy's production was of F4F's, PBY's, TBF's, and OS2U's. The OS2U was called upon to carry a heavier burden than had been anticipated, because its replacement, the SO3C (Seagull) did not meet navy requirements, and an improved design, the SC, was not yet being manufactured.

All the hard work of planning and sweating out the production details began to get results. Navy plane acceptances reached a total of 1,104 airplanes for the month of September 1942. The extent of the increase can be measured by the fact that for the same month in the previous year the Navy had accepted only 268 planes. There was another important trend to be noticed. Half the production of 1941 had been of trainers; by the fall of 1942 the emphasis on combat types was resulting in production at four times the 1941 rate.

By the fall of 1942, the F4F and the TBF were well established in the production line, the SBD was coming out at better than a hundred a month, and the PBY was being accepted in ever-increasing quantities. By July 1942, a newcomer that was to make an enviable name for itself, the F4U (Corsair), began to come off the production line in small numbers.

By the end of 1942, large-scale production of naval aircraft was well established. Planes were being made more than 2.6 times as fast as during the previous year while the "target" for production had been exceeded by 14 percent. A good deal of this success was due to the fact that a number of surplus trainers had been turned over to the Navy by the Army, making possible greater concentration on the production of combat aircraft. Only dive-bomber production failed to equal the number of deliveries called for. Since the automobile industry was converted to aircraft production during 1942, the prospects for the next year were excellent.

No account of aircraft production during the first year of war would be complete without mentioning the plans and designs of advanced and radically new aircraft. With an eye to the future possibilities of air war, by the fall of 1942 both the Navy and the Army had projects well under way for the design and manufacture of jet planes. Before the outbreak of war, in 1941, arrangements had been made for the General Electric Corporation to obtain information about the best English jet motors. After further experiments, the first

American jet plane, the P-59, designed by Bell Aircraft and sponsored by the Army Air Forces, made its test flights in October 1942.

Meanwhile, the Navy made arrangements with the Ryan Aeronautical Corporation to make experimental planes designated as XFR-1's, which used the General Electric I-16 jet engine in addition to a conventional gasoline engine. Anticipating our story somewhat, it might be noted that in the summer of 1943, after the completion of satisfactory tests, one hundred of these planes were ordered. These purchases illustrate the manner in which joint procurement facilitated experiment and new designs, since these jet engines were made in Army-sponsored plants — that is, finance, inspection, and production control were all handled by the Army.

An airplane grounded for want of spare parts is of little use to a fighting air force. Such a condition can easily develop, because a plane uses spare parts faster than any other piece of fighting equipment in the Navy. During the early months of the war there were times when the percentage of American naval planes grounded for want of repair material was a little too high for comfort.

As we shall see in another connection, maintenance forces through almost superhuman effort and superb ingenuity created spare parts from practically thin air or "cannibalized" planes to "keep 'em flying." Such procedures merit great praise, but more than improvisation was needed to win a war. An effective system of securing spares in sufficient quantities was absolutely essential. Our success in building up such a system in comparison with the enemy's comparative failure to develop one was an important factor contributing to victory.

Even before the attack on Pearl Harbor, the United States Navy had made a good start toward building an efficient system for the purchase and distribution of aviation spares, and the wartime history of the spare parts program bears out the farsightedness of the early measures. Two major developments were set in motion at the same time, in October 1941.

The first was the purchase of spare parts in quantities determined by the rate at which spares would be used. Previously the selection of spares had been left to the manufacturer, and a quantity equal in value to 20 percent of the contract had been ordered. The second development was the establishment of an agency to handle the ordering, stocking, and distributing of spares to the entire naval aviation organization.

This organization was known as the Aviation Supply Office. Established at Philadelphia, 1 October 1941, this agency eventually became the principal procurement office for naval aviation replenishment and maintenance material. The Bureau of Supplies and Accounts also entered the picture. Although the Bureau of Aeronautics finally acquired authority to make its own contracts, the handling of spare parts continued until the end of the war as a joint project of both bureaus.

There were several factors making this the best procedure. In the first place, a good deal of the replacement material was what the Navy calls "standard stock," or material in common use, such as paint or sheet metal. It was better to buy these items in general orders for the whole Navy through the Bureau of Supplies and Accounts than to duplicate the process in the Bureau of Aeronautics.

Then, as anyone who has ordered spare parts for his automobile knows, keeping stock bins full and getting the right part in the right bin depends upon a good system of stock numbers and on up-to-date inventories. This type of work was "right up the alley" of the Bureau of Supplies and Accounts, and since its people were handing the material over the counter at hundreds of repair bases it was logical to let this bureau to take care of all the details of delivering it.

On the other hand, the Bureau of Aeronautics, because of its technical staff and its connection with the manufacturers as well as its experience in operating planes, was in a better position to know the kinds and quantities of spares that would be needed. These special qualifications of both bureaus established the lines along which the machinery for ordering and distributing spares could be put together.

To carry out this task, an agency was set up that stood, as far as responsibility was concerned, halfway between BuAer and BuSandA. This was the Aviation Supply Office and its offshoot the Aviation Supply Depot. The roots of the latter went back to World War I, to a central supply depot for aviation material that had been set up at that time, and to the work of the supply officer at the Philadelphia Aircraft Factory.

To carry out the same task in World War II, the Aviation Supply Depot was built in Philadelphia. Material was received in its vast warehouses from the manufacturers and was distributed to the air bases and the fleet through supply annexes established at Norfolk, Virginia, and Oakland, California.

One of the major achievements of the Aviation Supply Office was the development of a standard numbering and inventory system. By renumbering the interchangeable parts for the different series of

Pratt & Whitney engines, for example, the ASO saved taxpayers not less than $30,000,000. To perform this service for all kinds of maintenance material a "Sears Roebuck" type of catalogue of aviation spare parts, complete with pictures and ordering instructions, was compiled.

Everyone who has worked in a war plant and has opened his lunch pail for inspection by the watchman at the gate, and the passing motorists who have wondered at the camouflage paint and nets used on aircraft plants, has been made aware of some of the details of the system of plant security developed by the War and Navy departments. To a people not accustomed to close governmental scrutiny of workers changing shifts, or FBI investigations of applicants for new jobs, plant security measures might seem to be an irksome and unnecessary detail thought up by busybodies.

Fortunately, the presence of enemy agents was detected so quickly by the FBI and other agencies that the elaborate security systems were never put to a real test. Had there been any organized sabotage, security measures would have been able to cope with the situation. That they were adequate is shown in part by the fact that there were no disasters comparable to the "Black Tom" explosion of the last war. There is no doubt that continual surveillance helped the morale of workers by providing visual and constant protection and by creating an atmosphere that made people "security conscious." If they did nothing else, security measures made some indifferent people know that "There's a war going on."

One of the most effective security measures was a check on employees. Authority to make these investigations was based on the provisions of the Aircraft Procurement Act of 1926 and the War Powers Act of 1940 that required employers to obtain the consent of the Secretary of the Navy to hire aliens for work in factories producing classified or aeronautical materials.

Much of the actual work of investigation was carried by the Federal Bureau of Investigation and the Office of Naval Intelligence, but Bureau of Aeronautics inspectors and other officials of the Security Section of BuAer were directly connected with the process.

Passive defense, as it was called, was also under the control of the Security Section of the Bureau of Aeronautics. This included such matters as blackout regulations and camouflage, and also the general problem of safety both for the plant and its personnel.

To secure these ends the Navy furnished funds for the construction of high steel fences around factories, and assisted materially in the serious matter of arranging for protection against fire. In view of the fact that the only attacks on our country came in the shape of a few shells lobbed into our western shores and paper balloons wafted our way through the stratosphere, some people may consider the expense of plant security measures unnecessary. A moment's reflection on postwar flights of four-engined bombers from Tokyo to Washington should, however, lead to a different conclusion.

Production of aviation materials made tremendous strides during 1943, a year which marked the achievement of the Navy's first production goal: the building of productive capacity. Expansion of plant facilities had been substantially completed by the end of 1943, concurrent with the mass production of new and improved fighting aircraft.

Early in 1943, the F4U (Corsair), the F6F (Hellcat), and the FM-2, an improved Wildcat, were the fighters upon which the Navy was relying. During 1943 Goodyear got into production of the Corsair under the designation, FG. A great increase in the number of "baby carriers" and the demands of anti-submarine warfare in the Atlantic gave rise to a special need for fighters of the Wildcat type.

Marine squadrons based ashore received the Corsair, while the Hellcat was used in ever-greater numbers by large carriers operating in the Pacific. Production of other types of combat aircraft continued at great speed, so that in 1943 a total of 23,144 aircraft of all types were accepted by the Navy. This was about two and a half times the 1942 rate, and it exceeded the estimate that had been established at the beginning of the year.

During the first year and a half of wartime production, the brunt of the engine production had been borne by Pratt & Whitney, the principal engine manufacturer allotted to the Navy in the division of industry between the armed forces. The success of this company in filling the great demand was the result of two factors: subcontracting and licensing other engine companies to manufacture according to the Pratt & Whitney design.

Four Pratt & Whitney engines were used: the 985, 1340, 1830, and 2800. Continental and Jacobs were brought in to make the R-1340, and Chevrolet and Buick helped to get the R-1830 past the critical point. Lycoming and Air-cooled made engines of similar size on licenses. Nash was licensed to produce the R-2800, 2-stage "B" engines for the Navy, and Ford produced the R-2800, 1-stage "B" engines for the Army.

Both companies were ultimately supplying all the service requirements for these engines, allowing Pratt & Whitney to concentrate on advanced models. A new Pratt & Whitney plant was set up at Kansas City to manufacture the improved R-2800 "C" engine, and complete success was achieved in producing an entirely new engine that had never been in production at other Pratt & Whitney plants.

The importance of these Pratt & Whitney engines to the Navy's air effort can be seen from their use in key aircraft. The F4U and the F6F used the R-2800-R engine rated at 2,000 horsepower. Late in 1943 the PBM came into production with the R-2600 engine, which gave place to the R-2800 single stage "C" in 1944. The PV-1 (Ventura), a land-based twin-engined bomber, used the R-2800-31. Navy Liberators, or PB4Y's, and the PB4Y-2 (Privateer), both used the R-1830 engine.

Other navy planes were powered by Wright engines. The FM-2, for example, used the Wright R-1820 which produced 1,350 horsepower at take-off. In the SB2C a Wright R-2600 engine was used, and it was also installed in the TBM, the General Motors version of the TBF.

Production of aircraft engines presented many problems both to the contractors and to the bureau, and at various times the ignition, bearings, piston rings, and other components threatened the schedules. By persistent work and diligent experimentation to develop more easily fabricated and longer enduring parts, the manufacturers kept the engine program moving along to equip the mountains of airframes being produced in other factories.

DURING THE FIRST YEAR OF WAR, as we have already seen, the assignment of finished aircraft was entrusted to the Munitions Assignment Committee (Air), which functioned under the control of the Combined Chiefs of Staff. Experience during this first year showed that more top direction was needed. There was need of an agency to direct actual production and settle matters of aircraft policy at the highest level.

As a result, the Aircraft Production Board (APB) was formed on 9 December 1942. This body provided a meeting place for high-ranking representatives of the armed forces and members of the WPB to discuss the joint aircraft program and thus secure the best balance in the war effort.

Since the Aircraft Production Board was a policy-forming agency, it needed an executive body to carry out its directives. Accordingly, on 19 February 1943, the Aircraft Resources Control Office, better known as ARCO, was brought into being. Essentially, the job of ARCO was to present the case of the aircraft industry before the War Production Board, or other agencies controlling the war effort, as far as manpower, materials, or machine tools were concerned.

Much of the field work and actual liaison with the Bureau of Aeronautics and the Army Air Force was accomplished by an older agency that was placed under ARCO. This was the Aircraft Scheduling Unit (ASU), established on 5 May 1941 to handle the problem of scheduling aircraft production for our own and the British armed forces, and, under ARCO, assigned to correlate the industrial needs of the Bureau of Aeronautics and the AAF.

If there had been plenty of material, these organizations might not have been necessary. However, with limited supplies of basic metals, such as copper, steel, and aluminum-magnesium, care had to be exercised to make the best possible use of this material from the standpoint of the war as a whole. To achieve this, the War Production Board set up what it called the Controlled Materials Plan to handle the allotment of basic war materials.

It was ARCO's task to present the claims for the Army-Navy program before the Requirements Committee of WPB. It is a tribute to the officers up and down the line in both services that the joint aircraft program worked so smoothly and well during the critical years of war.

At the start of the war, it was not immediately foreseen that a shortage of manpower would become one of the most serious production problems. Throughout the 1930's few American industrialists or labor economists had been obliged to think seriously about labor shortages. For this reason, the Bureau of Aeronautics before Pearl Harbor had considered manpower only in terms of labor-management relations.

For a time, this approach appeared to be adequate, since shortages of labor did not interfere with the first great expansion of the aircraft industry that began soon after the United States entered the war. Almost two years after the Pearl Harbor attack, however, when the national economy was straining to increase every kind of war production to equip armed forces that were expected to total more than ten million men, the attainment of maximum aircraft production began to be threatened by lack of workers.

This lack of manpower manifested itself with dramatic suddenness in the spring of 1943, first in the west coast aircraft plants and then in other parts of the country. Since there were other agencies, both in the Federal and state governments, that had a more direct control over the supply of labor, no immediate steps were taken in the Bureau of Aeronautics until it became evident that the aircraft production program was threatened.

The work with the manpower problem as developed by the Bureau of Aeronautics was tied closely to other Navy Department organizations and to ARCO. At first most of BuAer's work was carried out through the board. In order to deal adequately with manpower, BuAer then began in 1943 to build up its own staff. Officers with a background in labor relations work or with legal training were ordered to duty in the Production Division.

The organization was at first handicapped because its work was arbitrarily divided into manpower and labor relations aspects, which were handled pretty much as separate units. That these two phases of the labor problem usually flowed into each other was not fully appreciated until the crisis had become acute in the summer of 1943. When this fact became clear, the precedence of manpower activities was advanced, and a special section for dealing with interrelated labor problems in the aircraft industry was set up in the Production Division. Special labor advisers were added to BuAer offices in the field.

The policy and the activity of the Bureau of Aeronautics manpower representatives were determined by the integrated manpower and production control plan, known as the West Coast Manpower Program, which the Director of War Mobilization put into effect 4 September 1943.

After this date, operating machinery was set up for the continuous adjustment, by decisions made in localities themselves with occasional appeals to Washington, of the shifting needs of manpower to meet the military and naval production requirements. With various refinements and modifications, the West Coast Manpower Program was adopted wherever a critical labor shortage developed.

The War Manpower Commission was, of course, the top agency controlling manpower policies, but with the help of various area and coordinating committees, the labor needs of the aviation industry could usually be met without prejudicing other aspects of war production.

Foremost among the problems with which the Manpower Section of the Production Division had to deal was Selective Service. Without impairing the equal treatment of citizens upon which the functioning of Selective Service depended, the Manpower Section tried to protect the production of naval aircraft by obtaining occupational deferments or military furloughs for the key men. In this work close liaison was maintained with the Office of the Assistant Secretary of the Navy.

To assist the west coast aircraft plants at the time of the man-power crisis of August and September 1943, an arrangement was made whereby the Army and the Navy agreed to give six months' fur-loughs to key workers who had been drafted. Surveys were made of the factories, and lists were compiled of the number of skilled work-ers required to maintain output.

After much planning by the bureau representatives, the pro-gram fell down because the companies did not know which men were needed most. As a result of their inability to name the key men drafted into the Army or Navy, the bureau representatives were able to obtain but few releases.

A labor relations desk in the Manpower Section gave the BuAer contractors expert advice in the handling of wage problems and in the preparation and processing of the various applications at the stabilization agencies. Production of aircraft was promoted wherever possible by obtaining wage rates more favorable to the recruitment of additional labor.

Special assistance was also given BuAer contractors whose pro-duction was threatened by strikes. Because all aircraft production hinged upon the output of airframes, all disputes over wages paid by the airframe contractors were handled directly by the Bureau of Aer-onautics before the War Labor Board.

Manpower problems were more difficult to solve than those in-volving matériel, transport, or equipment. There was not only the human factor but there were also the laws of property and the con-stitutional rights of citizens. To attract people to war industry often required more than mere appeals to patriotism; the social customs; personal desires, and standards of living of the workers had to be recognized, and as far as possible in wartime, protected and nur-tured.

Considering the size of the industry that was developed, there was remarkably little governmental intervention in the matter.

It was not enough to get men and women to work in aircraft fac-tories. The vast majority of them had had no previous experience in this type of employment. The Bureau of Aeronautics played an im-portant part in helping to train these workers, and in presenting incentives that stimulated greater output. BuAer found that many of the techniques that were working out so well in the technical train-ing of its own men could be applied in aircraft factories.

Training aids of all variety, therefore, found their way into air-craft factories. Films, slides, posters, and other devices were used. Morale was built up by war films and by visits of war heroes. In addition, a monthly magazine, Wings, was published as a joint Army-Navy project. This magazine, which reached an important percentage of the aircraft workers, contained articles and illustrations on various phases of aircraft production. The publication stressed the conservation of materials and of time, and made suggestions for more effective contribution to the war effort.

Films were widely used, with effective use of the animated cartoon, and before the end of the war a catalogue of more than a thousand films of technical subjects was compiled. Another valuable type of visual aid was the graphic chart. One of the first of these was a set of charts prepared for the Goodyear Company in 1942, and from this original work a service was developed for analyzing management problems. Close relations between the bureau and the leading contractors were maintained by meetings such as the one held at Buffalo, New York, in March 1943, at which general problems were thoroughly aired.

Everyone who has worked in a plant on war contracts for the armed forces knows the sense of pride and achievement that comes with the award of an Army-Navy "E." This award originated in the Navy in 1906, when an "E" was given a ship for proficiency in gunnery, engineering, or communications. To give this award for industrial achievement was a frank recognition of the important role that American war workers were playing. At first only plants working on navy contracts were eligible, but in June 1942, it was retitled the Army-Navy "E." Every plant engaged in war production, whether private or government, then became eligible, with the exception of transportation, service, or utility agencies.

First consideration was given the quality and quantity of war materials produced; then the armed forces considered the record of the plant in overcoming stoppages, maintaining fair labor standards, training additional labor forces, and creating effective management. Awards were made on a plant rather than a company basis, and consistent performance was necessary to retain the award.

Whether the Army or the Navy made the award depended upon which service had the larger contracts in the plant, but confirmation of the award had to be made by both army and navy production boards. As evidence of the high performance necessary to win this award, it should be noted that only 30 percent of the plants working on war contracts had received it prior to V-J Day.

THE GREAT NAVAL AIR OFFENSIVE PLANNED for 1944 was made possible by the acceptance of more than 29,000 planes from the aircraft industry. This total may be compared with approximately 23,000 planes for 1943.

Production reached its peak in March 1944, when 2,831 planes of all types were accepted by the Navy. New fighters were being manufactured, such as the F7F (Tigercat), a twin-engined fighter designed by Grumman, the makers of the famous F4F and F6F series. Improved versions of these models were still being made and were giving an excellent account of themselves in air combat.

A replacement was at last available for the OS2U, and even the old SOC, which had been the work horses of the scouting fleet. This was the SC-1, a Curtiss model that possessed much greater speed than the OS2U. Still, the latter's exceptionally long duration of flight made it useful whenever long searches were required in areas where enemy fighters were not active.

By 1944 the PB4Y-2 was well established in production, and the old reliables, the PBY and PBM, were still coming off the lines. The Lockheed PV-2 (Harpoon) was also being produced in 1944, together with improved models of the SB2C and TBM.

The increasing emphasis upon the production of combat planes rather than trainers or utility types, was shown in the fact that almost 90 percent of the acceptances were for combat aircraft. There were two other trends during the latter period of the war. One was the fact that airframes were considerably heavier, a point that makes our production growth all the more remarkable.

In 1940 the average navy airframe weighed 2,740 pounds compared to an average airframe of 6,423 pounds in 1944-1945. This increased weight, of course, reflected improvements in range, speed and bomb load. Despite the trebling of output per employee, a second trend was to be noted, namely, that there were significant reductions in the cost of airframes to the Navy.

By 1944, the aircraft industry was on the whole a well-functioning machine. Production was keeping pace with operational needs. Efficient administrative organizations had been established to handle the complicated problems of labor, critical materials, machine tools, and co-ordination between the armed forces. The full might of our production forces was making possible the great military gains made on the battlefronts of the world.

While production remained at a very high level, 1945 in the aircraft industry saw the emergence of reconversion problems. Aircraft manufactures and workers are human. They were intensely concerned that implements of war should be made, but as the conflict appeared to be drawing to a close, it was only natural for them to think of the future. Factory owners began, therefore, to think about postwar work, and factory workers began to cast about for more permanent employment.

Since production might be affected by either of those tendencies, the Navy Department found it necessary to set up agencies to deal with them. Every effort was made to assist in the coming transition from war to peace by an intelligent policy of cutbacks. Contract termination services were developed, therefore, to make equitable settlements of war contracts and strike a balance between production for war and peace.

Lawyers who had been serving in a variety of billets throughout the Naval Reserve were transferred to Contract Termination in order to carry out a "man-to-man" type of negotiation with contractors who were seeking a release from war production. Close co-operation was the order of the day with the Army Air Forces, and since both services faced the same problems, common methods were worked out and agreed upon. While these adjustments were being made, the aircraft factories continued to produce war materials and a steady stream of planes and equipment reached the combat areas.

The aircraft industry had done its job well. From 1 July 1940 to 1 September 1945, over 83,000 planes had been accepted by the Navy. This figure in itself is eloquent praise of the labor of the men and women of America who left their normal way of life to make this achievement possible. Without their efforts, victory on the field of battle could not have been won.

30 – THE EXPANSION OF SHORE ESTABLISHMENTS

THE OUTBREAK OF WAR IN DECEMBER 1941, found naval aviation shore establishments getting under way on a huge program of expansion. The first year of war was to witness significant progress toward the achievement of this program. There were two main functions to be performed by these stations: They were to assist in anti-submarine warfare to protect our own shores and they were to provide space and equipment for the expanding training program for both pilots and enlisted personnel.

There is no space to trace in detail the story of the development of each of these air stations. A few examples, however, may give an indication of the problems that had to be solved and of the work that was accomplished.

One of the largest naval air activities that developed on the east coast was at Norfolk, Virginia. Like many another air station, it was built on what had formerly been mud flats or swamps. Air stations, in general, need room. The fewer the surrounding obstructions, either natural or man-made, the better could be the air approaches and the less possibility there would be of danger to the innocent bystander in case of training accidents. Where seaplanes were to be employed, access to the water was an obvious necessity.

At Norfolk, for example, Willoughby Bay was dredged to provide fill-in material for the surrounding swamps, with a result that over 350 acres of land were reclaimed to provide satisfactory beaches for the landing of seaplanes. The original sites proved to be too small for the expanding requirements made on the station.

Consequently, ten outlying fields, for the most part in low, swampy areas surrounded by level ground, were acquired in Virginia and North Carolina, and reclamation and construction were pushed during 1942. One of the main functions of NAS Norfolk was the training of carrier groups that engaged in landing and take-off practice and gunnery exercises on the outlying fields.

The outstanding naval air station in New England was at Quonset Point, Rhode Island. Situated between Providence and Newport, Quonset Point was constructed from low land and filled by pushing back the waters of Narragansett Bay. This huge station was used not only for training but for operational purposes. Patrols swept the seas off New England in the war against the German submarine.

At the same time advanced flight training was carried on at the base for carrier and other duty. As noted elsewhere, Quonset was the site of two of the Navy's outstanding schools for Naval Reserve officers, the indoctrination course for A-V (S) officers, and the ACI (Air Combat Information) school, and as the war progressed, other important training was also carried on at this base, especially in the field of anti-submarine warfare. Perhaps the best example of the creation of something from nothing was the construction of Floyd Bennett Field, New York. The original site was a spot in the bay aptly called Barren Island, whose only "improvements" consisted of garbage heaps and a glue factory that acted as a receiving station for the city's dead horses. The Navy entered the scene, removed the "improvements" and lifted some 6,000,000 yards of sand from the bottom of the bay. The field was raised to a level of 16 feet above high water and had adequate layers of subsoil and topsoil above the sand base. The result was that Barren Island ceased to be barren — and ceased to be an island, for filled land now connected it to Flatbush Avenue and Brooklyn. During the war, this field was used as a base for anti-submarine warfare and also as a center in which naval aircraft were assembled, serviced, and delivered to the fleet.

The three largest naval aviation flight training centers that were to develop during the war were at Pensacola and Jacksonville, Florida and Corpus Christi, Texas. The expansion of Pensacola was natural in view of its pre-eminence in the prewar period. Facilities were expanded, new fields were added to make this one of the leading training stations in the country. In Texas, along the southern Gulf coast, Corpus Christi carried on functions similar to those at Pensacola. NAS Jacksonville became the hub of a network of stations devoted to operational training. The spot had certain advantages that caused it to be selected over other suggested sites. It possessed a superior strategic location, with good operating conditions. The site was immediately available, and was suitable for immediate construction. Transportation facilities were good, there were no local objections to the proposed development, and the region was relatively free from possibility of hurricane damage.

In 1942, in addition to the development of important stations along the coast, there was the beginning of a new trend, the establishment of large training stations in the heart of the country. As early as February 1942, the Secretary of the Navy had approved a recommendation that Naval Reserve aviation bases be established at Memphis, Tennessee, and Norman, Oklahoma, for primary training. These bases were also designated for technical training, and it was through this function that they acquired their greatest significance.

These bases were huge inland enterprises, on which only the training smacked of the sea. Such huge bases as these naturally took time to construct. On 15 September 1942 the first class of twenty cadets took off at Memphis in Piper Cubs. At that time there were no outlying fields, the main base mat was only 20 percent complete, and the base itself was still a huge construction camp, lacking in permanent heating, telephone service, water, electricity, and even a sewage disposal system. In contrast, it should be noted that this station was to develop a capacity of nine hundred cadets or aviation pilots. The construction of naval air stations was a co-operative work of various bureaus in the Navy Department. The Bureau of Aeronautics and the Bureau of Yards and Docks working together formulated policies on construction. BuAer and the Bureau of Supplies and Accounts co-operated on matters of storage, and BuAer and the Bureau of Ordnance on the establishment of arsenals and magazines.

In addition, there was close liaison with two agencies outside the Navy Department. Close contact was maintained with the Army Air Force, and a mutual exchange was made of information regarding air facilities. The most important relationship, however, was with the CAA (Civil Aeronautics Authority). Many of the fields used by the Navy were municipally owned airports that normally were under control of the CAA, but were leased to the Navy for the duration of the war. While the Navy was developing such stations as Norman and Memphis, the CAA was also constructing fields at the Navy's request and then turning them over for use upon completion. This action was made possible by an authorization of Congress under a program known as the DLA (Development of Landing Areas for National Defense).

Under this program airfields were chosen by the CAA, upon the approval of either the Army or the Navy, in locations that would be useful to civil aviation after the war. Under the usual procedure, local governments acquired these sites, and the CAA allotted funds for the construction of appropriate runways, lighting, fencing, or the clearance of normal approach zones. Fields of this sort usually had few other improvements, since they were employed as auxiliary landing places. NAS Quonset Point, for example, made frequent use of this type of facility. Another valuable contribution of the CAA was in the construction of navigational aids throughout the country, such as airway beacons and localizers. They also carried on important experiments on landing devices, and conducted a number of helpful aeronautical surveys.

The second year of war saw the greatest actual increase in new construction facilities, since much of the half billion dollars' worth of such work that had been contracted for in 1942 was completed and made available in 1943.

During 1943 an ambitious project for the construction of a string of lighter-than-air bases was completed. A great deal of oil and ship production was being carried on in southeastern Texas. German submarines had found their way into the Gulf of Mexico and had inflicted considerable damage on shipping in that area. Part of the Navy's program to protect shipping from the Gulf ports to the east coast was to provide blimp coverage over this route.

Accordingly, bases were constructed at such spots as Hitchcock, Texas; Houma, Louisiana; Richmond, Florida; Glynco, Georgia; and South Weymouth, Massachusetts. The technique of construction was essentially the same in each of these, with the exception of Richmond, which was designed and constructed as a sizable assembly and repair base to service blimp squadrons that were beginning to dot the Caribbean and Latin American areas. The stations were relatively small, self-contained bases, with huge wooden hangars, large enough to hold six blimps, and even a rigid airship, if and when such a craft should again be constructed.

As in the case of Norfolk and other large air stations, most were built upon marshes. The usual inconveniences were encountered. Irritated personnel often complained that they could walk knee-deep in mud and at the same time have the dust swirling along the surface. Mosquitoes, the indigenous occupants of most of these sites, carried on a bitter struggle against the intruders. Station newspapers were filled with boasts concerning the size and potency of these insects. It was asserted that six flying in formation could bend a prop, and that one could bowl over a seaman if he did not brace himself for the attack.

Adding these LTA bases to auxiliary air stations, air facilities, and outlying fields, by the end of 1943 naval aviation had constructed a combined total of 148 aviation shore activities. Against this figure should be placed the total of 41 in use on December 7, 1941, and 78 a year later. These 148 activities were able to take care of 16,000 aircraft. All this growth during 1943 testified to a truth that the layman had found hard to grasp during peacetime, that navy planes, whether based ashore or afloat, needed shore facilities in the United States.

In addition to these physical conditions, there were important developments in the use of naval aviation shore establishments. There was a tendency to make more use of smaller auxiliary fields for primary training. Another trend was the sharp increase of operational training. The addition of new carriers had laid an emphasis upon the need for land-plane training. Coastal bases increased their capacity, and interior bases, such as Traverse City, Michigan, and Clinton, Oklahoma, were shifted to this type of training.

As 1943 drew to a close it was seen that submarine warfare was on the decline. Bases along the Atlantic coast, especially, began to turn from anti-submarine operations to operational training, and several LTA bases were closed when squadrons were shifted to more active regions. Marine Corps training during 1943 also made added demands, and facilities were provided for this expansion. Outstanding in this connection was the establishment of imposing Marine Corps air stations on the west coast.

New fighting techniques created a demand for specialized stations. Toward the end of the year, more attention was paid to the development of night fighters. Separate fields were designated for this training, since natural hazards to flying had to be reduced to a minimum. Furthermore, men who fought at night had to sleep in the daytime, and it was therefore unsatisfactory to have planes carrying on daylight operations at these fields. A specialized use of a naval aviation base was that of commissioning and delivering new aircraft. Increased production placed a heavy burden on the aircraft delivery unit at NAS New York (Floyd Bennett Field). To relieve the pressure, a similar unit was established at Trenton, New Jersey. The function of this unit was to deliver to the fleet aircraft received from east coast manufacturers.

During its first eighteen months of operation the Trenton Naval Air Facility equipped, flight-tested, and delivered to the fleet 9,183 fighters and bombers. A somewhat similar unit was set up in the west. In the Arizona desert an Auxiliary Acceptance Unit accepted PB4Y aircraft from the Consolidated-Vultee Aircraft Corporation, San Diego, and other sources, tested this aircraft and installed radio and radar equipment. The main development in naval aviation shore establishments during 1944 was the co-ordination and consolidation of administration. Such a development was natural and essential, for the heavy work of construction had been completed. Building was to continue, but the high mark had been reached, and the problem for the remainder of the war was to make the best possible use of the facilities at hand.

Needs were continually changing. Lighter-than-air bases had passed their real usefulness, for example, and bases created to handle this type of craft could be and were shifted to other uses.

Sometimes this shift was gradual, and for a time blimps and heavier-than-air aircraft operated from the same bases. Newer techniques of fighting, such as rocket warfare and night fighting, continued to become more and more important and required special facilities. Naval Air Transport was rapidly expanding, and additional bases were assigned to this activity.

As a result of these and similar problems, late in 1944, a general order of the Secretary of the Navy set up three major types of air base commands. The most inclusive of these was known as the Naval Air Bases Command, which was an outgrowth of an earlier administrative organization, the Naval Air Center. With few exceptions, these commands corresponded to the naval districts, and gave the commander of the naval air bases administrative control over the naval air bases in the district.

Normally, Marine Corps air bases were included under this control. In the case of Cherry Point, North Carolina, however, a separate Marine Corps Air Base Command was established, with functions similar to the naval air bases. The training bases were already well organized, and rather than disturb this organization, naval air training bases were exempted from the naval air bases organization.

In 1945 naval aviation shore establishments underwent continuous changes to meet fluctuating needs. As the war drew toward its close, certain activities became less important. With the passing of the German submarine, lighter-than-air activity became less important. More and more LTA bases were turned to other uses, and huge hangars that had once housed operating blimps were, in 1945, converted to storage purposes.

Some of the bases were shifted to heavier-than-air use, and others maintained curtailed blimp operations and assumed other duties in addition. NAS Houma, for example, in the swamps of Louisiana, devoted itself to fleet training. NAS Santa Ana, in the orange groves of southern California, on the other hand, continued blimp operations but was assigned a special project (Drones).

Primary training in 1945 became less important than operational training, and as a result there was a good deal of shifting about to meet these changing training needs. At the beginning of 1945, the Navy had 171 air facilities. It was able to refrain from further construction by two means. The first, as already indicated, was to convert stations from one use to another as rapidly as possible. The other was to reassign facilities between the Army and the Navy. Through joint committees, the Army transferred a number of facilities that it no longer needed for the use of the Navy in critical areas. This work had begun in 1944 but was carried on into 1945, and by the use of army facilities the Navy was saved the expense and effort of new construction.

We have noted that in 1944 new techniques, such as night fighting, had imposed a burden upon shore establishments. These burdens were increased in 1945. Rockets and jet-propulsion affected not only the problems of training but of research. One of the outstanding developments in this connection was the growth of the Naval Air Test Center at Patuxent, Maryland. Authorized shortly before our entrance into the war, this center was of immense value to the prosecution of the war.

In view of the emphasis upon scientific developments during the war, it is likely that this activity will be of great assistance in the future in keeping naval aviation in the fore in technical equipment. Some sixty miles from Washington, in a rural community, Patuxent covers a large area, some six miles across. Its main functions have included flight, radio-radar, and armament testing, as well as experimental work. In addition, the station had functioned as an eastern terminal for NATS.

In April 1945, it was decided to make important changes on the west coast. Most important of these was the decision to make improvements at Moffett Field, originally devoted to lighter-than-air aircraft, to carry on NATS overhaul of 4-engined transports. A novel feature of this plan was to use one of the 1,000-foot hangars for the creation of a "production line" type of preventive maintenance, which on a mass-production scale would apply maintenance to engines before rather than after repairs actually became necessary.

The end of the war in Europe naturally reduced the pressure on certain shore establishments, and by V-J Day more than twenty stations had been reduced to a caretaker status or otherwise taken off the "active" list.

The influences of the naval aviation shore establishments have been manifold. Without them, of course, the tremendous training program could never have existed, nor would anti-submarine operations from our shores have been possible. As in the case of other production, these establishments stand as witnesses to the ability and determination of a united people.

31 – "KEEP 'EM FLYING!"

THE IMPRESSIVE GROWTH OF NAVAL AIR POWER cannot be explained simply in terms of the number of first-line fighting planes which contributed so effectively to the surrender of the enemy. The factors that made this air power such a potent instrument for war can be listed ad infinitum. Such a listing, however, would not in itself tell the whole story. Building air strength was one thing; keeping it powerful was quite another. Once created, naval air power had to be continuously maintained.

A combat aircraft is considerably more than a flying mechanism. It must be equipped with such things as guns, bombs, navigational devices, special instruments, cameras. It requires fuel, lubricants, and painstaking mechanical checks, not only to keep it flying, but to keep it in fighting trim. The relative importance of these factors to the ultimate fighting performance of the aircraft can hardly be exaggerated. Without any item of equipment, the possibility of combat inefficiency or downright failure is evident. Without proper and adequate maintenance, the best military aircraft in the world would soon become as useless as a rusty shotgun on a wall.

No plane can fly long without upkeep. Combat aircraft require even more specialized upkeep or maintenance than types engaged in routine operational flying. Not only must normal engine overhaul be attended to, but battle damage repaired. These planes had to be restored to fighting condition, serviced and rearmed by speedy and efficient maintenance crews, and returned to action in a minimum of time.

The standard naval aircraft radial engine is a complex unit of over 14,000 parts. That it must not fail in flight is the obvious axiom on which maintenance is focused. It can function only within specified time limits before it requires routine overhauls. The longer it flies, the more intensive and extensive becomes the necessary maintenance until after 500 to 600 flying hours or approximately 90,000 flying miles it must be completely disassembled and rebuilt or even replaced altogether.

It is an obvious but important fact that an airplane is not like an automobile that can be pulled over to the side of the road for minor repairs by the driver; nor is it like a ship whose power plant can be maintained while under way. Even though emergency repairs can sometimes be made in the air, the airplane must return to its base, carrier, or tender for proper service and maintenance by a highly skilled crew of properly equipped specialists.

In the days before the war, maintenance was a relatively simple matter when confined to well-equipped naval air stations with permanent facilities and established sources of materials, personnel, power, communications, and transportation. Major overhaul and assembly were performed customarily by large assembly and repair (A&R) shops located at permanent air stations in the United States or in territories owned or leased by this government.

Routine maintenance, short of major overhauls or structural repairs or changes done by these A&R shops, was usually performed by each pilot for his own plane, with the help of a small ground crew that traveled with the squadron for this purpose.

The exigencies of combat made this old system as obsolete as many of the aircraft it maintained. A war fought over such vast distances in so many parts of the world required that maintenance function where aircraft were in combat. Furthermore, the complexities and number of new types of aircraft which were constantly being developed meant that maintenance personnel had to be not only experts in aircraft types but specialists as well in aircraft component parts.

In the early part of the war it was necessary to return aircraft engines to Hawaii or the United States for major overhauls, a process which kept each engine out of action for approximately six months. As the war progressed into the southwest Pacific and then northward into the Philippines, this problem was rapidly overcome by the establishment of large semi-permanent A&R shops together with supply depots of aircraft parts on island bases. These shops were not more than a day's flying time from combat areas, and overhauled engines were never out of action for more than twenty-one days.

It was estimated that A&R facilities built in three months at Noumea, New Caledonia, saved the Navy and our allies over $12,000,000 in investment for new engines and by serving army, navy, marine, Australian, and New Zealand aircraft of all types, made immediately available some 45,000,000 additional flying miles.

The Integrated Aeronautic Program, an outgrowth of the recommendations made by a board headed by Rear Admiral (later Vice Admiral) Arthur Radford and charged with developing an Integrated Aeronautic Maintenance, Material and Supply Program, in 1944 relegated all major overhaul and repair to the continental United States and provided for replacements of completely new aircraft rather than the return of renovated aircraft to combat areas. Thus, the entire maintenance problem was greatly lessened, because new aircraft which required a minimum of operational maintenance were always on hand. Service units thereafter devoted their efforts to routine maintenance and repair of battle damage rather than to fighting normal wear from hard use.

The most significant departure from traditional maintenance procedures were the new policy of keeping maintenance crews permanently attached to a ship or shore station instead of accompanying squadrons as they alternated between ships or shore establishments for training or combat duties. Even before December 7, 1941 it was recognized that in the event of war, shore establishments would be required to perform a larger volume of maintenance, leaving carriers and tenders as free as possible of routine upkeep in order to concentrate their efforts on the kind of maintenance that kept a maximum number of planes constantly ready to fly.

But as the first trickle of naval aircraft into combat areas swelled into an unprecedented flow, it became obvious that existing maintenance organizations were antiquated and that the need for an entirely new system was acute. This problem was resolved by the creation of several kinds of naval aircraft service units designed to meet the over-all maintenance requirements of naval aircraft of all types and especially "tailored" to support specific models.

The common denominator of all these service units was that they were completely integrated mobile organizations trained to maintain naval aircraft from ship or shore bases anywhere in the world. Like any other naval vessel or activity, each service unit was usually organized into several divisions; engineering, flight operations, ordnance, communications or electronics, first lieutenant, supply and disbursing, personnel, and medical. Under each division were the various departments and shops which performed the actual maintenance work on aircraft.

There was usually a shop for each type of upkeep and repair: machine, metal, hydraulic, instrument, electronics devices, carpenter, paint, propeller, photo, CO-2 and oxygen, procure rigging, instrument, tire. Because these service units were created for the sole purpose of aircraft maintenance, they were relieved for the most part of any additional duties not strictly related to support of their own personnel and the aircraft assigned to their care. For that reason, these units seldom operated independently. They were always based on a naval air facility in the United States or its territories or upon advance base units in forward areas.

At advanced bases these service units usually operated with an ACORN, the Navy's term for an organization of men and equipment designated to construct rapidly and operate an air base or to repair and operate a captured enemy airfield. In effect, the ACORN built and operated a forward-area naval air station on which the service units maintained aircraft.

In those few cases where major repairs were performed in combat areas, Aircraft Repair and Overhaul units, AROU's in naval parlance, handled the job. Such units operated from our bases at Noumea, Guadalcanal, and Espiritu Santo. By far the most numerous and versatile of these units, however, were the Carrier Aircraft Service units, commonly called CASU's.

A few weeks after the Japanese attack on Pearl Harbor the first four CASU's were commissioned began to operate in the Hawaiian area. Because CASU's maintained most types of combat aircraft including fighters, fighter-bombers, torpedo planes, dive bombers, and night fighters in the later phase of the war, they were redesignated Combat Aircraft Service units, or CASU (F)'s (the [F] indicated Forward Area).

During some of the Pacific island campaigns they supported marine, army and Allied combat aircraft as well as their regular navy charges. While all CASU's were theoretically stamped from the same mold, much of their value stemmed from their flexibility. Those units which operated in combat zones concentrated their efforts on keeping their squadrons in the air, while CASU's based in the United States or Hawaii assumed the added responsibilities of receiving new planes, training personnel in their duties and commissioning men and planes as squadrons.

This function of the CASU's operating from permanent naval air stations was especially important because they were responsible for organizing and training the service units that maintained aircraft aboard aircraft carriers, the Carrier Aircraft Service divisions, or CASD's. Quite often experienced CASU personnel became the nuclei for new CASD's in process of formation.

Scout Observation Service units, or SOSU's, were similarly created to service scout observation planes which operated from battleships, cruisers and, in a few instances, from destroyers. Patrol Service units, PatSU's, were established to maintain patrol aircraft, usually the large land-based Liberators or Venturas as well as the "Cats," Catalina flying boats. PatSU's were usually part of the Headquarters Squadron (HedRon) of a fleet air wing. Since all these service units were basically the same in organization and function, with suitable training a CASU could be redesignated a PatSU, or a SOSU could become a CASU to meet the requirements of changing squadron types under their care.

Lighter-than-air (LTA or blimp) maintenance was generally organized along similar patterns, Blimp HedRon's were usually responsible for airship maintenance which differed little in most cases from that required by airplanes, except for specialized problems such as fabric damage and helium control. Major repairs and engine overhauls were normally performed at permanent air facilities in the United States, while interim checks and minor repairs were affected at advance LTA bases in the South Atlantic and Caribbean areas.

Aside from that performed aboard carriers, aircraft maintenance afloat was confined primarily to two types of vessels, aircraft tenders and aircraft repair ships. Aircraft tenders were well-known ships of the train long before the outbreak of war. Equipped with complete facilities for maintaining all types of seaplanes, they operated as floating bases for as many as twenty aircraft and their crews at a time.

On occasion they served to transport air groups and aviation supplies in addition to their normal maintenance duties. They were attached to fleet air wings and often served in the Atlantic and Pacific as flag ships of fleet air wing commanders. Aircraft repair ships converted from both Liberty ships and LST's were developed during the war to supplement shore-based maintenance units. They were completely equipped with the same kind of shops and facilities as shore-based units and provided emergency aircraft maintenance offshore until a beachhead was secured and the CASU was in full operation. During this period their work was largely limited to maintaining component parts of aircraft brought from the beach in small boats.

The Japanese publicly attribute their defeat to American scientific skill and mechanical "know-how." It was a combination of these two factors, especially under the severest combat conditions, that kept our naval aircraft, sometimes inferior in performance and oftentimes inferior in numbers in the early phases of the war, at the peak of fighting efficiency until our superiority in all categories was overwhelmingly and permanently established.

Improvisation was a keynote throughout all the American armed forces. Seldom did it show up so markedly as in the aircraft service units which moved with the fleet and operated from our successive "steppingstone" advance bases. This American "know-how" produced remarkable results in operational maintenance of naval aircraft.

One of the first and perhaps most famous is the story of how a few Marines at Wake Island in the first days of the war kept their pitifully few planes flying against the enemy by determination and ingenuity in patching, repairing and "cannibalizing" parts of one damaged plane to keep another in the air. Even if not against such hopeless and overwhelming odds as those faced by the Wake garrison, this same story of keeping planes in condition to strike the enemy was repeated many times over as naval aircraft spearheaded and then supported the difficult assaults and landings of combined American and Allied forces from Guadalcanal to Okinawa.

In the early phases of each of these campaigns, while a beachhead was being secured and before adequate facilities were built, aircraft maintenance units were invariably hard pressed to keep the first planes operating from captured bases in flyable fighting condition. Usually there were planes awaiting battle-damage repairs, refueling, and rearming as soon as the maintenance units arrived. Wrecked or damaged planes not worth repairing served as a steady source of supply for replacement parts for other aircraft until such time as regular maintenance supplies were brought in. Many naval aircraft served important combat duties as "victims" of such "cannibalism."

Maintenance cannot function without materials and supply. It was evident in the very nature of many landing operations that during the initial phase maintenance supplies might be inadequate. The immense quantity of supplies needed for simple maintenance precluded its total arrival on the beachhead simultaneously with the service units.

By improvisation and cannibalism plus actual fabrication of needed parts, maintenance units managed to keep planes operational until a normal supply flow was established. Even then these units sometimes continued to improvise, for aside from expected parts and structure failures, battle damage demanded the greatest amount of replacement spares and no matter how carefully anticipated, battle damage could on occasion exceed available supplies of parts. Planes were often forced to jettison valuable and scarce equipment, thus placing unusually large demands on supply.

Complete sets of radio, radar, and ordnance gear were "ditched" on occasion to insure a plane's safe return to its base. While such items were always replaced by supply, there were many instances in both the Atlantic and Pacific war theaters in which the Army Air Forces provided badly needed maintenance supplies in the interim. Similarly, the Navy supplied maintenance services and parts to the

Army, Marines, and Allied aircraft when requested. In some of these instances the appearance of obsolete or unfamiliar aircraft types for which supplies were not immediately available challenged but never defeated the ingenuity of maintenance and supply men.

Almost everywhere in the world normal problems of aircraft maintenance were complicated by enemy action, geography, and the elements. In addition to regular maintenance instruction, service units in the Pacific were trained in Commando tactics and jungle warfare as well as foxhole and slit-trench "architecture."

Some had many opportunities to use such training. CASU (F) 17 went ashore with the Marines on Tarawa, "a square mile of hell," on November 24, 1943, four days after the first landings and immediately began to maintain aircraft on the captured Japanese airstrip on Betio Island. Since this landing was the first American attempt to take a small, low-lying coral island, there was no definite precedent to follow in planning operations. A CASU was not supposed to carry its own tools and equipment but to rely upon its supporting ACORN for them.

Nevertheless, it was realized that in the confusion of landing it might be impossible to locate urgently needed crated equipment among the tons of supplies put ashore. It was the CASU's primary duty to service planes immediately to insure a continuous Combat Air Patrol (CAP) over the newly won beachhead. Therefore, each man in the first landing wave was given basic maintenance hand tools to carry in with him. This was probably the first time any CASU went ashore with every possible piece of equipment lashed to gasoline and oil tanks, and to every other available mobile unit. These vehicles served not only to carry equipment ashore but as markers indicating its location.

CASU (F) 17 personnel demonstrated their resourcefulness as improvisers in using captured enemy equipment. Tents, walls of damaged Japanese buildings, scrap lumber, packing cases, and salvaged tin were used to build shops. For four weeks a reconditioned Japanese generator was the only means for charging aircraft batteries and the only available transportation was an abandoned enemy sedan. Even Japanese medical supplies provided valuable ointments for treating lip blisters and infected sores.

32 – Naval Air Transport Service

Railroad timetables and other travel schedules have always possessed a peculiar fascination for the American people. In peacetime no dream has been as persistent among both young and old as that conjured up concerning distant places. It is regrettable that wartime security prevented the arm-chair explorer from reading the schedules set up by the Naval Air Transport Service, or as it was familiarly but not disrespectfully called, NATS.

Without moving from his chair or shifting his pipe in his mouth, the vicarious traveler could have planned a trip from an airport outside Paris, France, to the east coast of the United States, across the continent to Oakland, and thence to such distant places as Okinawa or the Philippines. There were side trips that could be taken, perhaps a jaunt down through the Caribbean to Rio, possibly a trip to Australia by way of the Hawaiian Islands, or conceivably to the remote reaches of the Aleutian chain.

These schedules were there, it is true, but of course it is equally true that the trips were not pleasure cruises, as will be sworn to by anyone who has sat for hours on end, literally, in the uncompromising bucket seats of the navy transports. These flights were a vital part of the war effort. If aircraft could be used as a potent fighting weapon, it was found that it could also be employed as a reliable and speedy method of transportation for essential personnel and equipment.

The story of NATS in this war, like that of any other major development in naval aviation, is account of a hard-won progress from small beginnings to a huge and highly organized system. In general, the development falls into three major categories—NATS in the Atlantic theater, NATS in the Pacific area, and the ferry service that took planes to the combat regions.

Even prior to the entrance of the United States into the war, the experience of the British had demonstrated a need for an air transport service run by the military forces; yet in this country both the Army and the Navy were slow to realize the importance of aviation for the rapid distribution of men and supplies.

The Navy, for example, in this period, depended upon a few utility squadrons to meet its requirement for air transport. There were no schedules, no established routes, when the acquisition of new bases as a result of the destroyers-for-bases deal with England made it clear that utility squadrons, already flooded with other duties, were inadequate for the task of supplementing the slower movement of goods by sea through submarine-infested waters.

The Navy was the first major service in this country to take steps to fill this gap in transportation facilities. Some time in 1941 plans for a regular, scheduled air service began to take shape.

The first proposals came from Mr. C. H. Schildauer, a civilian employed by the Glenn L. Martin Company as sales manager of marine equipment. As a graduate of the Naval Academy, as a former employee of Pan-American Airways for whom he had helped investigate both transpacific and transatlantic routes, and finally as a man experienced in flying boats, Mr. Schildauer had both great technical capacity in questions relating to air transport and close connections with the Navy.

As a result of the work of Mr. Schildauer, it was decided to inaugurate an air division within the Naval Transportation Service, which was to function under the Chief of Naval Operations. Accordingly, on 1 December 1941, Mr. Schildauer was called into the service as a commander (later captain) to commence plans and studies for the establishment of such a service. The advent of war less than a week later speeded up the program, and on 12 December 1941 the President authorized the Naval Air Transport Service.

The early months were hectic. Personnel had to be secured, and like Captain Schildauer, other men were drawn from the commercial airlines. Not too many could be brought in from these lines, however, for fear of paralyzing the commercial air transport system. These men who joined NATS were trained and thoroughly schooled in airline procedures.

This fact had its obvious advantages, but it also had disadvantages, for the airline procedures of maintenance and operation in many cases differed considerably from those of the Navy. The Navy was willing to compromise; from this nucleus of personnel the vast NATS organization was built up, pilots and maintenance crews assigned by the Navy were trained in approved airline methods, and NATS routes were pushed into areas where no commercial pilots had penetrated and under conditions that they had not faced before.

The beginning of operations was dependent upon the acquisition of planes as well as of personnel, and in this connection difficulties appeared immediately. In the first place only sixteen planes capable of the extended overwater flights contemplated by NATS existed in the United States, and they were in private hands.

In the second place, new plane construction of the standard Douglas two-engined transport, the DC-3 (termed R4D by the Navy and C-47 by the Army), and the four-engined transport made by the same company, the commercial DC-4 (called R5D by the Navy and C-54 by the Army) were both in that sector of the aircraft industry that came under army supervision. New land planes for NATS could, therefore, be obtained only through the Army, which was struggling with its own problems and, concentrating on heavy bomber development, was not inclined to encourage the building of transports.

It was not until June 1942, that the Army set up its own military transport service, the justly celebrated Air Transport Command.

A possible source for seaplanes existed in the patrol boats already being used by the Navy or under development. Although these had not been designed with the thought of use as transports, it was found that they could be adapted for such use by making changes in fittings and equipment. The Martin Mariner (PBM) and the Consolidated Coronado (PB2Y), therefore, came to be widely used by NATS.

Another possible source of planes was to use the craft of commercial airlines. In this case, however, instead of commandeering these planes, in actuality, the Navy reached a two-way arrangement whereby NATS used commercial planes, and the commercial lines used some navy seaplanes in connection with their contract work with NATS.

With a minimum of personnel, much of it inexperienced, and what planes it could obtain from the airlines, adapt from combat types, and beg from the Army, NATS began operations. On 29 January 1942, the first plane, an R4D was received at Norfolk, and on 3 February the first flight was undertaken. For a month, pilots and crews were given their initial training for the great event, and on 2 March 1942, the first regular flight took off from Norfolk for Squantum, Massachusetts.

A couple of days later another flight set out from Norfolk for Jacksonville, Florida. One week after the first R4D took to the air, the first NATS squadron was formally commissioned, and it might be said that NATS was really under way.

During 1942 the main emphasis was on development on the North American continent. In April, operations were commenced on the Pacific coast, and a squadron was set up at Kansas City to make possible more coordinated transcontinental flights.

New Squadrons were needed and commissioned, with the emphasis on the Caribbean during the first year of operation. One seaplane squadron was established at Dinner Key, Florida, and served the dual purpose by carrying men and materials to the Caribbean and of being a training center for new pilots. Another squadron, using land planes, operated out of Miami to points in the Caribbean and ultimately as far south in Latin America as Montevideo. Despite the rise in importance of the Caribbean area, the pressure became increasingly heavy on Norfolk, which was burdened by other aviation and naval problems. Consequently, much of the NATS activity was transferred to Patuxent, Maryland. This was a new station, devoted primarily to testing, but it had a great deal of available space, and Patuxent became important for NATS, both from the standpoint of operations and of maintenance.

Until December 1943, NATS did not attempt direct transatlantic flights, but left this phase of operations to contract operators Pan-American Airways, and American Export Airlines. These lines maintained services to the United Kingdom by way of Newfoundland and Ireland, and from the latter point connections were maintained to Port Lyautey in French Morocco and to Lisbon. Pan-American also operated south of Port Lyautey as far as Leopoldville in the Belgian Congo. At times, also, both lines operated shuttle services from North Africa to Natal, Brazil, from which point it was possible to transfer to planes operating south from Florida.

Although with the assignment of a Skymaster (R5D) to a NATS squadron a trial flight was made to Scotland as early as May 1943, no regular service was initiated until the end of the year. On 20 December 1943 the first R5D began flying the route from Patuxent River to Port Lyautey.

By May 1944, these big planes had added to their itinerary a run from the United States to Scotland.

With the great increase of equipment and personnel during 1944, it became advisable to release the commercial airlines from their contracts and permit them to return to their more usual tasks of providing commercial transportation for representatives of nonmilitary departments of government, of Allied and neutral nations, and of civilians having legitimate reasons to travel in wartime. The two airlines had contributed heavily to the Navy's war effort not only by maintaining essential routes when NATS possessed neither the equipment nor the personnel but also by showing the infant NATS the way in many phases of operation. By late 1944, however, the child had outgrown its teachers and was able to go it alone.

Atlantic operations reached their peak during 1944. During that year NATS maintained routes all the way from the United Kingdom to Uruguay, and over these routes flowed a steady stream of high-priority mail, cargo, and personnel. No matter how frequent the flights or how large the planes, there never seemed to be quite enough service to meet the demand. Unlike much combat flying, air transport was at its best when it was least exciting — when schedules were maintained, accidents nonexistent, and routine flight uninterrupted.

In addition, however, to the regular schedules that formed the backbone of the work, NATS was frequently called upon for emergency deliveries. In May 1943, it was learned that a carrier, badly needed in the Pacific, was being held at the Canal Zone pending the arrival of droppable gas tanks for its planes.

A NATS plane took off from Norfolk, despite the fact that the field was closed by weather, went to Chicago and picked up the tanks. From Chicago it went to Long Island for fittings for the tanks and then proceeded by way of Miami to Panama, arriving about noon on the day following the receipt of the order.

The following February, a French destroyer was desperately short of ammunition unobtainable in the Mediterranean area. Two R5D's arrived in Bizerte, each carrying about 10,000 pounds of the needed material only 34 hours and 40 minutes after leaving Floyd Bennett Field, and one of the planes performed the added service of spotting and reporting a German submarine en route.

The most important of these special missions occurred in connection with the Normandy invasion, when, on 14 May 1944, the order came to move approximately 112 tons of special equipment which according to the official report "was of the most urgent nature and one of factors upon which the success of the operation hinged."

NATS employed eight R5D's to transport 165,250 pounds of cargo in sixteen round trips between 15 and 23 May, while the Air Transport Command moved the remaining 57,205 pounds in eleven trips.

The successful prosecution of the war in Europe enabled NATS to turn its attention more and more to the Pacific. One after another, Atlantic operations were curtailed and then canceled. On 12 February, even the central command of NATS moved west as Rear Admiral J. W. Reeves, Jr., took up his post at Alameda, California, as head of a fleet command.

Because of the vast distances involved, a naval air transport service was even more vital to the prosecution of the war in the Pacific area than in Europe. The early problems were much the same. As one of the NATS pioneers phrased it, "We started with a headful of ideas and a handful of equipment and personnel." Fortunately, in the Pacific as well as in the Atlantic, there was the Pan-American Airways to fall back on.

Prior to the war, the Pacific Division of Pan-American Airways had maintained a flying boat service from San Francisco to Honolulu, and on to the Far East. This service had been terminated by the outbreak of war, but within a week the shuttle service to Honolulu had been resumed, with Pan-American's planes under navy ownership and operated by Pan-American pending a formal contract with the Navy Department.

These contract operations with the Pan-American Airways were the beginning and foundation of naval air transport in the Pacific. As the pressure for an increase in the shuttle service became greater, the Navy and the airlines company at first worked in co-operation to meet the demand. Operations by the Navy itself began on 1 April 1942 with the establishment at Alameda of a NATS squadron. By the middle of May routes had been opened to Corpus Christi, Texas, the great naval air training base, and to Seattle to link with the Alaskan Division of Pan-American.

With one exception the planes flown were land planes. The exception was a seaplane, an XPBS-1, that was used on the flight to Hawaii. During the spring and early summer, this plane shuttled between San Francisco and Hawaii, but at the end of that period it crashed, killing one of the copilots. One of the passengers, who was uninjured, was Admiral Chester W. Nimitz.

By the end of the summer, not one but many seaplanes appeared to replace the XPBS-1. Coronados were used for the trip to Hawaii, and Mariners were used for island hopping in the South Pacific.

The opening of scheduled flights to the far Pacific was preceded by survey flights. One of these was made late in August and left small groups of key personnel to welcome a larger survey flight that followed the first. As a result of these flights, bases were established from Palmyra to Brisbane, Australia, and by the end of 1942, NATS service to Australia was in full swing.

The following year saw a great rise in the NATS activity and organization in the Pacific. Young NATS officers were sent out practically single-handed to obtain a seaplane anchorage and secure necessary equipment. One of the most commonly told yarns is of Lieutenant Dobbins, whose special assignment was to find and procure a boat to meet incoming planes at Espiritu Santo. The naval officer found a man who had an extra boat, but would release it only for some wire.

Lieutenant Dobbins scurried about and found another man who had some extra wire, plus an insatiable thirst. The officer found a means with which to quench the thirst, received the wire in exchange, bartered it for the boat, and the planes were met at Espiritu Santo. A difficult portion of the Pacific venture of NATS was the Alaskan theater. Planes were valuable in evacuating wounded from Attu. In addition, a listing of the cargoes of NATS planes in the Alaskan area would include refrigerated whole blood, Russian fliers, penicillin, the Truman Committee, and a 4,844-pound pump for a submarine.

Meanwhile, the expansion of NATS in the whole Pacific area was phenomenal. A few figures will demonstrate the point. In the last half of 1944, NATS Pacific flew 18,323,409 plane miles as compared with 5,030,829 plane miles in the first half of 1943. Total ton-miles show an even greater growth, with 68,329,587 flown in the last half of 1944 as compared with 14,153,182 in the first half of 1943.

It was one of the anomalies of NATS operations that the facilities were rarely adequate until after the operational peak had passed. Thus, the South Pacific bases had no sooner been improved to a point at which they provided reasonable facilities and living accommodations than the war moved ahead toward Japan, and these accommodations had been left for more primitive conditions farther along the road to victory. One of the outstanding developments of NATS in the Pacific was the inauguration of a service, described elsewhere, to evacuate wounded from the advanced area.

Two new planes were designed for NATS use. One, the Conestoga (RB-1) a stainless-steel plane that was developed during a period when aluminum was critical, proved unsatisfactory for naval use. The other, however, was the huge Mars (JRM-1) originally designed as a patrol plane but found much more adaptable to transport work. The original Mars did yeoman work in the Pacific from November 1943 until the spring of 1945.

Before the outbreak of hostilities, planes had been constructed at the factory and picked up there by pilots from the squadron to which they had been assigned. Under war conditions, it would have been impossible to continue this practice that would have created bottlenecks at the plants and delayed prompt delivery to combat units. In order to meet this problem, commissioning units were set up to install special equipment and armament. This action still did not solve the problem of getting the planes to the operating units. At first the burden fell upon the naval air stations in New York and San Diego and the naval air facility at Columbus, Ohio, since they were situated near the principal manufacturers of naval combat planes.

At these locations aircraft delivery units were established as part of the air-station organization. Originally, these units relied for fliers upon pilots with civilian experience whose age prohibited their use in combat. All sorts of fliers, amateurs, ex-stunt pilots, crop-dusters, former combat airmen with the Navy and the RAF, were enrolled and set to work. Training consisted mainly of beginning with ferrying primary trainers and progressing through more advanced trainers to combat types.

Because no regular routes existed, ferry pilots chose their own, obtaining gas and service from army, navy, or civilian fields along the way, keeping in touch with the home unit by sending a telegram indicating where they intended to spend the night. It was rough and rugged, but it kept planes moving. By autumn of 1943, however, it became apparent that such a haphazard method of ferrying vital war planes needed to be junked for something more appropriate to the demands. The net result was the commissioning, on 1 December 1943, of the Naval Air Ferry Command, with Captain John W. King at the head. Organized as part of NATS, the Ferry Command underwent few organizational changes, but did undergo remarkable development.

The most important features of the new command were the establishment of definite ferry routes with competent service units along the line. Formerly, if a plane broke down along the line, parts and repairmen had to be dispatched from either New York or San Diego. The setting up of an organization, of course, did not solve all problems. The question of personnel was a critical one. Some of the old-timers objected to the somewhat binding rules of the new schedules and controlled conditions of flight. For a time, pilots from operational squadrons were assigned to ferrying duties, but this worked with indifferent success. Eventually, special schools were set up to train ferry pilots, and this eased the situation. All in all, the ferry service performed well a vital function in the work of the Naval Air Transport Service.

The termination of the war did not bring to an end the work of NATS, and in fact in the Far East the routes were extended for the use of the occupation forces. The work of NATS will live in another way. Starting with a heavy debt to commercial airlines, after the war these airlines will in turn profit immeasurably by the wartime experiences of NATS.

33 – NAVAL AIR SERVICE

THE USE OF THE ATOMIC BOMB at Hiroshima and Nagasaki impressed upon the world with dramatic and explosive suddenness the major role that science was playing in World War II. At the same time the atomic bomb, with its still incompletely studied implications for the future, has tended to blind people to an appreciation of the other contributions of science in the war.

The scientists who were fighting on our side were opposed, especially in Germany, by technical experts of rare skill and ingenuity. The enemy, who for years had been preparing for war, had sponsored research and experimentation in the tools of war without regard to cost.

On the other hand, our scientists were more concerned with inventions for a peacetime existence, and at the same time funds were limited for research for war.

Once the war started, our science demonstrated that it had two assets that enabled it to surpass the enemy. In the first place, our scientists showed under pressure that their abilities were fully as great as those of the enemy and often exceeded them. Coupled with these skills was the fact that American industry, and particularly the aircraft industry, developed successfully the technique of introducing modifications (or slight changes in design) without disrupting the system of mass production.

Thus, it was possible for American engineers and scientists to maintain continuous superiority in quality as well as quantity.

In an account such as this, there is no place for a detailed description of the work of science in naval aviation. This work extended to all fields of aviation activity, and in many cases was exceedingly complex. On the other hand, the results of many scientific developments are observed in the combat narrative, as, for example, the introduction of a new plane, the use of radar, or the value of radio to certain operations.

In addition, it is the purpose of this chapter to note in a very general way some of the major scientific problems that confronted naval aviation and to indicate the solution to these problems that helped build naval aviation to its high peak of efficiency.

A dominant problem, of course, was to create planes superior to those that would be met in combat. Not one, but a variety of planes had to be produced — trainers, fighters, dive bombers, scout and observation planes, torpedo planes, patrol craft, and transport and cargo planes.

In the building of carrier-based planes, we had special problems to solve. Compared to airfield runways, the largest carrier deck is extremely short, and methods had to be devised, therefore, to get planes into the air with shorter take-off runs. Two solutions to the problem were suggested by the engineers and scientists. One was the improvement of catapults and their use on carriers as well as on other ships of the line. The other solution was the development of what was called JATO, or jet assisted take-off.

Engines were also built which added a jet unit to the standard reciprocating aircraft motor. This type combined the longer range of the ordinary engine with the great speed resulting from jet propulsion.

Carrier craft, also, must normally have stronger under carriages and fuselages than land-based planes in order to withstand the shock of arrested landings on the carrier deck. The scientists in this case were confronted with, and satisfactorily solved, the problem of providing strength without unduly increasing weight or sacrificing performance. The result was one of the most amazing developments of the war, namely, that carrier planes were able to compete successfully with land-based aircraft.

A military plane consists primarily of three components, the airframe, or body of the plane, the motive power, and the accessories, such as armament, navigational gear, radio, and radar equipment.

In the matter of airframes, the Navy's problem was doubled by the fact that both land and seaplane types were needed. Several types developed prior to Pearl Harbor were found adaptable to combat operations. The PBY's, for example, were found to be a satisfactory type of seaplane for many functions, and aside from the addition of an amphibious version, its production continued without too much variation.

The PBM, on the other hand, underwent a number of modifications before the top-performing PBM-5 was produced. Constant improvement was made on fighters and dive bombers. The durable F4F gave way to the superior F6F, and had the war continued, the F7F and F8F might have made equally enviable names for themselves.

Experimentation was continuous on such matters as flaps on the trailing edge of wings, and brakes to check the speed of a dive bomber; tests and studies continued throughout the war on problems of strain and stress.

On lighter-than-air aircraft, the K-type ship superseded the smaller L-ship, which was not suitable for long patrols. By 1944 a third type, the M-ship, considerably larger than the K-ship, was introduced, but did not have enough advantages to displace the K-ship.

In the field of motive power, science achieved outstanding successes. For a variety of reasons, the Navy has tended to concentrate on the production and perfection of the air-cooled engine. Among the results have been its light weight in relation to its power, and, most important of all, its great durability.

In the early stages of the war, for example, at Midway, high-altitude fighters were necessary. As the war progressed, we turned to the offensive, and low-altitude aircraft became more important.

The result was the development of two main types of engines. For altitudes above 25,000, two-stage engines were developed, embodying superchargers and other equipment for producing maximum performance at high altitudes. For medium- or low-altitude flight, the additional stage was not only unnecessary but actually impeded the performance of the plane. Consequently, as the war progressed, more and more attention was devoted to the single-stage engines, and by the end of the war the Navy had top-performance engines of both the single- and two-stage types.

In addition, naval aviation made use of devices such as water injection. With engines operating at continued high output, there was danger of cylinders becoming overheated and the combustible mixture detonating. It was found that with a proper injection of water or a special coolant directly into the cylinder, the cylinder could be cooled sufficiently to remove this danger. As a result, the speed of a plane could be considerably increased. This added spurt of power at the right time saved more than one pilot's life.

Although in the closing months of the year, increased attention was being given to the jet engine, it could be said that in general, throughout the war, it was found that the greater durability and all-round performance of the reciprocating engine made it still the best type to use. Because the account herewith is limited to the war record and does not include predictions of the future, it is perhaps enough to state that the Navy continued to conduct careful tests and experimentation on all kinds of motive power.

The air-cooled engine brought with it special problems. One of the most difficult was that of properly cooling all parts of the engine. In the early days, the engine was placed where it got the full force of the wind. The parts exposed to the wind were cooled, but portions of the engine not reached by the wind became overheated.

A long process of experimentation and adaptation produced a number of answers to this problem. In the first place, a covering or cowl was placed over the engine, and flaps or "gills" were hinged into this cowl to make possible a regulated access of air to all parts of the engine, and, in addition, special baffles conducted air to vital parts, such as spark plugs.

In the main, this use of cowl flaps has been the most widely employed method. There was, however, one difficulty with flaps: they created a drag that in some cases slowed up the plane more than was desirable. Consequently, experiments were made with other cooling devices. In the South Atlantic, where engines were consistently overloaded and had to contend with tropical weather, fans were used to direct the air to different parts of the engine. In later experiments, some fans were powered, jet-like, from the exhaust gases of the engine itself.

A great deal of work went into the types of metals to be used in engines. In this connection, naval aviation had to contend with problems of corrosion from sea air and sea water, of deterioration in tropical areas, and damage as a result of cold in the North Atlantic and Aleutian regions. Combat operations, obviously, created a great strain on engines.

It was found, for example, that dive bombers burned out bearings in the pull-out from a dive; the Navy found a solution in the production of a silver-surfaced bearing that could withstand a temporary loss of lubrication.

Fuels presented another difficult problem that resulted in a great deal of research. One of the Navy's contributions in the field of high gasoline performance was the establishment of a dual standard for aviation gasoline — a minimum requirement for a "rich" mixture and another for "lean" mixtures. As a result of the extreme variation in temperatures which developed in aircraft engines, great attention had to be paid to lubrication.

It was found, for example, that oil returned from the engine was filled with air and could cause trouble if reused in this condition, especially at high altitudes. Consequently, means were developed to de-aerate the oil. Another problem, especially in cold weather, was that of maintaining the proper temperature of lubricants.

Aviation engine oil is more viscous than automobile lubricants, having the consistency of toothpaste at 32° F, and when below 0° F being almost as solid as wax. Quick warm-up devices were necessary, therefore, and were developed after considerable study and testing.

A great deal of effort was expended on the subject of starters. The question of weight was very important in this regard, as a starter is just so much dead weight once it has completed its function of starting the engine. On airfields, portable starts could be used, making use of the electrical power of the station, but the aircraft had to have its own starting mechanism for use away from a well-equipped station.

For a time, the Navy made use of the cartridge type starter. A cartridge, similar to a blank shotgun shell, was fitted into a chamber and the engine was turned over by firing the cartridge. This had the advantages of light weight and low energizing power, since the cartridge could be operated from two flashlight cells. It had the disadvantages of being somewhat unreliable, and of requiring careful maintenance.

The Navy had developed and used as standard equipment the so-called inertia type starter, which made use of energy created by a rapidly spinning flywheel, turned either by hand or by electricity. Generally speaking, however, efforts have been concentrated on the development of a direct cranking starter similar to those used on automobiles. The main drawbacks were problems of weight and power, but as the war progressed, satisfactory answers were found.

Closely allied to engine development was the improvement of propellers. The casual observer might think that there is nothing unusually complicated about a propeller. On the contrary, the study of this important airplane part has taxed the ingenuity of the best scientists.

Questions of air density, the pitch, shape, radius, and number of blades, all combine to create questions difficult of solution. The blades of early propellers were at a fixed angle to the hubs and merely whirled about. By the end of the war, however, naval aviation had light-weight propellers of hollow steel of great strength, that constantly changed the pitch to give the maximum performance for the particular requirements of the plane.

On take-off, for example, the blade setting was adjusted to speed up the ascent; on level flight the angle was shifted, analogous to "shifting into high" in a car; and at high altitudes where air density decreases, the blades were set at a greater angle to take larger "bites" of air. If an engine gave out, a propeller could be "feathered" and stopped to reduce its drag, and also prevent damage to the engine which might result from the propeller's windmilling and turning the motor over in flight.

It was also found that correctly designed propellers could be used to check the too rapid descent of a dive bomber and relieve the pressure on the wing flaps. On the M-type airships and on large seaplanes the propellers could be reversed, so that the airship could go backward, and the seaplane could reverse while taxiing on the water.

AMONG THE MOST IMPORTANT ACCESSORIES of naval aviation were those operated by electricity. The significance of radar and radio in modern warfare is difficult to exaggerate. Once again, the problems connected with electronics are far too complicated and technical to be taken up in any detail in this report. It may be of some value, however, to point out a few highlights along the road, since these developments did play such a major role in gaining victory.

Naval aviation first came into contact with airborne radar through the work of the British in anti-submarine warfare a considerable period before the attack on Pearl Harbor. Our patrol planes operating from Iceland in 1941 made use of a relatively crude form of radar secured from the British.

After we entered the war, the United States developed an American version, and by the end of 1942, hundreds had been installed in patrol craft both in the Atlantic and Pacific theaters of war. It was this model that made possible the depredations of the "Black Cats" in the New Guinea area, and saved pilots' lives in the Aleutians. This type of radar was too heavy for use in fighter aircraft, however, and one of the outstanding developments of naval aviation was the introduction and successful use of a light-weight radar in fighter planes. At the same time, bigger and better radar equipment was developed for use in the larger planes.

During most of the war, information regarding radar was highly classified, or "hush-hush." Since the termination of hostilities, however, the restrictions have largely been dropped, with the result that the average American probably has a fair idea of the nature of its operation. At the risk of pointing out that which is generally known, a brief description of the operation of radar might be in order.

Radar is an instrument that sends out a series of electrical impulses. When these impulses hit an object that stands out from the surface of the land or sea they bounce back and are reflected on a cathode ray screen. This screen is graduated so that one can tell the distance the object is from the radar. In a rough way, the image, or "blip," on the screen resembles the object that is indicated.

The contour of a coastline being approached from the sea will appear on the screen, and if, for example, there is one mountain that stands out, it can be used as an aid to navigation. A conning tower protruding from the surface of the sea could be picked up on radar, thus poor visibility did not stop the search for submarines.

From relatively crude beginnings radar has developed into a great variety of specialized instruments. It has been used not only for search and navigation, but for aiming bombs and torpedoes. It has been outstandingly successful in night fighters; in fact, it is safe to say that night fighters could hardly have operated so efficiently without this equipment.

As we have noted, radar is of no use below the surface of the water. Scientists, therefore, tackled the tough problem of attempting to develop a device that could detect objects under the water. Two devices were tried with a certain amount of success. One of these is known as MAD (magnetic airborne detection) gear. This equipment, as its name indicates, locates underwater objects that are con-structed of metal. Installed on K-type airships, the MAD equipment has been fairly successful in covering limited areas in search of ei-ther submarines or mines.

The other device was a type of sonar equipment. This consisted of an expendable buoy that was dropped from the plane into the wa-ter. Within this object was a mechanism that picked up underwater sounds, such as the noise of a submarine propeller, and transmitted it to the search aircraft.

Another use of radar was in the field of recognition. In the dark days of the Battle of Britain it was essential to know the identity of planes flying across the Channel. The British, therefore, developed a system, known as IFF (identification, friend or foe), that was adopted by the United States. By setting up a standard recognition device, it was possible to recognize both friendly ships and aircraft. Installed on all our ships and planes late in 1943, this equipment was of great service throughout the duration of the war.

Another vital electronic device was, of course, the radio. More was known about radio than radar at the outbreak of the war, but there was still a great deal of work to be done. What was needed was a light-weight instrument that could be tuned quickly and that would be thoroughly reliable. During 1942 a satisfactory long-range radio was produced and installed on planes. It was found, however, that long-range transmitting had one serious defect. The enemy could listen, too, and in fact assigned men to monitor our circuits.

An alternative was suggested in the development of VHF (very high frequency) radio equipment which transmitted over a much shorter range and could reach planes from a carrier, for example, but would not normally be picked up by the enemy. This type of radio was rapidly improved to include more channels and to have more reliability of operation.

Radio was of great value in air-sea rescue work. One of the most interesting "gadgets" was the so-called "Gibson girl." This was a little transmitter with a pinched-in middle portion that was held between the knees of the life-raft survivor while he cranked enough energy to send out a message for help. Later the VHF "Handy Talkie" was developed; it was but little larger than a flashlight. This was even more satisfactory for the downed pilot since he could communicate directly with his rescuers.

Electronics came to the aid of aircraft navigation, or avigation, as it is sometimes called. Both radar and radio were widely employed in this connection, and there were a number of variations that increased the security of air travel. One of these variations was called LORAN (long range navigation). Using what might be termed a combination of radio and radar, LORAN transmitters were set up at strategic spots throughout the world. By means of the proper equipment a navigator within approximately fifteen hundred miles of one of these transmitters could fix his position at any point in that area. This is one wartime device that will be of inestimable value during peace.

A good deal of attention was given to the development of radio and radar countermeasures. The enemy was carrying on scientific work, too, and if his radar or his radio could be rendered useless, a great advantage could accrue to our side. Both sides made frequent and effective use of "window" or "chaff." These were thin metal foil strips that were dropped from a plane as it approached the target. These strips had a tendency to clutter the radar screen with so many "blips" that the plane itself could not be isolated in time to prevent the attack.

The question of lighting had a great many ramifications. A great deal of study and experimentation had to go into the kind of lighting, for example, to be used on the panel of a night fighter, so that the pilot could see the instruments without impairing his night vision. Some work was done, also, on the use of very powerful searchlights to blind the defenders during night attacks on submarines.

The field of electronics in war is relatively new. Electronics played an important part in determining the character of World War

II, and our successful use of electronic devices aided materially in achieving victory.

When war appeared imminent, it became painfully obvious that we knew next to nothing about Japanese installations in the islands of the Pacific. For years, the enemy had excluded all prying eyes from his possessions and from military areas in the home islands, yet exact knowledge was necessary to any amphibious landings, indeed was vital even to the success of air strikes.

The intelligence information so urgently needed was in a large measure supplied by the marriage of the airplane and the camera, whose offspring in the form of thousands of pictures were subjected to analysis by a new art — that of photographic interpretation. As in so many other respects, we learned our first lessons from the British who had of necessity gone about developing the art after being driven from the continent in 1940.

When the first United States naval officer to visit England and observe their methods returned late in the summer of 1941, plans got under way for the establishment of a school at the Naval Air Station, Anacostia, D.C., and the first students reported in January 1942. Before hostilities came to an end over 800 received basic instructions in photo interpretation and 650 were given further training in special aspects of the subject. Because of the close proximity of the school to the Navy Department, photographs were constantly being referred to it for expert analysis, and early in 1943 an interpretation unit was set up at Anacostia to handle the growing number of requests.

By November of the same year, the value of this activity had become so great that a Photographic Interpretation Center (redesignated as the Photographic Intelligence Center in January 1945) was set up not only to include the school but also to provide services to the Navy Department, to maintain a central file of reconnaissance photographs, to develop new interpretation procedures, to carry on liaison functions with the Army and other government agencies, and to have in readiness photographic interpretation teams for the use of the fleet.

Another activity of the center was training in methods for constructing three-dimensional terrain models, which had been begun early in 1943. Made from aerial photographs and other such sources of information as might be available, these models gave an exact picture of terrain features and were of inestimable value in planning amphibious operations and briefing pilots. Not only was instruction given at the center, but it also built over five thousand rubber models for the use of the armed forces.

As the students of the school permeated through the fleet, there was a growing realization of the need both for photographic planes in existing fleet air wing and carrier organizations and also for long-range photographic squadrons to which were assigned specially equipped Liberators.

The first of these squadrons commissioned in October 1943, served in the South Pacific where in the hard school of experience many additional lessons were learned and applied. Most interesting of these was the centralization of trained personnel in a photographic interpretation squadron, which when joined to one of the specially equipped aircraft squadrons became a photographic group.

The first of these groups was created in August 1943, in the field at Guadalcanal. Even so brief a word as this indicates that photography was becoming a big business; it was also one in which techniques were constantly changing and developing; and it was considered important enough so that the Navy built a special laboratory for it at Anacostia.

At the same time that the Navy was building up its work in this line so were the Army and our allies and there existed a constant exchange not only of intelligence information but also of techniques and procedures. Joint manuals were issued and the experience of one branch of the service proved invaluable to the other. For example, when our carriers moved against the Japanese homeland, photographic interpretation techniques developed by the Army Air Forces and the RAF for industrial Germany were taken over by the Navy and applied to the similar problems encountered at Kobe or Nagasaki.

THE IMPROVEMENT OF WEAPONS KEPT PACE with other aviation developments during the war. At the outbreak of the conflict, the most generally used weapon was the .30-caliber machine gun. As the war progressed it became necessary to replace this gun with weapons of greater fire power. The introduction of armor plating and self-sealing fuel tanks rendered the .30 caliber gun more or less impotent, and it was replaced by the more destructive .50 caliber machine gun, the 20mm gun, and, in the PBJ, the 75mm cannon.

Some of the most important changes made during the years from 1941 to 1943 were those in weapon accessories. Hydraulic chargers were developed, as well as pneumatic and electric chargers, and electric remote-trigger controls were installed. During the next two years, improvements were made on the .50 caliber and 20mm guns. The barrels were improved, for example, to permit more extended firing.

The dropping of bombs involved a number of problems. Strong and foolproof bomb racks had to be constructed. Because effective aiming was necessary, various bombsights were utilized, and toward the end of the war radar was used with success in securing hits through clouds and overcast. Bombs not only have to be constructed to explode but must detonate at the right time.

A depth charge, for example, has to explode at a certain depth and proximity to cause maximum damage to a submarine. A bomb dropped from a slowly moving plane at low altitude must not explode too soon or the plane itself would be a casualty. Different types of bombs were needed for various objectives; for example, fragmentation bombs were more effective against personnel, heavy bombs were needed for large ships or huge gun emplacements. The Navy Bureau of Ordnance worked on these and many other problems and produced superior fuses and bombs for all occasions.

In the main there were two types of gun installations on planes, fixed and flexible. The former were directed by aiming the entire plane. Planes having a number of these guns installed and synchronized in their operation, presented a terrific fire power over a restricted area, capable, for instance, of cutting an enemy aircraft into fragments. The fixed fire power of navy planes was enormously increased.

In 1935 the Navy's standard fighter, for example, carried two .30 caliber machine guns; the F7F, on the other hand, had four 20mm guns and four .50 caliber guns and could fire 22 times as much weight of projectile per second as the older aircraft. The free firing guns were on flexible or movable mounts, that sometimes were and sometimes were not installed in turrets. The advantage of a turret, which could be power driven, was that it could be extended a short distance from the body of the plane and thus made possible a wider range of fire.

One of the outstanding weapons developed in this war was the rocket. Like most ordnance developments, this was the product of work by all military services and allied scientific agencies. A rocket was essentially a missile that carried its own motive power (jet-like in form) within it. Its main advantage was its combination of light weight with a terrific impact effect. With an installation of rockets, a fighter plane could become virtually a bomber, and the advantage of this development to carrier warfare is of course obvious. Like all other innovations, the use of rockets required a great deal of experimentation.

At first, long runners or launchers were attached to the underside of aircraft wings. These tended to create too much of a drag, and it was with considerable relief that it was discovered that a "runway" was unnecessary, and that rockets could be started on their way from "zero length" launchers that weighed little and did not materially affect the aerodynamic performance of the planes. A great deal of study also went into the problem of making rockets more accurate, and significant gains were made in this direction.

An outstanding rocket developed for aircraft use was the "Holy Moses." This was a five-inch rocket that was especially valuable for strafing and pin-point attack, in view of that fact that it had greater accuracy of aim than a bomb. It also proved to be especially effective against lighter shipping. Need for a more powerful weapon led to the development of an 11-inch rocket named "Tiny Tim." Although it delivered a blow nearly equal to that of a torpedo, it could be carried and launched by navy fighters.

Another important offensive weapon was the aerial torpedo. The Japanese had demonstrated the power of this weapon in the attack on Pearl Harbor, and, once again, we improved it, and turned it back on the enemy with devastating effect. As in the case of other weapons, a great deal of experimentation was made on the torpedo. One of the major problems to be solved was that of launching the device successfully from a plane traveling at high speed. There was a danger that the torpedo would hit the water with such an impact that its course in the water would be deflected. A solution to this problem was found in a device that protected the vital control surfaces and still permitted the torpedo to run "hot and true" through the water to the target.

As naval aviation improved its offensive weapons, it also gave attention to protection of the personnel in naval aircraft. Some sort of armor had long been thought of, but before 1940 it was out of the question because of the excessive weight that would be involved. The development of lightweight metals and other materials of great strength, however, made armor possible.

At first, main consideration was given to protecting the pilot; but later, as materials were improved, the gunners and other personnel were afforded a certain amount of protection. As we have already noted, the introduction of armor was a major factor in causing a shift from the .30 caliber machine gun to weapons of heavier caliber.

By 1944 heavier armor became necessary and was installed because of the enemy's use of heavier guns. There was danger not only from enemy aircraft fire, but also from antiaircraft fire, commonly known as "flak." It was found that anti-flak suits constructed of laminated layers of nylon and other synthetic materials could be of considerable protective value. Flak curtains of nylon were also discovered to be an effective substitute for fixed metal armor plate.

RANKING HIGH IN THE SCIENTIFIC AIDS to naval aviation have been those in the field of medicine and surgery. The work that surgeons, doctors, and nurses have done for the wounded in all the fighting forces is an indispensable contribution that lies outside the scope of this account.

In addition to these services, however, medical science has brought its special talents to bear on a number of problems that deal specifically with aviation.

In recognition of the fact that such problems existed, a separate classification of Flight Surgeon was used to designate a group of doctors who flew and who were aware of the particular problems of the pilot and other aviation personnel.

One of the most important functions of this group of men was to see that only those physically and mentally fit to fly became aviators. Carefully designed examinations were prepared for the candidates for flight training. The medical officers' functions, however, did not cease with picking out the right men.

Their task also was to see that these men, once they passed their training and were engaged in naval aviation operations, maintained their fitness. Periodic tests were given to make sure that personnel did not slip below the physical requirements.

Mental conditions, likewise, were watched. Combat fatigue and excessive tours of duty under unfavorable weather conditions were observed by flight surgeons with a view to keeping men in good mental as well as physical shape. The flight surgeon had the last word on whether or not a man was fit to fly and could remove men from flight lists until he considered that they were in condition.

In other ways medical science concerned itself with naval aviation. One of the important developments of the Navy's air war was dive bombing. As planes were improved to stand the strain of pulling out of a dive, something had to be done to enable the pilot to resist this same strain. It was found, for example, that as a plane pulled out of a dive, the blood in the pilot's body rushed toward his feet. If the dive and pull-out were sufficiently strong, enough blood would be forced away from the aviator's head to cause him to "black out," and the chances were strong that he might not recover his senses in time to prevent a crash.

Aviation medicine attacked this problem from two angles. One was to attempt to educate fliers so that they would know their limits in such matters. The other was to attempt to devise a so-called "Anti-G" suit that would prevent or lessen the possibility of black-outs. Roughly speaking, this was a suit that could be pumped up so tightly about the lower portion of the body that blood could not force its way down and away from the brain. Although these suits were not the most comfortable apparel in the world, pilots soon found that wearing them gave an added edge over the enemy.

The Bureau of Medicine and Surgery made many studies of the problem of maintaining life at high altitudes. The use of oxygen had made possible flight operations up to about 43,000 feet. These altitudes, as well as flight conditions in the Aleutians and in other cold climates, brought about, in addition, a great deal of experimentation on various types of flight clothing. As a result, a number of different flight uniforms were devised, from lightweight summer equipment to electrically heated clothing for the coldest weather.

Medical science performed a great service in helping the development of night fighting. Careful studies of night vision were made, and tests were conducted on proper lighting. Conversely, daytime flying was made easier as a result of the development of high-grade sunglasses and goggles.

Both the Bureau of Aeronautics and the Bureau of Medicine and Surgery were interested in all types of safety devices. For example, originally the safety belt extended merely across the waist. It was found, however, that in crash landings, the occupant of the plane was often thrown against the instrument panels. A scientifically devised shoulder harness was developed, therefore, that saved many lives. The effects of parachuting were also studied; parachute harnesses were improved, jumping techniques were developed, and once again a saving of life resulted.

Another problem connected with safety was that of survival after crashes or forced landings. Studies were made of compact foods that would sustain life over an extended period. In addition, there was the question of providing an adequate water supply. Outstanding in this connection was the development of a kit that produced a drinkable water from sea water. Its main importance was that it permitted a seven days' supply of water in the space that formerly would have held but a single day's supply. For protection against the sun, special sunburn lotions were developed, and exposure suits of light weight were made the subject of experiment.

Toward the end of 1944, a new service was offered by medical science. At NAS Alameda, a school was started for nurses and corpsmen for the "air evacuation of casualties." From the graduates of these schools, together with flight surgeons and others, so-called VRE squadrons were set up. Wounded had been evacuated by air from combat areas as early as 1942, but in these new squadrons the problem was tackled by a well-trained team of experts.

The nurses had to learn how to administer oxygen, they had to be prepared for "ditching" procedures in case the plane should be forced down in flight, they had to be able to deal with air sickness, as well as with the various injuries of the patients who were being evacuated. By a careful screening process, the wounded who most needed to be flown back to this country made the trip — and made it under the expert care and men and women specially trained for their duties.

EPILOGUE

WITH THE COLLAPSE OF THE JAPANESE EMPIRE, naval aviation completed its wartime mission. In conjunction with the other American armed forces, it had been assigned the task of beating the enemy into submission. As we have seen, this had been a long and difficult process. Fields had been scraped clean of cornstalks or drained from swamps to make way for huge training and operational bases. To these stations had come young persons from all walks of life to learn their specialized jobs for war. Their training had been lengthy and often arduous and had continued in the combat areas. Practice and more practice, coupled with certain innate characteristics of American youth, provide a fundamental explanation of American victory.

With this highly trained fighting group had gone the finest equipment. Here again, the process of building up superiority in both quality and quantity had been slow. Our nation had first to get into a fighting frame of mind, and second to turn the great wheels of American production and the sweat of American manpower to the manufacture of the best planes and equipment that could be devised by our scientists and engineers. This machinery, once well under way, provided a second explanation for the success of naval aviation.

These two factors, highly trained men and excellent planes and equipment had been vitalized by a third element that completes our understanding of victory. This was organization, or the "know-how" of warfare. Once again, this was a factor that had been fused in the crucible of wartime experience. Gradually, naval aviation had built up the technique of warfare, from the co-operation of two planes on patrol to the carefully worked-out plan of a great task force.

An important feature of this organization had been close correlation between the sea and the air forces, and the best evidence of this fact was to be found in the successes of the fast carrier task force.

Naval aviation had completed its wartime mission with the strikes over Tokyo. It had conquered the Jap air force. Together with the submarine service, it had accounted for 90 percent of the losses of the Japanese navy, 78 percent of the losses of the Japanese merchant fleet. It had thus contributed handsomely to the economic strangulation of the empire. Naval aviation is now confronted with a new mission — to aid in the maintenance of world peace. The new task is a difficult one; it will take a different type of work, and naval aviation must reorganize to carry out the mission.

It is not within the scope of this report to suggest the nature of this reorganization, but it is hoped that certain lessons of the past will not be forgotten. Two points, especially, should be kept in mind. In the first place, this nation must be more careful than it was in 1919 and 1920 in throwing away the tools with which the war was won.

In the second place, if the operations of naval aviation in peacetime are to equal in quality its work in World War II, naval aviation must remain a part of the framework of the Navy as a whole.

46970561R00231

Made in the USA
Middletown, DE
03 June 2019